Ecological Studies
Analysis and Synthesis

Edited by

W.D. Billings, Durham (USA) F. Golley, Athens (USA)

O.L. Lange, Würzburg (FRG) J.S. Olson, Oak Ridge (USA)

H. Remmert, Marburg (FRG)

Volume 67

Ecological Studies

L. R. Pomeroy   J. J. Alberts
Editors

Concepts of Ecosystem Ecology

A Comparative View

Contributors
J. J. Alberts, R. T. Barber, D. C. Coleman, K. W. Cummins,
C. F. D'Elia, J. K. Detling, E. C. Hargrove, P. F. Hendrix,
K. H. Mann, S. L. Pimm, L. R. Pomeroy, W. A. Reiners,
P. H. Rich, P. G. Risser, J. E. Schindler, H. H. Shugart,
D. L. Urban, W. J. Wiebe, R. G. Wiegert

With 93 Figures

Springer-Verlag
New York Berlin Heidelberg
London Paris Tokyo

Lawrence R. Pomeroy
Institute of Ecology
University of Georgia
Athens, GA 30602
U.S.A.

James J. Alberts
Marine Institute
University of Georgia
Sapelo Island, GA 30602
U.S.A.

Library of Congress Cataloging-in-Publication Data
Concepts of Ecosystem Ecology.
  (Ecological studies ; v. 67)
  Bibliography: p.
  Includes index.
  1. Ecology. I. Pomeroy, Lawrence R., 1925–
II. Alberts, James, 1943– . III. Series.
QH541.145.E845 1988   574.5   87-32210

Typeset by Impressions, Inc., Madison, Wisconsin
Printed and bound by Edwards Brothers, Inc., Ann Arbor, Michigan.
Printed in the United States of America.

9 8 7 6 5 4 3 2 1

ISBN 0-387-96686-2 Springer-Verlag New York Berlin Heidelberg
ISBN 3-540-96686-2 Springer-Verlag Berlin Heidelberg New York

Preface

This volume derives from a conference honoring Eugene P. Odum on the occasion of his retirement. Professor Odum has been influential in ecological thinking through his focus on the ecosystem as a functional unit of nature. In recognition of his influence, he has received all of the major awards for which an ecologist is eligible: the Mercer Award, the Tyler Award, le Prix de l'Institut de la Vie, and the Crafoord Prize. He has been elected a member of the U.S. National Academy of Sciences. At the University of Georgia, he is the only professor to have performed the feat of occupying two chairs at once, that of Alumni Foundation Distinguished Professor and that of Fuller E. Callaway Professor. In his case, retirement is a relative term, since Odum continues to be as active as ever in both the pursuit of fundamental principles in ecology and their application to problems of human impact on our environment. Therefore, it seemed appropriate to honor him with something other than the usual *Festschrift* of otherwise unpublishable offerings from former students. Instead, we assembled leading experts in the functional analysis of ecosystems and some related topics to consider two questions:

1. What is the proper study of ecology, and are we doing it?
2. What have we learned about ecosystem function?

Clearly, these are ambitious questions to which no final, totally definitive answer is forthcoming. Yet, there is a need to step back periodically from our narrow pursuits to look out over a broader scope of science, hoping to get some perspective on where we have been, a glimpse of where we are headed, and some sense of where we are now. We had naively hoped that each participant would consider both questions. While that did happen to some degree, the contributions tended to divide into those mainly concerned with philosophy and methodology and those mainly concerned with the status of our knowledge of a specific system. So we have arranged this volume accordingly, beginning with a series of chapters which consider how we go about studying ecosystems. This is always a somewhat controversial topic, and we hope the present treatment is no exception, for we believe that more careful consideration of the question of methodology is needed. The two central sections consider the state of our knowledge of some terrestrial and aquatic ecosystems. The final section considers problems of current or future concern in studies of ecosystems, terminating with an opinionated summary by the editors.

Terrestrial and aquatic ecologists stopped talking to one another more than a decade ago. This was evident at the conference, where the two groups came together as strangers, both personally and professionally. There was a clear dichotomy in viewpoint, although some instances of parallel evolution in the two areas could be seen. This is also evident in the respective sections of this volume, with often different emphases. Although a comprehensive coverage of all major ecosystems was beyond the scope of both the conference and the book, we believe that a number of conclusions emerge which bring together concerns shared by investigators of both terrestrial and aquatic ecosystems:

• The ecosystem is a distinct and important hierarchical level of organization of life on this planet.

• Ecosystems are remarkably stable and resilient, but these are attributes that emerge in large part from the nature and interactions of the component species populations. Therefore, ecosystem scientists must be conversant with population biology, if not the reverse.

• Ecologists have until recently tended to focus on macroorganisms as ecosystem components. Several chapters in this book point out major roles of microorganisms in ecosystem functions. The disparity of treatment of microorganisms from chapter to chapter shows that this concept is moving through the field of ecology.

• Ecosystem studies tend to be descriptive, but this is not necessarily a fault. In Chapter 1 we suggest that this is a natural result of the short history of ecological studies—hardly 100 years—together with the inherent complexity of such a high level of biological integration. Of course, experimental science is stronger science, and we see experimental approaches emerging rapidly.

• As ecology is a young science, it follows that there is still contention over most of those few general principles to have emerged in ecology, including Liebig's "law," Redfield's ratio, the stability–diversity relationship, the trophic level, ecological efficiency, and so on. In a young science this is not necessarily a bad thing.

Because this is a collection of contributions by leading ecologists, we have not attempted to shape the volume into a unified view of ecology. The differences between marine and terrestrial, as well as empirical and theoretical, specialists are preserved, and there are indeed contradictions between chapters. These serve to emphasize subjects of contention, usually worth future consideration. We have, however, subjected each chapter to rigorous external review, and we thank the following for their diligent and timely assistance in that endeavor:

John D. Aber, Forestry, University of Wisconsin, Madison
Karl Banse, Oceanography, University of Washington, Seattle
Steven M. Bartell, Oak Ridge National Laboratory, Oak Ridge
A. Joy Belsky, Cornell University, Ithaca
Stuart F. Chapin, III, Institute of Arctic Biology, University of Alaska, Fairbanks
Scott D. Cooper, Biological Science, University of California at Santa Barbara
Zdenek Filip, Institute of Water, Soil, and Air Hygiene, Langen, West Germany
Thomas Fontaine, NOAA-GLERL, Ann Arbor, Michigan
L. R. Ginsburg, Ecology and Evolution, SUNY, Stony Brook
Robert Goldstein, EPRI, Palo Alto, California
Gary Grossman, Forest Resources, University of Georgia, Athens
Eugene C. Hargrove, Philosophy, University of Georgia, Athens
Rodney Heitschmidt, Texas Agricultural Experiment Station, Vernon
Charles S. Hopkinson, University of Georgia Marine Institute, Sapelo Island
Robert E. Johannes, CSIRO Marine Laboratories, Hobart, Australia
Dennis H. Knight, Botany, University of Wyoming, Laramie
W. K. Lauenroth, Range Science, Colorado State University, Fort Collins
William M. Lewis, Jr., Evolutionary, Population and Organismal Biol., University of Colorado, Boulder
J. M. Melillo, Ecosystems Center, Marine Biology Laboratory, Woods Hole
Judith L. Meyer, Zoology, University of Georgia, Athens
James Morse, Biology, University of South Carolina, Columbia
Robert N. Muller, Forestry, University of Kentucky, Lexington
John Ogden, West Indies Laboratory of Fairleigh Dickinson University, Christiansted, St. Croix, Virgin Islands
Michael Pace, Institute of Ecosystem Studies, New York Botanical Garden, Millbrook
Robert P. Patterson, Crop Science, North Carolina State University, Raleigh
Gary A. Peterson, Agronomy, Colorado State University, Fort Collins
H. Ronald Pulliam, Zoology, University of Georgia, Athens
G. Phillip Robertson, W. K. Kellogg Biological Station, Michigan State University, Hickory Corners
David W. Schindler, Freshwater Institute, DFO Canada, Winnipeg
Steven V. Smith, Oceanography, University of Hawaii, Honolulu
Walker O. Smith, Graduate Program in Ecology, University of Tennessee, Knoxville
Douglas Sprugel, Forest Resources, University of Washington, Seattle
Robert Strickwerda, University of South Carolina, Aiken

Richard Waring, Forest Science, Oregon State University, Corvallis
Robert G. Wetzel, W. K. Kellogg Biological Station, Michigan State University,
  Hickory Corners

We also thank Edward Chin, Director, Marine Sciences Program, University of Georgia, for initiating the idea of a conference and for financial support of the conference. The text was edited by Patricia Freeman-Lynde. Illustrations were prepared by Michelle Taxel and Terri Ainley.

Lawrence R. Pomeroy
James J. Alberts

Contents

Contributors

ALBERTS, JAMES J.

Marine Institute, University of Georgia, Sapelo Island, Georgia, U.S.A.

BARBER, RICHARD T.

Monterey Bay Aquarium Research Institute, Pacific Grove, California, U.S.A.

COLEMAN, DAVID C.

Department of Entomology, University of Georgia, Athens, Georgia, U.S.A.

CUMMINS, KENNETH W.

Center for Environmental and Estuarine Studies, Appalachian Environmental Laboratory, University of Maryland, Frostburg, Maryland, U.S.A.

D'ELIA, CHRISTOPHER F.

Chesapeake Biological Laboratory, University of Maryland, Center for Environmental and Estuarine Studies, Solomons, Maryland, U.S.A.

DETLING, JAMES K.

Natural Resource Ecology Laboratory and Department of Range Science, Colorado State University, Fort Collins, Colorado, U.S.A.

HARGROVE, EUGENE C.　　　　Department of Philosophy, University of Georgia, Athens, Georgia, U.S.A.

HENDRIX, PAUL F.　　　　Institute of Ecology, University of Georgia, Athens, Georgia, U.S.A.

MANN, KENNETH H.　　　　Department of Fisheries and Oceans, Marine Ecology Laboratory, Bedford Institute of Oceanography, Dartmouth, Nova Scotia, Canada

PIMM, STUART L.　　　　Department of Zoology and Graduate Program in Ecology, University of Tennessee, Knoxville, Tennessee, U.S.A.

POMEROY, LAWRENCE R.　　　　Institute of Ecology, University of Georgia, Athens, Georgia, U.S.A.

REINERS, WILLIAM A.　　　　Department of Botany, University of Wyoming, Laramie, Wyoming, U.S.A.

RICH, PETER H.　　　　Department of Ecology and Evolutionary Biology, University of Connecticut, Storrs, Connecticut, U.S.A.

RISSER, PAUL G.　　　　University of New Mexico, Albuquerque, New Mexico, U.S.A.

SCHINDLER, JAMES E.　　　　Department of Zoology, Clemson University, Clemson, South Carolina, U.S.A.

SHUGART, H. H.　　　　Department of Environmental Sciences, University of Virginia, Charlottesville, Virginia, U.S.A.

URBAN, D. L.　　　　Graduate Program in Ecology, University of Tennessee, Knoxville, Tennessee, U.S.A.

WIEBE, WILLIAM J.　　　　Department of Microbiology, University of Georgia, Athens, Georgia, U.S.A.

WIEGERT, RICHARD G.　　　　Department of Zoology, University of Georgia, Athens, Georgia, U.S.A.

1. The Ecosystem Perspective

Lawrence R. Pomeroy, Eugene C. Hargrove, and James J. Alberts

1.1. The Concept

Although Tansley (1935) is credited with the name, and often with the concept, of ecosystems, the concept of a higher level of organization had been forcing itself to the attention of biologists for many years (Möbius 1877; Forbes 1887; Lotka 1925). Another manifestation was the concept of the superorganism, which was given various interpretations, proponents applying it to anything from a termite colony to the landscape. What survives is the idea of a hierarchy of levels of integration. The term, superorganism, turned out to be an unfortunate one, as it implied specific kinds of organization that are not present beyond the species population. Controls that involve population genetics and natural selection end at the species level. This spills over into community organization, in a sense, through coevolution, but the genetics are still those of interbreeding populations of the individual species. We observe significant homeostasis at the ecosystem level, but it must be understood in terms of the processes of the component populations in their physical and chemical environment. A major thrust of ecosystem studies at this time is to understand in an integrated way the response of communities to their environment, and how this results in the resilience shown by most ecosystems.

It may not seem necessary to begin by asking if ecosystems are an imaginary construct of the human mind. Yet, some other well-developed and much used concepts in ecology appear to be just that, and some population ecologists assert that ecosystems are mere random assemblages of species populations, subject to change without notice (Ricklefs 1987). Ecosystems were defined by Tansley (1935) in physical as well as biological terms, and to the extent that they are delimited by physical features, they are distinct. At the same time, they are open systems, so even though the margin of a lake or of the ocean is a clearly recognizable boundary, there are important fluxes of materials and even organisms across it. In other cases, boundaries are more subtle combinations of physical and biological factors, producing broad ecotones between more clearly defined ecosystems; forest to savanna, for example. Yet the distinctions between forest and savanna are clear, useful, natural ones. Because of their usually large size, encompassing whole landscapes, the boundary questions do not seriously compromise quantitative approaches to ecosystem study. In fact, ecological gradients have been treated quantitatively (Curtis 1959; Whittaker 1960).

What are perhaps less clear are the distinctions between several terms for ecological units. The community is a biological unit that is readily distinguishable from the ecosystem. However, as Barber (Chapter 9) points out, ecologists may see a given landscape, or seascape, as one ecosystem or a set of related ones, and the relation of ecosystem to community may vary according to individual definitions. The community, on the other hand, is usually interpreted as the local assemblage of species populations, without specific regard for the physical environment (Krebs 1985), while the term biome is reserved for the assemblage of species in the ecosystem as a whole. This difference in definitions reflects the fundamental difference in perspective between studies of ecosystems and studies of communities.

1.2. Homeostasis

The stability of ecosystems, despite the lack of strong internal controls, is remarkable. Ecosystems are resistant to change at more than one level. Studies of succession have shown that there is typically an orderly progression of species characteristic of the climate and geological substratum in any ecosystem. If the successional clock is set back by some catastrophe, the same orderly progression begins again, assuming that the catastrophe has not altered the climatic or geologic regimes. A more modern view of this process would overlay the simple concepts of succession with those of island biogeography: the movement of species into an ecosystem from outside and the extinction of species within. However, the latter processes may occur on a longer time scale than successions. The existence of succession in marine and aquatic plankton communities is debated. The changes in planktonic populations that might be identified as successional are more rapid than terrestrial succession, but there is a repetitive cycle of appearance of dominant species populations, which is often a predictable one. It may be an annual cycle, or one in response to less frequent phenomena, such as the El Niño events off Peru and comparable ones off South Africa. Moreover, most plankton are microorganisms. When we know enough about soil microbial ecology, we may find better terrestrial analogs of the plankton.

The property of resilience may be presumed to be a property of the resident species, both those that are current residents, and those that are dormant or potential residents in the form of spores, seeds, or other propagules. However, the idea that stability is a function of species diversity has promoted more controversy than consensus. One problem is the definition of stability. Tropical forests and coral reefs are very dynamic systems on both short and long time scales. Yet, on time scales at least up to the span of human investigations of them, and within the limits of those investigations, these large, productive tropical ecosystems have the appearance of maintaining a large number of species. The frequency of disturbance is cited as a factor promoting diversity (Connell 1978; Huston 1985; Loucks 1970). How, then, does one explain salt marshes of the lower intertidal zone which contain only a few—and often only one—dominant species of higher plants? They are stable systems of high resilience; salt marshes replace themselves rapidly after perturbation. This resilience has been attributed to the reserves of nitrogen and phosphorus available in the sediments (Pomeroy 1975), while Brown (1981) attributes the low number of species to a very low rate of colonization in the face of high environmental stress. While there appear to be demonstrable relationships between stability and diversity, they are not the simple, elegant ones we might prefer. Processes at both population and ecosystem levels may be involved in system stability.

1.3. Aquatic Versus Terrestrial Ecosystems

Ecologists have assorted themselves into aquatic and terrestrial specialists. Is this because there are fundamental differences in the systems themselves, or because studying them requires different equipment, and funding comes from different sources? Probably the societal constraints have more to do with this separation than do the fundamentals of ecology. However, the isolation of the specialized groups of ecologists is real, and one of the goals of the conference on which this volume is based was to bring together people who do not regularly communicate through meetings or journals. A perusal of the membership lists of the Ecological Society of America (ESA) and the American Society of Limnology and Oceanography (ASLO) reveals that about 7% of ESA members also belong to ASLO, while 10% of members of the smaller ASLO belong to ESA. These statistics probably are a conservative measure of the disciplinary separation, because scientists tend to publish and attend meetings most frequently in association with organizations most closely representing their interests.

Putting scientific specialization aside, what are the real distinctions between aquatic and terrestrial ecosystems that are important to their structure and function? The obvious one is the presence of a liquid phase in the aquatic ecosystems which forms something functionally analogous to the gaseous atmosphere of terrestrial ecosystems. The analogy has been well exploited by physical oceanographers, who recognize that the same physical laws govern the circulation of oceans and the atmosphere, and thus one finds comparable processes at work. The aquatic biologist finds that most of the biota are in, rather than at the bottom of, the liquid phase, because light does not reach the bottom, except around the margins of lakes and ocean basins. This leads us to a second, related distinction. The presence of an illuminated solid substratum has influenced the evolution of the biota in terrestrial systems more than it has in the sea and other smaller aquatic systems. It has resulted in a highly divergent evolution of autotrophs of different mean size, with implications for food chains, which remains a perplexing subject. This is touched on in several of the chapters of this volume.

While air is more transparent to sunlight than is water, its low viscosity has limited its penetration by organisms. Water is a more supportive medium, both for microorganisms and macroorganisms. Moreover, at least for the microorganisms, diffusion of materials in water solution is a significant source of essential materials. Another significant distinction is the oxygen tension and diffusion rates in air and water. The lower rates of diffusion produce a refuge for primitive, obligately anaerobic microorganisms in the hypolimnia of lakes and in aquatic sediments, which often are anaerobic. The same properties make anaerobic sediments a relatively stressful environment for rooted vegetation, although the trees of river flood plains and the herbaceous plants of marshes have evolved remarkable means for coping.

Yet, there is more similarity than difference between aquatic and terrestrial systems, because they are constrained by the same physical and chemical laws and are populated by related biota evolved from common ancestry. Trophic structure appears to be different in detail, although similar in function, with sometimes remarkably convergent equilibrium states evolving. This suggests that factors other than the contents of the collective community genome constrain ecosystem development. One feature that is just making its impact on the community of ecologists is the omnipresent and nearly omnipotent microbial community. Microbial metabolism is a significant part of the total metabolism of all ecosystems. The energetic implications are just becoming apparent, and they are major ones. In terrestrial soils, natural waters, and aquatic sediments, there is a large secondary production of microorganisms, often consuming 25–50% of the energy available from net community photosynthesis. Microbial food webs may take various forms, and may be energy links to metazoans or energy sinks operating in direct competition with metazoans. Where they are energy sinks, microorganisms are nutrient remineralizers. They are especially important in nutrient-limited systems, such as the open sea, where most nutrients are supplied by recycling. Much of the nutrient flux is attributed not to bacteria, the classical decomposers, but to protozoa that consume bacteria and release excess ammonium and phosphate into soils and waters (Clarholm 1984; Linley and Newell 1984; Pomeroy et al. 1984; Pomeroy and Wiebe 1987).

1.4. Emergent Properties

On strictly philosophical grounds, the existence of emergent properties is questionable. Edson et al. (1981) cite Hempel and Oppenheim (1948), who point out that unpredictable phenomena emerge from insufficient theories about system behavior. Thus, if we know enough about a system, we should be able to predict all of its properties and behavior in a strictly reductionist way. Einstein, in his early work on Brownian motion, stated that if we know enough about the state of every molecule in a system, we can predict the state of the system. However, in a footnote he says, "Dear reader, do not believe that you can do that." From this realistic viewpoint, it seems that we may have with us for a long time properties of ecosystems that appear to be emergent. These probably should not be cited as evidence that ecosystems are real entities. They are, however, reason to study ecosystems as entities, because such properties are best studied by holistic, "global" methods rather than by the summation of the components.

The identification of properties of ecosystems as emergent ones has proven to be difficult and controversial. Regrettably, the example chosen by Odum (1977) of high primary production at Enewetak Atoll as an emergent property has not withstood the test of time (Chapter 10). Although these are what Salt (1979) terms collective properties, experience shows that we cannot add them up from components. In the real world, there are cryptic components, components with a high variance, and other operational impediments. When we measure them synoptically at the ecosystem level, we find that processes such as photosynthesis, assimilation, and respiration do go on at predictable and often remarkably stable rates in spite of successional changes in the system. This we now attribute to the existence of guilds of species which perform a common function. Competition for niche space within a guild often results in its being dominated by one or a few species. Through time, the dominant species may change with little accompanying change in the rate of the processes which are controlled by other, often physical or chemical, factors such as light intensity, nutrient supply, or temperature. Thus, ecosystems exhibit higher stability than do species populations and often are little affected by the results of continuously occurring competitive shifts in populations. This stability is not a property of the species populations, but of the ecosystem, although it is conferred by the interaction of species populations, competing to fill specific niche dimensions.

Other emergent stability characters are structural. Natural forest succession involves competitive replacement of tree populations with only minor loss of canopy, as well as continued stable rates of functions. In mature forests in both temperate and tropical regions, blowdowns are quickly replaced, but by individuals of species determined by complex species interactions and involved reproductive strategies. In grasslands, grazing and fires may influence species dominance, usually with no loss of continuity of plant cover or loss of soil, except when driven to extremes by human exploitation or major natural catastrophes. It is interesting that emergent properties of both structure and function at the ecosystem level—first recognized by ecosystem ecologists—are frequently the result of population level processes. This is a theme repeated throughout this volume: the truly holistic study of the ecosystem brings together ecosystem and population processes as a continuum of functional response to changing conditions. Full understanding of ecosystem structure and function cannot be attained simply by grinding up samples, nor by gathering data from satellites, but neither can it be understood by a simple summation of population processes.

Ecology is indeed a holistic science, as Odum (1977) strongly stated. But perhaps we have reached a stage of sophistication where we can step back from old battles with reductionist cell smashers and specimen collectors to ask not for holism or reductionism but for realism. One feature of the biosphere that clearly emerges from ecological studies is the range of scales of space and time over which events occur. Each kind of event may have a characteristic time or half-time, ranging from nanoseconds for intracellular processes (which are highly relevant to ecology and even to biogeochemistry) to millions of years for the evolution of continents and ocean basins and their contained ecosystems. We are really dealing with a continuum of functional response over a range of time scales, and it is often necessary to think about a substantial part of that range in order to properly describe and understand ecological processes.

Dealing with a wide range of time scales continues to challenge ecologists. Improved instrumentation has made it possible to measure events on very short time scales. However, integrating those measurements over ecosystem-scale units of time and space is tricky and has led to problems. One that is currently contentious is the rate of photosynthesis of the ocean. Integrative measurements and spot measurements do not agree, although the methods themselves seem to be valid (Jenkins and Goldman 1985). At the other extreme, and on a more positive note, palynology of long cores from lakes and former lake beds has given us a perspective of population changes and ecosystem evolution over hundreds of thousands of years (e.g., Hooghiemstra 1984). On that time scale, communities do not reach a climax state, but are continually evolving through the processes of invasion, extinction, and competition, while events such as climatic change have demonstrable impacts.

There is contention among ecologists, and between ecologists and their financial sponsors, over the value of long-term studies of ecosystems. Ecologists understandably wish for long-term records of ecosystem properties. Where these can be mined from natural repositories in glaciers and sediments, they have proven valuable. However, augmenting them with systematically recorded data is not so easy as we might think. Quite aside from the cost and dedication necessary to create long data sets, we have to look into a crystal ball to predict what sorts of data will be valuable in future decades or centuries. Some kinds are easily identified; others are not. We can point to both valuable and nearly worthless sets that have been accumulated to date. Recognizing this need presents a challenge to ecologists (Strayer 1986). Conferences such as the one on which this volume was based may help us define tractable long-term programs.

1.5. Scientific Methods In Ecology

In the hierarchical arrangement of biological studies, ecology occupies more than one level. Autecology and the study of individual species populations border on other more traditional branches of organismal biology. Such studies involve large inputs of systematics, behavior, and physiology. The viewpoint is still a reductionist one. Studies of communities take various forms and, in principle, should bridge the gap between studies of species populations and attempts to understand the function of ecosystems. Such studies are less reductionist in their methods than are studies of species populations. Between community studies and ecosystem studies there is an important transition in viewpoint and method. We move from something strictly biological to something almost akin to geology. And in the ultimate extreme, studies of the biosphere, we often adopt the methods and viewpoint of biogeochemistry, a science founded by a geologist (Vernadskii 1926). Therefore, it should not be surprising that there are strong and even polemical discussions of the validity of methods and viewpoints within this range of investigators. One approach will not serve all of ecology, let alone all of science. Our goal here is to examine specifically the situation for the study of ecosystems, but it must be put into a perspective that recognizes biogeochemistry on the one hand and population ecology and population genetics on the other.

1.5.1. Ecological Explanation, Prediction, and Knowledge

Most scientists tend not to be self-conscious about what they do. In spite of talk about *The Scientific Method*, most scientists "know" how to conduct proper studies and how to judge the studies reported by other scientists. Many of them are not even aware that those standards of judgement are unique to each subdiscipline, and are in fact subject to considerable variation, resulting from the nature of the materials and the state of scientific development of the subject. Successful practitioners often are given more latitude in their methods and pronouncements (Price 1986). Dubious though this may seem, it does permit recognized innovators the opportunity to share potentially useful but untested theories with others.

While scientists are working, philosophers of science are observing their antics and developing formal statements about how knowledge is advanced through science. Most of the time we are as unaware of this as is a baboon troop unaware of being observed by naturalists through binoculars. That situation is preferable to the one that exists today in ecology, where we are suffering from a casual reading of a few philosophers of science and are seeing attempts to impose on ecology as a whole a single, highly circumscribed set of rules defining what constitutes "science." In order to understand what has happened, it is necessary to review the status of the philosophy of science and ascertain to what extent it can be applied to the practical, day-to-day activities of ecologists.

We may think of ourselves as pure scientists engaged in the most basic research, but from the viewpoint of the philosophy of science our activities seem pragmatic and practical. Our methods and decisions result more from the workings of human nature, or "common sense," than from formal rules of evidence. Reichenbach (1951) says, "Scientists often have strange beliefs, and make fallacious inferences with good results. The logician is not interested in copying the scientists' mistakes." Would it not be more rigorous to pursue science "correctly," according to procedures established by the philosophers? So it might seem, if we can find the procedures appropriate for our particular scientific endeavors in the literature of the philosophy of science.

1.5.2. Laws and Paradigms

Most philosophers of science have focused their studies specifically on physics and, to a lesser degree, chemistry. While understanding physical laws is a very difficult enterprise, the interactions in physical systems tend to be few, and single processes can be isolated experimentally. Characteristically in modern physics, mathematical models come first, and observations and experiments come later to test the hypotheses emerging from the models. Some dramatic instances of predictions from mathematical models have come from research on subatomic particles. We should remember, however, that those models were not naively created. They were the result of many years of empirical experimentation and the testing of many hypotheses. Hanson (1965) says, "Physicists do not start from hypotheses; they start from data. By the time a law has been fixed into an hypothetico-deductive system, really original physical thinking is over."

Most philosophers of science reserve the term law for the most fundamental laws of physics and chemistry. By this criterion, there are no laws in biology. So our first warning to ecologists delving into the philosophy of science is to understand that philosophers are not necessarily generalizing about all kinds of science, and they have their own arcane terminology, which, like the terminology of science, includes many seemingly everyday words that have taken on new and restricted meaning. Some, however, do have wide application. When Kuhn (1962) developed his concept of paradigms in science, he chose examples from physical sciences, but he specifically stated that examples could be found in the biological sciences. Many of us have used the term paradigm in the Kuhnian sense and find it useful.

In the absence of universal laws comparable to those of physics, biologists must seek other means to validate their work. Popper (1957) recognizes two approaches to knowledge, theoretical and historical: ". . . while the theoretical sciences are mainly interested in finding and testing universal laws, the historical sciences take all kinds of laws for granted and are merely interested in finding and testing singular statements." For example, continental drift is not a universal law, but a singular statement about an ongoing historical event (Kitts 1977). In these terms, ecology is historical science. It is true that ecologists accept the laws of physics and chemistry, as well as the body of knowledge of biochemistry and biology, as given. Otherwise we would be physicists, chemists, or whatever. The part of Popper's remark dealing with singular statements may seem like an unfair criticism, and to see this in a better perspective, we need to consider how Popper and others view the quest for knowledge. This is where ecologists and some other scientists have sometimes gone astray in attempts to apply Popper's philosophy to empirical science.

1.5.3. Theories

It should be noted at the outset that there are significant areas of disagreement among leading philosophers. Indeed, there are two major schools with almost diametrically opposed views. The logical positivists have developed a formal system of stating hypotheses, the so-called Ramsey sentence, which they believe "proves" them, if all of the prescribed elements are present in the statement. In practice, this is said to work only for a limited number of physical theories, and it has found little application among biologists (but see Hulburt 1983, 1985). The essential elements include what are called descriptive and theoretical elements. The former includes strictly those elements discerned by the human sensory apparatus, but a list of examples includes "blue" and "wood" (Suppe 1977). The latter are those outside sensory experience, and the list of examples includes both "electron" and "virus" (ibid.). This contrast, which may seem nonsensical to a scientist, highlights a schism among philosophers of science regarding the validity of indirect evidence, not based on direct sensory experience, that is ultimately rooted in the conflicting rationalist and empiricist foundations of modern science. While one school has attached special importance to direct sensory experience, another school, associated with the later Wittgenstein (1953), recognizes that sensory experience is not all that direct and not really that different in content or validity from data of chart recorders and oscilloscopes. Scientists may be impatient with this line of debate, but it should not be dismissed too lightly. After all, much time in empirical science is spent examining and questioning the validity of observations and experimental methods. Carnap (1966) says, "The line separating observable from nonobservable is highly arbitrary." Hanson (1969) cites the zipper analogy of Braithwaite (1953): the directly observable and the indirect observations must be closely interdigitated and firmly attached at the end. These problems do not exist for the realist philosophers.

The realist school, known to its detractors as the irrationalists, was founded by Karl Popper. They do not see science, even at its best, as a means for finding truth in any absolute sense. Indeed, the search for absolute truth is like the search for the Holy Grail. It is a pleasant, genteel occupation, not likely to be terminated by success. Therefore, in the serious pursuit of scientific knowledge we are left with varying degrees of practical assurance short of "truth," and there are several roads we may take in seeking assurance. Of these, the most rigorous is said to be through the testing of theories. Theories can be tested and falsified, but not proven (Popper 1957). Thus, the technique is to keep erecting and falsifying theories through experiment, thereby narrowing the range of acceptable possibilities. Platt (1964) proposed that this system, which he called "strong inference," be made the universal criterion for progress in science:

> . . . we have long needed some absolute standard of possible scientific effectiveness by which to measure how well we are succeeding in various areas—a standard that many could agree on and one that would be undistorted by the scientific pressures and fashions of the times and the vested interests and busywork that they develop . . . I believe that strong inference provides this kind of standard . . .

The strong inference approach is a reductionist approach that requires experimental falsification of hypotheses. Here is where we have gotten into trouble in ecology.

Population ecologists have embraced the teachings of Popper, and for them this sometimes works reasonably well. The investigator considering the individuals within a species, or several species populations, is presented with a finite number of plausible interactions to test. Experimentation is often possible, and many useful natural experiments occur, such as those of island biogeography. Population ecologists can sometimes erect opposing theories and falsify one or more of them. Where it is workable, this line of evidence is powerful and appealing. We should, however, take note of the comment of Hanson (1969, p. 255) regarding experiments.

> Crucial experiments are out of the same logical bag as pure observations and hard, uninterpreted facts. They are philosophers' myths that many scientists spend a good deal of time telling and retelling, not because they properly characterize the real situation in experimental science, but because they trip lightly off the tongue and make laboratory work seem a frightfully objective business. But talking in this manner is rather like an Englishman eating spaghetti. Slicing and cutting things up in this way may make for easy oral manipulation, but it is only a travesty of the genuine scientific situation in which theories, hypotheses, experiments, and facts are interlocked and intertwined in a fearfully complicated way. It is the task of philosophy of science, not to scissor all these elements into tidy and discrete packets, but to try to give an account of how, why, and where these aspects of science conflate and mingle. (Reprinted from *Perception and Discovery*, by N. R. Hanson, with permission of the publisher, Freeman, Cooper & Co.)

This account of scientific reality should be recognized readily by ecologists. We should remember that ecology is not physics. In ecology, theories rarely precede experiment, if the truth be told. Strong inference is not as strong as in physics. The isolation of single factors for testing is usually difficult (Quinn and Dunham 1983). We are, after all, dealing with populations of organisms in a changing physical environment. We have to deal with all of the levels of variance of the beasts, bugs, or plants laid over the variances of chemistry, physics, geology, and meteorology, embracing a wide range of scales of space and time. These are problems characteristic of the higher levels of hierarchical systems (Allen and Starr 1982). Moreover, experiments in ecology commonly involve constrained, somewhat artificial systems, and the results may require observational validation in the field (Salt 1983).

Trouble and strife enter ecology when influential investigators and grant administrators try to tell us that *only* through falsification experiments is *science* accomplished, and *only* when a theory is explicitly stated is an investigator engaged in valid, useful ecological research. Rules that barely work for physicists are being applied naively to ecological studies. There is a significant dichotomy in the hierarchy of ecological studies that we must recognize here. One often perceives, as indeed we did during our conference, a disparity between how people say they should conduct research and how they actually do it. This is not unique to ecology, and in part it is a failure to recognize the distinction between long-term goals and short-term operational needs. Ecosystem studies, because of their inherent complexity, have been largely inductive (Salt 1983). Conclusions are sought from accumulated data. This is a philosophically weak approach, one Popper dismisses out of hand. Yet, the fact remains that studies on imperfectly described systems will be flawed and useless. Complex systems require complex description, and, for better or worse, ecosystem research is still largely in a descriptive stage. The transition from description to experiment was made in organismal and cellular biology 100 years ago (Cohen 1985). Population and community studies, which are, like cellular biology, reductionist in approach, involve less complex sets and can be approached by deduction. The dichotomy is well-illustrated by the models produced. Ecosystem models are complex, descriptive, and primarily heuristic. Population models are simple and sometimes predictive. Obviously, the latter represent a higher level of assurance and utility. But, with a few exceptions, simplified, excessively condensed ecosystem models have proven to be neither heuristic nor predictive. While Platt's (1964) advice to study the simplest system you believe has the properties that interest you is very well-taken, oversimplification of complex systems through condensation produces garbage. Given this situation, what should ecosystem investigators do? Philosophers of science have not yet considered a scientific method for ecosystem studies to be worth their attention, so we do not have the benefit of their opinion on how we should proceed. If we must seek a model from the literature, physics is clearly at the wrong end of the spectrum of sciences. Strong inference may be a desirable goal, one we can achieve only after other preliminary approaches. Other, more routinely applicable models exist. Indeed, we have a choice.

1.5.4. Normic Explanations

Much of the work to date on ecosystems has been descriptive. While this is often dismissed as "descriptive biology," or simply as "not useful," it is a necessary beginning to the study of complex systems. Descriptive science is not simply concerned with reporting observations. An ecologist's description of an ecosystem is not the same thing as a description of a stroll through the forest. Rather, it is an organized, and more or less formalized, conceptual model of structure and function of a complex system. That kind of "description" must exist before more rigorous and incisive investigations can proceed. This volume is replete with citations of attempts at explanations that preceded sufficient knowledge of the structure of the system and their failure. It also includes appeals for more data (e.g., Chapter 13), as does other recent literature (e.g., Smetacek and Pollehna 1986).

Because of the very high level of complexity of ecosystems and the very short history of true ecosystem studies, we are still far from the point at which we should reasonably expect to establish universal laws, if indeed they can be established in ecology. Rather, we have what Scriven (1959 a) calls normic statements. Kitts (1977) shows that normic statements are the common currency of geology, which is in this respect a better model for ecosystem studies than is physics. These are statements of what usually or normally happens. They are not statistical statements and do not deal with probability. Rather, they attempt to state what will happen in a particular case, given a series of conditional statements about the circumstances. Much of this volume is constructed around such statements. It is difficult to dismiss them as "not useful." Kitts (1977) says, "Essentially generalized universal statements are not as common in science as some accounts would lead us to believe. They are found in the most rigorously formulated physical theories and seldom anywhere else." Normic statements are a practical solution intuitively adopted by working scientists. Unlike a statistical statement which says nothing about the individual case, a normic statement states what will always happen in a defined set of circumstances. In ecosystem studies, as in geology, the conditions may be so complex as to create a degree of uncertainty in the outcome. They are therefore less predictive than physical laws, where the conditions are few and are rigorously defined. This is, in effect, the price paid for dealing with higher levels of integration, where the potential interactions are many, complex, and variable.

In all fields of science, prediction and explanation are closely related. Ideally the same general statements (universal or normic), combined with appropriate singular statements, ought to provide both (Hempel 1942)—that is, whether one is predicting or explaining should depend only on whether the event explained or predicted has occurred. In practice, however, especially in those cases depending on normic statements, scientists are often able to provide explanations that will not serve equally well as predictions. The reason for this asymmetry is that normic statements frequently provide conditions that are necessary, but not sufficient, for an event to occur. The standard example concerns the relationship of syphilis to paresis (Scriven 1959 b; Kitts 1977). Although it is necessary that someone have syphilis before contracting paresis, syphilis alone is not a sufficient cause of paresis, and the other causes are imperfectly understood. As a result, even though a physician can appropriately explain to a patient who has the disease that his paresis was caused by syphilis, he cannot predict in advance that a particular patient with syphilis will contract the disease, since paresis is extremely rare. The situation is much the same with lung cancer and smoking, since many people smoke throughout their whole lives without any problems. Faced with intricate webs of conditions that are necessary but not sufficient, in ecosystem studies, as in medical studies, ecologists often need to turn to probabilistic observations of particular events to bolster their predictive capability.

1.5.5. Probabilistic Explanations

Ecologists increasingly depend upon statements to which we assign a formal predictive value on the basis of the analysis of variance. This is now so institutionalized that several scientific journals employ statisticians to scrutinize routinely all manuscripts accepted for publication. Ecologists accept the proposition that statements have greater value when we know the size of the envelope of uncertainty around them. A majority of philosophers of science ignore or gloss over the significance and application of probability theory, probably because it involves inductive reasoning. Here again, we have to distinguish between the requirements of a search for absolute truth and the search for knowledge at the more mundane, but achievable, level of science. While induction is not the road to universal laws, it is the technique by which we get out of bed in the morning and put on our shoes. It also serves important functions in empirical science, and Reichenbach (1949) has shown how the analysis of variance can be used to strengthen induction. It may be well to recall that the rejection of induction by philosophers originated in large part with Hume's skeptical problem about the future (Hacking 1984). Hume's point was that we have no basis other than our sensory experience for statements of causality. Although this is not the central issue it was in Hume's day, philosophers of science are still troubled by the inherent circularity of inductive reasoning. As a practical measure in science, we have to accept these limitations and proceed (Gould 1965).

Underlying all explanations and predictions is the principle of uniformity: Similar conditions always and everywhere yield the same results in those cases of true cause and effect, but not in cases of empty correlation. This assumption underlies everyday behavior and all of science (Gould 1965). However, any test of this principle or of particular cause and effect events must involve induction, the inference about the whole population from the sample. Such a test is not a logical proof. Indeed, the realist philosophers do not accept the principle of uniformity (Toulmin 1953). Science is thus relegated to the level of everyday experience, albeit more formally carried out. The search for perfect proofs and ultimate truths is limited to logic and mathematics. This does not diminish the real value of scientific research, but it defines the limits on procedures appropriate to research. We are not constrained to eschew all inductive reasoning, which would include the analysis of variance. We accept the fact that there is an envelope of uncertainty around our conclusions. We are not limited to falsification of hypotheses, a procedure poorly suited to many scientific inquiries, but perhaps a goal to be achieved. There is no single high road to absolute truth—indeed for questions about the real world there is no road at all. There are, however, several low-lying paths, all without exception through swamps and thickets of uncertainty, leading to results which have proven useful in practice but always are tainted with the inductive character of the principle of uniformity. Science, even at its best, is not on the highest level of assurance, which is reserved for mathematics and logic, endeavors that are independent of the physical universe.

For the practical, working ecologist, a rigorous probabilistic statement will do very nicely. It is a higher level of description that brings us one step closer to elusive generalizations about ecosystem structure and function and is an essential precursor to testable hypotheses. Probabilistic statements are the bread and butter of current ecosystem studies. They tell us, under a defined set of circumstances, what we can expect to happen most of the time, at least in that small corner of the universe we have observed. The search for methods that tell us what happens *all of the time* is a wholly different business, one we have likened to the search for the Holy Grail. However, we should not assume that probabilistic statements can always be reduced in the future to deterministic statements or laws. Some aspects of nature are not deterministic (Braithwaite 1953).

We should not confuse the existence of a variance around observations of natural events with the randomness of natural events. Some ecologists suggest that all nature is stochastic, a reverse twist on Einstein's famous quip, which would now say, "God always plays dice." When we measure events in the real world, there is often a heavy background of random noise. One thing ecosystem ecologists have learned that seems to have eluded some of their colleagues, is that the noise is often a signal waiting to be read. That signal may, in fact, be a function of events in the physical environment. Ecology is, as originally defined, inclusive of physical and chemical interactions. For example, rare and seemingly random observations of anomalously high productivity in the waters of the southeastern U.S. continental shelf are now known to be associated with specific, intermittent upwelling events which do not reach the sea surface and may have an obscure surface signal, even with the use of satellite information (Atkinson, Menzel, and Bush 1985). Once the physical regime was described and the upwellings were located in space and time, the variance around biological events was reduced, and predictive ability increased. Thus, what may appear to be God playing dice is usually an ecologist trying to describe an inadequately defined system.

Nature can, however, be stochastic. Examples of randomness in the succession of tropical forests and coral reefs appear valid, as may some aspects of island biogeography (Connor and Simberloff 1977). While the arrival of new species on an island may be influenced by the ability of potential colonists to cross the intervening space successfully, as well as the numbers available and the distance to be traversed, stochastic or quasistochastic events, such as weather, influence the success of the process. Similarly, the timing of the opening of blowdown spaces in tropical forests, relative to the long, irregular reproductive cycles of tropical trees, may make some aspects of succession in tropical forests a lottery (Bazzaz and Pickett 1980). The same might be said of tropical reefs.

The observation that nature can be stochastic—from the fall of a tree to quantum mechanics—places limits on determinism. Anyone who has played a pinball game knows that the existence of unvarying physical actions and reactions does not necessarily produce a result that is deterministic insofar as the player is concerned. The effect of one small variable spread among many dichotomies in a long causal chain is the basis for the game.

1.5.6. Revolutions in Science

The falsification of theories is the most rigorous way to improve our assurance, but it is the reverse of the normal psychology of scientists. In the real and sometimes naive world of practicing scientists, people tend to seek what is "true," rather than what is false. Scientists typically seize on an appealing theory and never let go. Kuhn (1962) suggested that science progresses only through the death of scientists, whose outmoded theories and paradigms die with them. His description of the constant testing of paradigms through the practice of what he called "normal science," followed by increasing instances of falsification and resultant dissatisfaction , followed in turn by the emergence of a new paradigm, has been appealing to ecologists. The Kuhnian terminology of paradigms and revolutions is frequently encountered in the ecological literature. In fact, Kuhn's description of major changes in theoretical physics seems to ring true for biology in general and ecology in particular. The processes that work on a large scale also work in microcosm. Small concepts also grow, are tested through the processes of normal science, and are overthrown by teapot revolutions. All the while, the world looks the other way, and only a few interested specialists note and record the relatively minor events in the progress of science.

Many philosophers of science reject Kuhn's (1962) paradigm of scientific progress. They complain of its generality or vagueness and of the fact that it portrays science as a discontinuous process that does not build, so they allege, on former work. What few of them assert, but seems to be on their minds, is Kuhn's description of science as a social process conducted by people. Scientists are, after all, people, and they share the failings common to the species. If there is any perfection in science, it is in the product, not the practitioner. Science is successful precisely because no one individual or group can lay down indisputable dogma for all time. The "truth" is always subject to revision and, while building on the past, we do throw away what no longer works. Chemists have, for example, thrown away alchemy and phlogiston theory. Ecologists have thrown away the superorganism paradigm. The sociological realities described by Kuhn seem not to have been assimilated into the formal structure of the falsification of hypotheses or vice versa. In one sense, however, there is agreement. Popper (1972) says, "Should anyone think of the scientific method, or of *The Scientific Method*, as a way of justifying scientific results, he will be disappointed. A scientific result cannot be justified. It can only be criticized, and tested." This is a good description of what happens during the cycle of a Kuhnian revolution in science. However, what is going on in the literature of science and in the minds of scientists during that cycle may be strikingly different. Again, Popper (Ibid.) says, "Whenever a theory appears to you as the only possible one, take this as a sign that you have neither understood the theory nor the problem which it was intended to solve." How many ecologists can pass that test of rigor in their thinking?

1.5.7. Methods for Ecosystem Study

A candid analysis of the methods of ecosystem study should not lead to the conclusion that the situation is hopeless. In all humility, we must continue with descriptive studies where these are needed to define a system. While further descriptive work must be justified in terms of what we need to know for higher scientific purposes, it is unrealistic to believe that descriptive work can always be directly and immediately linked to the needs of testing some specific hypothesis. Long-term studies of all major ecosystem types are still a necessity. Hopefully, these will involve interactions among many investigators and will encompass many approaches, including the experimental. Indeed, there are now many opportunities for experiments and hypothesis testing in ecosystem research, as the chapters that follow will show. We must not, however, succumb to physics envy and try to force experiments for which we lack the basic descriptive foundation.

Some experimental approaches to ecosystems are now commonplace. Often these involve the isolation in some manner of a small, manageable segment of the system. This can vary from a vial of water or a core of soil to a hectare or more of landscape. While isolation involves simplification, this is completely within the spirit of the strong inference mode of research. Falsify first with the simplest system that will work, then test further with other cases. For ultimate validation, if there is such, we should take note of how astronomers have to work. Natural events perform experiments for us, and if their coming can be predicted and put to use, this is a sign that progress is being made. We see real and potential examples of this in Chapter 9. Modeling also serves an important function as an adjunct means for experiment and prediction. In a complex system, modeling can save years of manual labor by providing not only theories but some initial falsifications. In this respect, ecosystem models are, and should be, unlike typical population models. Indeed, good ecosystem models may be more likely to generate theory than many population models. Moreover, as Shugart and Urban demonstrate (Chapter 14), modeling provides a short-term approach to long time scales, improving our understanding of events that occur over more than one human lifespan.

Much ecological modeling and some hands-on field experimentation has been flawed by insufficient descriptive knowledge of the system. You must know all of the parts before you can understand or manipulate the mechanism, be it a pocket watch or an ecosystem, and dumping the parts on the floor is not a good beginning. A case in point is the elucidation of the roles of microorganisms in ecosystem structure and function. In the original Lindeman trophic hierarchy, microorganisms were simply decomposers. They were not assigned a trophic level. Consequently, they were not included in early models of ecosystems, conceptual or mathematical. The recent work has shown that microorganisms are the major movers of energy in most ecosystems, both terrestrial soils and natural waters and sediments (Azam et al. 1983; Hobbie and Melillo 1984; Elliott et al. 1984; Pace et al. 1984; Pomeroy and Wiebe 1987). Bacteria are frequently successful competitors with metazoans for mutually available energy sources. Because microorganisms are small and ubiquitous, their natural populations are more easily manipulated than are those of metazoans. As a result, we have seen a flowering of experimental studies with natural microbial populations. These are not a revisitation of the classical population studies of Lotka, Volterra, and Gause, although they are built on those foundations. Instead, they are ecosystem-oriented measurements of the flux of energy and essential elements. Questions about the linkage of microbial processes to metazoan food chains are being debated. So, through the incorporation of microbial processes into our concept of ecosystem function, we are seeing a shift toward experimental approaches to the study of ecosystems.

Investigators of ecosystems have reason to approach their subject with humility. It is not quite as unreachable as the stars or the atoms, but it is orders of magnitude more complex than cellular biology. Ecosystem studies are in a transition between description and experiment, where organismal and cellular biology were 100 years ago (Cohen 1985). Yet, even the experimental, "hard" sciences contain a strong undercurrent of paradigm–myth, bordering on acts of faith. When our cellular–molecular colleagues suggest that we ecologists really do not achieve "scientific" results (as, for example, Davis 1987), ask them if they have read Gilbert and Mulkay (1984).

Although much of science in general is technique limited, there are occasions in every field when important advances are made with very simple instrumentation. At the same time, it is important to any field to have at its disposal appropriate equipment. We are fortunate in ecology to have at our disposal satellites, automated terrestrial systems, and marine buoys for gathering data, a wealth of physical and chemical instrumentation, and computing systems to process the data and model ecosystem function. As we shall see in the chapters that follow, this technology is already making biosphere level studies a reality in what is now a quantitative, analytical science.

2. The Origin of Ecosystems by Means of Subjective Selection

Peter H. Rich

"Ecosystem" was defined in 1935 by Sir Arthur Tansley as:

> the whole system (in the sense of physics) including not only the organism-complex, but also the whole complex of physical factors forming what we call the environment of the biome—the habitat factors in the widest sense. Though the organisms may claim our primary interest, when we are trying to think fundamentally we cannot separate them from their special environment, with which they form one physical system. It is the systems so formed which, from the point of view of the ecologist, are the basic units of nature on the face of the Earth (Tansley 1935).

Tansley's concept of ecosystem was an alternative to the Clementsian doctrine of "monoclimax": the insistence that in time the species composition of every hectare of a regional climax succeeds to homogeneity.

Raymond Lindeman, in 1942, published the actual research upon which the science of ecosystem ecology was founded (McIntosh 1985). Based on biomass data from three lakes, his paper described the flow of material and energy from the environment through the predatory hierarchy of plants, herbivores, carnivores, and decomposers (Lindeman 1942). Lindeman's ecosystem receives energy from solar radiation captured by plants. The captured energy is used immediately by plants to transform inorganic constituents and energy into organic matter. Energy and material are returned to the environment by the process of respiration. Energy is made useless by respiration, but the inorganic constituents of organic matter become available again for incorporation by plants. The transmission of material and energy through a particular level of the predator hierarchy in a unit of time is called productivity, and is the basic manifestation of "process" of ecosystems. Lindeman outlined the trophic dynamics of ecosystems clearly; all the parts and their interconnections were elucidated. He seriously underestimated non-predatory losses, but that was discovered and corrected by Odum and de la Cruz (1963). By his own account, Lindeman believed his work on lake ecosystems explained the existence of Eltonian pyramids (Elton 1927) and the progress of hydrarch succession (Cooper 1926). Lindeman died in his 27th year, before his momentous sixth paper (Lindeman 1942) reached print.

Missing at the founding of ecosystem ecology was an explicit statement of a fundamental and compelling question: *Do exchanges of matter and energy described by trophic dynamics evolve because excess fertility is an integral property of living organisms? Or does excess fertility evolve because exchanges of material and energy are an integral property of living organisms?*

In more technical language: *Do organisms create order out of chaos (locally reverse entropy) by evolution which causes energy and material to be pulled through trophic levels by predation? Or does order emerge from (is entropy locally reversed by) material and energy exchanges which give rise to order, organisms, evolution, and predation?*

2.1. So What?

For a scientist, neither alternative can be proved. The basic question concerns the origin and propagation of causation in ecosystems. Why, then, should a scientist worry about a question for which there is no scientific answer? At the very least, the question is someone else's problem; for instance, a philosopher's. Scientists *are* philosophers, to the extent that most hold the degree "Doctor of Philosophy". But many scientists, ecologists included, fail to understand that every science has metaphysical foundations (Burtt 1954). Despite sturdy denials by advocates of the Scientific Method, science makes such metaphysical assumptions as the aforementioned existence of cause and effect in ecosystems. My topic is the different assumptions underlying ecosystem ecology and their effects upon ecosystem theory. My thesis is that the scientifically unanswerable question of who does what to whom to create an ecosystem has been answered, not once, but twice! The two answers have historical origins in two competing philosophies of nature; one accepted and associated with 18th and 19th century science, the other controversial and identified with intellectual revolutions in 20th century science.

2.2. Deism And The Malthusian Axiom

The first and traditional answer to the question of who does what to whom to make an ecosystem is that excess fertility is an integral property of organisms, as is predation an integral property of predators (Rich 1984a). Living organisms cause the exchanges between the biotic and abiotic worlds observed by Linde-man. That answer is found in 19th century deism. Deism is the belief in God the Creator solely on the evidence of design in nature, rejecting supernatural revelation. In the static, 18th century version of deism, God created a law-bound system of matter in motion and gave it a certain quantity of momentum which He sustains from moment to moment, leaving the system to produce the material universe in the course of time. In the evolutionary 19th century version described by Lamarck and Darwin, God was far more remote from the universe, and the proof of the Creator's existence was not the static design of nature but rather the evidence of progressive adaptation and improvement in the organic world. Deism reinforced science by making it the study of God's works, whether those works were conceived statically or in terms of evolutionary processes (Greene 1981).

In Darwin's theory the origin of adaptation and improvement in organisms and, therefore, nature, is the Malthusian axiom: organisms produce more young than the environment can support, precipitating a struggle for existence. Natural selection of heritable variations among the offspring then produces new species, genera, etc. (Lewis 1980). The source of assumptions underlying evolutionary theory is apparent in two sentences in the next to last paragraph of *The Origin of Species* (Darwin 1859):

> To my mind it accords better with what we know of the laws impressed on matter by the Creator, that the production and extinction of the past and present inhabitants of the world should have been due to secondary causes, like those determining the birth and death of the individual . . . And as natural selection works solely by and for the good of each being, all corporeal and mental endowments will tend to progress towards perfection.

In the next (last) paragraph of the first edition of the *Origin*, Darwin rhapsodized about ". . . life, with its several powers, having been originally breathed into a few forms or one." In the second and all subsequent editions he added to that expression the phrase "by the Creator."

All of which suggests that life originated and gradually improved by processes ordained by an intelligent Creator. However, elsewhere in his writing Darwin wondered if the patterns apparent in nature constituted a design worthy of a Designer. For instance, how could God authorize the cruel predicament proposed in the Malthusian axiom? Those expressions of skepticism have led some Darwinian scholars to interpret the quotations above as attempts by Darwin to assuage his essential agnosticism and disarm hostile reaction. Contrary to that interpretation, Darwin's published references to God as the author of nature are consistent with those in his unpublished notebooks and with what is known of his character (Greene, pers. comm.). In the final analysis, regardless of his own doubts, Darwin's published appeals to deism were part of his impact and authority in society and science.

Evolution exists, then, because the design of nature makes excess fertility an integral part of life. A mechanical analogy of fertility in organisms is the alternator, an integral part of an automobile. So long as the engine is running, the alternator generates excess electricity for storage in the battery. The arrangement makes energy available for restarting the engine. Like fertility, the alternator is designed for excess output. Insufficient output is useless. Actual starting effort realized depends upon designed output, actual operation, and environmental factors, where starting effort is analogous to reproduction and designed output to fertility.

2.3. Herbert Spencer

The second and alternative answer to the question of who does what to whom to make an ecosystem is found in the philosophy of Herbert Spencer, a contemporary of Darwin but "never a Darwinist" (Kaye 1986). Initially also an evolutionary deist, Spencer's interests drew him to the physical sciences, especially thermodynamics. At one point he stated his ultimate goal as "the interpretation of all concrete phenomena in terms of the redistribution of matter and motion" (letter to John Fiske). By the time his *First Principles* was published in 1861, Spencer had abandoned deism in favor of the idea that the processes of nature were manifestations of an "unknowable" force which produced the heterogeneous from the homogeneous. In Spencer's cosmology, nature assembled itself in a continuous progression from the simple to the complex. Properties inherent in the most basic configurations of matter and energy gave rise to processes which produced new, more complex configurations, which caused the processes leading to the next stage of evolution, and so on. Foreseeing effects discovered later in quantum theory, particle physics, and disequilibrium thermodynamics, Spencer outlined the evolution of stars from hydrogen, chemical elements from stars, life from chemistry, and evolution from life. Considered by many the founder of sociology and a cofounder with Freud of psychology, Spencer also wrote on the evolution of personality and society. For Spencer, evolution itself, evolved.

Anticipating Ilya Prigogine's "dissipative structures" (Blackburn 1973; Nicholis and Prigogine 1977; Odum and Pinkerton 1955; Prigogine 1978; Prigogine and Stengers 1984), Spencer argued that evolution develops mature levels of organization which are so profoundly integrated that they are exempt or immune from the destructive interactions of nature emphasized by Darwin. Thus, nature evolves advanced, inherently positive and cooperative properties in addition to its primitive, negative, and chaotic ones. Spencer based his argument on the dependence of plants upon the nutrients and carbon dioxide provided by animals and the dependence of animals on the organic matter and oxygen produced by plants. He theorized that Darwinian evolution best describes the early stages of an undifferentiated system progressing toward specialization and integration. At first there is so little interdependence that any part of the system can suffer extinction without affecting the overall system. But evolution leads to more specialization of function and division of labor. Finally, truly functional differentiation appears which is exempt from extinction because the system, itself, would become extinct if one of its necessary functions ceased. So long as the system persists, the specialized species remain competitively superior.

Ironically, it was idealistic and uncompromising adherence by Clements to an extension of that point; i.e., insistence that evolution and succession should lead to perfectly homogeneous regional climaxes of uniform species composition, which triggered Tansley's criticism of Clementsian dogma. Comparing his ecosystem with Clements' succession, Tansley states:

> The gradual attainment of more complete dynamic equilibrium ... is the fundamental characteristic of (ecosystem) development. It is a particular case of the universal process of the evolution of systems in dynamic equilibrium. The equilibrium attained is however never quite perfect ... (Tansley 1935).

Clements, as a graduate student at Nebraska (Worster 1977), read Spencer's book, *Principles of Biology*, at a time that Spencer was far more popular than Darwin in America for "... his ability to make what appeared to be a scientifically based agnosticism sound like theism ..." (Kaye 1986). Tansley was employed by the aging Spencer in the revision of the final (1899) edition of *Principles of Biology*. Thus, the ecosystem concept appears to have been born of an argument within Spencerian theory, not between Spencerian and Darwinian theory. However, elements of Lamarckian inheritance, underlying much evolutionary and ecological thought of that period, probably were involved as well (McIntosh 1985). Spencer had no formal scientific education, and he blundered badly attempting to defend Darwin's theory of pangenesis, seriously compromising his reputation among later biologists. Today, Spencer's influence is more apparent in ecosystem theory than in philosophy. On the other hand, Spencer's philosophy anticipated the existence of natural phenomena which are among the most significant discoveries of 20th century science.

2.4. The Ecosystem Axiom And Redox Potential

The ecosystem axiom (Rich 1984b) depends on Spencerian cosmology the way the Malthusian axiom depends on deist cosmology. The ecosystem axiom states that the energy captured by photosynthesis is not stored *in* the products of photosynthesis but, rather, as environmental redox potential *between* the products of photosynthesis. The separation of photosynthetic products described by the ecosystem axiom also creates a predicament for living organisms, but not the predicament Malthus observed.

Redox potential simply refers to the affinity of an atom for the electrons in its outer orbits. Different kinds of atoms have different affinities. When two different kinds of atoms or molecules are placed together, electrons on the atoms with less affinity for electrons will be attracted to the atoms with greater affinity. Redox potential is the potential energy of that attraction. In practice, redox potential is measured with a hydrogen electrode, which itself contains an oxidation-reduction reaction:

$$2H^+ + 2e^- \leftrightarrow H_2$$

Redox potential, called *Eh* when measured with a hydrogen electrode, is positive when the environment surrounding the electrode has a net affinity for electrons and is pulling them out of the electrode, causing the reaction in the electrode (as shown above) to go to the left. A negative redox potential indicates that the environment has enough electrons to push them into the electrode, causing the reaction to go to the right.

In photosynthesis, electrons are removed from water, which has a strong affinity for electrons, and added to carbon dioxide, which has a weak affinity for electrons. The oxidized product, from which electrons have been removed, is oxygen, which is gaseous and enters the atmosphere. The reduced product, which has gained electrons, is organic matter which is particulate or soon becomes particulate, and enters the lithosphere. That separation creates a thermodynamic predicament for heterotrophs whose existence depends upon undoing by respiration what photosynthesis produced. The return flow of electrons (from organic matter to oxygen) must overcome physical segregation (between lithosphere and atmosphere) of the two products by dissimilar phase (solid and gas) and by their chemical isolation behind high activation energies characteristic of organic matter. Thus, photosynthesis creates an electron disequilibrium; i.e., a redox potential, between organisms made of electron-rich organic matter and the environment (atmosphere) from which the electrons were removed. However, the redox disequilibrium created by photosynthesis also is the environment in which life evolved "its several powers" (Rich 1984a). In thermodynamic terms, the co-evolution of environment and metabolism increased the intrinsic power of living processes by increasing the redox potential between the two ends of phosphorylation: potential $\times$ capacity = power.

Life evolved on the Earth during the Archean Eon (ca. 3.6–2.6 billion yrs. BP). The first protobionts probably were assembled from scraps of organic matter produced by UV radiation or lightning discharges. The biochemistry of photosynthesis and respiration, microfossils from preCambrian rock, volcanic geochemistry, paleometeorology, and data from other planets in the solar system returned by NASA probes indicate that little redox potential existed between organic matter and environment when life originated. That potential coevolved with prokaryotic bioenergetics during the Proterozoic Eon (ca. 2.6–0.6 billion yr. BP), which followed the Archean. For two billion years, as photolithotrophy evolved into photosynthesis, electrons were transferred from the environment to organic matter, and the organic matter transferred to the lithosphere by sedimentation (Broda 1975; Folsome 1979; Lovelock 1979; Schopf 1983). As a result, the concentration of oxygen in the atmosphere rose from an unknown but low concentration to about 20%, aerobic phosphorylation evolved, and the redox potential between the reactants and products of metabolism approached 1.5 volts, probably 10 times higher than at the origin of life.

The exact mechanisms which produced the modern redox environment and the actual sequence of events in the Proterozoic Eon are controversial. But, without knowing precisely *how* redox potential evolved, an explanation of *why* it evolved is found in modern disequilibrium thermodynamics. Disequilibria, such as that proposed in the ecosystem axiom, evolve "dissipative structures," so-called because they exist at the expense of dissipating energy; i.e., increasing entropy (Blackburn 1973; Gladyshev 1982; Gochlerner 1978; Prigogine 1978; Wicken 1979). Total entropy, like power, is the product of energy quality (intensity) times energy quantity (capacity), and is equivalent to power dissipated in biological (isothermal) processes. Although dissipative structures exist by increasing entropy, their structure represents a local entropy minimum. The reduction of entropy appearing in one part of the disequilibrium system depends upon an increase in entropy elsewhere. Life is a dissipative structure which dissipates solar radiation and lowers the temperature (increases the entropy) of energy reradiated by the Earth. Environmental redox potential is the intensity factor of the power and/or entropy transformed in the biosphere's dissipative structure, and will tend to increase until the product of intensity times capacity of energy dissipated (entropy) matches the product intensity times capacity of energy available (power) in solar radiation (Rich 1984a; 1984b). Disequilibrium thermodynamics implies that order in nature results from transformations of energy initiated by an innate propensity of energy to become entropy; i.e., order is a means to create entropy. Thus, *objective* recognition of Spencer's "unknowable" and disputed principle of evolving evolution involves realignment of *subjective* implications and assumptions underlying thermodynamic theory.

2.5. Effects Of Assumptions On Ecosystem Theory

The Malthusian axiom is both subjectively profound and profoundly subjective. It is the point of view of the too numerous progeny of an embattled species preoccupied with their own struggle for existence and salvation. Inevitably, the subjective axiom produced an evolutionary theory which cast nature in subjective terms: individual survival, struggle, selection, fitness, etc. The science of ecology founded on evolution (McIntosh 1985) was predisposed to discover the fundamental niche and the competitive exclusion principle. That world view also compromised Raymond Lindeman's (1942) elaboration of Tansley's ecosystem concept into a science (Rich 1984a; 1984b). Lindeman's hypothesis that biological productivity is what predators do (i.e., predation) is predicated upon the Malthusian view that the source of order in nature is higher organisms (i.e., predators). Underlying the conventional world view is the finite universe of the Creator; i.e., equilibrium thermodynamics. Essentially, environment is an arena in which abiotic environment (physical–chemical resources) was created once, and has thereafter been in the possession of the biotic environment (competitors and predators) or chaos. Hence, order in nature comes from what predators do to secure material and energy (predation), not what material and energy do to produce predators (evolve dissipative structures). Thus, Lindeman was permitted to define the ecosystem in terms of demand, from the top predator backwards to the primary producers (Rich 1984a). Supply of resources may be an *object* of study in conventional ecosystem ecology, but not the *subject*, per se. Ecology is a biological science because its subject is organisms, and the ecosystem is a branch of ecology because it takes organisms' point of view.

The enduring significance of Tansley's ecosystem concept is its challenge to the subjective predisposition of ecology, not its resolution of Clementsian dogmatism. Ecosystem ecology is no more (or less) a biological science than it is a chemical or physical science, because the subject is the "whole system," a unified, general, and compelling scientific enterprise. Coevolved between metabolism and environment, redox potential resolves the predicament implied in the Ecosystem axiom and endows living processes with the "several powers" attributed to the Creator in *The Origin of Species*, including excess fertility and predation. The role of redox potential in resolving the Ecosystem predicament complements the role of fitness in the resolution of the Malthusian predicament.

Power is the product of a quantity (capacity factor) of energy multiplied by a quality (intensity factor) of energy. Excess power had to become available through the evolution of environmental redox potential (intensity factor) before excess fertility evolved in organisms (i.e., before evolved fertility could become excess). Returning to the automobile alternator analogy, in an atmosphere of insufficient oxygen the engine and alternator of an automobile will not produce excess power, regardless of design. Darwinian theory, based on the Malthusian axiom, explains only half the phenomenon of evolution: the evolution of fitness (capacity factor) in organisms. The other half of evolution is the evolution of potential (intensity factor) between metabolism and environment. Power brings together in one concept the two phases of evolution (intensity and capacity) which produced the modern ecosystem: the "one physical system" defined by Tansley. Power also validates Tansley's statement that ". . . the systems so formed . . . are the basic units of nature on the face of the Earth." However, Tansley's ecosystem has yet to become ". . . the point of view of the ecologist . . ."

The profound subjectivity, implied equilibrium thermodynamics, and comfortably traditional metaphysics of the Malthusian axiom have seriously inhibited the development of ecological theory by isolating ecology from the mainstream of 20th century science. The scientific objectivity of the Ecosystem axiom, modern disequilibrium thermodynamics, and informed appraisal of subjective assumptions underlying ecology provide a path to unity, generality, and wider acceptance of ecological theory and practice. The Malthusian axiom depends upon deist assumptions which, strictly speaking, cannot be disproved. Similarly, the assumptions underlying disequilibrium thermodynamics cannot be proved. In the widest sense, we should expect that neither is correct. Thoughtful scientists understand that more and better approximations of reality await them in the future. Similarly, perceptive scientists know they make assumptions, that those assumptions affect their science, and that their assumptions require careful consideration. Despite widespread recognition of the conceptual problems in ecology (Reiners 1986), the assumptions underlying the Malthusian axiom and their effects remain undetected by a majority of ecologists.

Acknowledgements

I am grateful to Prof. John C. Greene for calling my attention to Spencer's writings and helping me understand deism. Prof. Greene, Drs. E. H. Jokinen, M. W. Lefor, S. K. Lehman, and an anonymous reviewer provided thoughtful and constructive comments on the original manuscript.

3. The Past, Present, and Future of Ecological Energetics

Richard G. Wiegert

In this chapter I provide a brief review of the historical development of ecological energetics, some cogent discussion of theoretical constructs and some suggestions for future work. To accomplish this, I have organized this chapter into three sections. First, I discuss historical developments, focusing on the earliest beginnings of ecological energetics in the 18th and 19th centuries, the pioneering work of Lindeman, and the expansion of energetics studies after World War II. The latter quickly branched into somewhat separate areas emphasizing, respectively, population, ecosystem, and physiological energetics. The exponential rise in the number of studies of ecological energetics in the 1960s culminated in the development of energy flow models in the 1970s and 1980s. This section concludes with a brief discussion of the most important technological improvements in methods of measuring energy parameters. Second, I discuss three important areas of past theoretical activity: ecological efficiencies, the thermodynamic basis for the ecological energy budget, and the principle of maximum power. The third section, current and future directions, comprises an introductory explanation of energy flow in inanimate versus animate systems plus discussion and evaluation of (1) resource foraging theory, (2) energy quality analysis, (3) energy flow versus cycling, (4) trophic levels, (5) energy storage, and (6) energy flow models, particularly as the latter are used in the development of theory and as they intersect with the problem of the control of energy flow in ecosystems.

3.1. Historical Perspectives

3.1.1. Early Beginnings

The first scientist to attempt seriously quantification of (or perhaps I should say identification of) the loss of energy associated with the work of staying alive was Antoine Lavoisier (1777). He first showed that living organisms consume energy as food and lose energy in the form of heat, and that a quantitative relationship exists between the loss of energy of the food and the heat produced. Indeed, Brody (1945), in his classic work on animal energetics, credits Lavoisier with conceiving the First Law of Thermodynamics. Lavoisier was a nutritionist (perhaps the first such), and Brody remarks on the curiosity that the First Law, one of the most fundamental generalizations of physics to this day, was conceived by a nutritionist, formulated by a physician (Mayer) and a physiologist (Helmholz), and experimentally demonstrated by a brewer (Joule).

The two principles (or Laws) of Thermodynamics most relevant to biology, and thus to ecology, are:

First Law	$\Delta H = Q_p + W_p$	(1)
Second Law	$\Delta H = \Delta G + T\Delta S$	(2)

where ΔH is the change in system enthalpy;

Q_p and W_p are the net heat, or net work, respectively, exchanged between the system and its surroundings, at constant pressure; ΔG is the net change in the Gibbs free energy of the system; T is temperature in °k; and ΔS is the change in the entropy (mean heat capacity) of the system.

The assumption of constant pressure is the Gibbs formulation (the most useful form of the Laws to biology), which factors out the work of expansion and contraction (see Wiegert 1968).

Equation 1 is a statement of the conservation of energy, that is, a given unit of energy in any one form is equivalent to the same unit of energy in any other form. The energy entering a system is equal to the energy stored minus the energy leaving, corrected for the energy equivalent of mass exchanges. Thus we are able to construct energy budgets for systems.

Equation 2 states the relation between forms of energy and their conversion from one form to another. In the *spontaneous* conversion of any form of energy into any other form, except heat, the conversion can never be 100%. Some energy always remains as the "bound" heat energy of the system represented by the product of the system temperature times the change in mean heat capacity of the system. The latter is always positive in spontaneous reactions.

Lavoisier's demonstration of the connection between animal work and heat was not immediately applied to populations. The first reference I have found to the measurement of energy in a population was the study of silkworm energetics by Hiratsuka (1920). Bornesbusch (1930) was the first to describe the energetics of a community assemblage, the floor of a birch forest. This early attention to forest soils is ironic in view of their later neglect by 20th century ecologists working in ecological energetics. Energy flux also began to be used in descriptive ecology by aquatic ecologists, culminating in the trophic dynamic principles put forward by Lindeman (1942) in what became a powerful influence and probably the most quoted paper in the field of ecological energetics. Lindeman synthesized a number of different views of how ecosystems were organized, the main theme being that the dynamics of a system could be represented by a description of amounts, efficiencies, and rates (added by later workers) of energy transfer and transformation. For a more recent discussion of Lindeman's world view, see Rich (Chapter 2). Perhaps partly because of Lindeman's early death (even before his major paper was published), but mostly because of the disruption of World War II, ecological energetics did not really begin to move beyond the trophic dynamics era until the 1950s.

3.1.2. Population vs. Ecosystem Energetics: The Beginning of Experimentation

After the resumption of work in ecological energetics in the early 1950s, the field became divided along the lines of population energetics vs. the energy flow of whole systems. Perhaps this split was more apparent than real, because throughout the period when the University of Georgia was seen as the focus for work on the energetics of systems, E. P. Odum was conducting some very elegant work on the ecological significance of energy accumulation and use in populations of migrating birds (Odum et al. 1965). Conversely, at the University of Michigan, thought of at the time as a center of population ecology, Smith and Slobodkin and their students were studying laboratory ecosystems comprised of two or more species. F. C. Evans and his group at the University of Michigan were really taking an old field ecosystem perspective and at the same time studying the component populations (mid-1950s to early 1960s).

This period produced a number of useful descriptive studies of ecological energetics, some of them embodying either natural or controlled experimentation. At Michigan, Slobodkin (1959) and his student, Richman (1958), were examining the energy efficiencies of *Daphnia*. Engelmann (1961) reported the energetics of soil and litter populations of an old field. Golley (1960), at Michigan State University, was bridging the gap between population and ecosystem energetics in the field by measuring the energy flow in a grass–grazer–predator food chain. Golley's major professor, Don Hayne, was a University of Michigan Ph.D. and the person who guided my M.S. degree work on the laboratory energetics of Golley's grazer, the vole *Microtus pennsylvanicus*. His guidance and advice led eventually to my work with F. C. Evans at the University of Michigan on the energetics of the meadow spittlebug (*Philaenus spumarius*) (Wiegert 1964).

Following the classic study of both Odums on the energetics of a coral reef at Eniwetok (Odum and Odum 1955), total system energetics descriptions were published on Silver Springs, Florida (H. T. Odum 1957), a cold spring in Massachusetts (Teal 1957), and the Georgia salt marsh (Teal 1962).

3.1.3. Exponential Rise Period (Early 1960s to mid-1970s)

By the mid-1960s, there were enough studies of ecological energetics to permit some generalizations. In the physical sciences, open system thermodynamics had become respectable, and ecologists were much interested in more general and theoretical questions about energy transformation and transfer and their associated efficiency ratios. Already in the previous decade we had discussed the more controversial aspects of efficiency ratios, stimulated by the papers of H. T. Odum and Pinkerton (1955) and Slobodkin (1959). Now was the time for more carefully controlled and experimentally oriented work.

Two studies had special impact on my own thinking and on that of my colleagues at Georgia. The first of these investigations was the elegant work on the energetics of yeast populations by Battley (1960a, 1960b, 1960c; 1971). For the first time with any population, Battley measured all parameters of the energy budget, and computed the accompanying entropy changes. The latter was possible because the chemical compositions of both the media and the yeast were known, as were the overall chemical reactions involved. His data permitted computation of, among other things, the free energy changes accompanying anaerobic as opposed to aerobic growth.

Perhaps more significantly, Battley was also able to decompose the overall growth reaction of yeast into a nonconservative "work" part and a conservative part in which new cells were produced. In essence, the change in Gibbs Free Energy (ΔG) of the non-conservative reaction minus the ΔG of the conservative reaction estimated the amount of substrate that could be converted (maximally) into yeast cells. Battley (1960c, 1971) found that both anaerobic and aerobic growth, on glucose, ethanol or acetic acid, produced an estimate of ~35% for the maximum conversion of substrate to new cells. This may turn out to be of general application to all living populations, but the same rigorous approach to the energetics of other populations has yet to be applied.

Exceptions to the maximum 35% growth efficiency limit have been noted. Certain high yield bacterial growth experiments were reported in which conversion efficiencies of substrate to cell material greater than 60% were found (Bauchop and Eldsden 1970). Embryonic growth efficiencies between 60–70% were cited by Needham (1931). In the former instance, the greater yields may have contained material absorbed but not yet incorporated into new cellular material, thus bypassing the work of conversion and increasing the apparent efficiency. In the latter case, the same explanation is possible, but there may also be some fundamental difference between the growth efficiency of populations of freeliving cells and multicelled embryos. In his recent monograph, Battley (1987) concludes that conclusive evidence for efficiencies greater than 35% for conversion of substrate to cellular material still does not exist. The difficulty lies in showing that all substrate absorbed into the cell is actually converted.

Beyers (1963) worked at almost the opposite end of the ecological energetics spectrum. His interest was focused on the energetics of whole ecosystems encompassed within sealed glass chambers. He demonstrated that a stable assemblage could indeed persist for long periods (relative to generation times) as a thermodynamically closed system; i.e., exchanging only energy with its surroundings. One of the many interesting findings was that net production rates in such systems were independent of temperature. This may be the only example yet known of a phenomenon known as "congeneric homotaxis", that is, instead of a feedback control regulating the metabolism of a population in the face of changing temperatures and enzyme catalyzed rates, the system has several populations conducting the same process, but with optima at different temperatures (for discussion see Hill and Durham 1978; Hill and Wiegert 1980).

By the 1970s, energy was being used as a conserved, "bookkeeping" unit in the simulation models developed in response to the increased interest in ecological modeling as a tool to generate theoretical constructs and to guide research programs. I return to this topic later.

3.1.4. Methodology

An historical review of ecological energetics requires brief mention of some landmark developments in equipment and methodology. Because questions of energy content, photosynthesis, and respiration are all of major concern, most of the big jumps in ecological energetics have occurred concurrent with increased ability to measure energy content of smaller and smaller samples of biological material and increases in the ease, speed, and/or sensitivity of measuring the rate of energy flow as carbon is either fixed or degraded. Because these latter flows are usually measured indirectly as oxygen produced during net photosynthesis and oxygen and/or CO_2 released by respiration, the improvements in methods have often involved better equipment to obtain measurements of these gases.

To measure energy content, the traditional macro and micro "Parr calorimeters," which required 50 mg to 1 g of sample, were supplemented in the 1950s by a unique calorimeter which could measure the energy content of samples as small as 1 mg, although 10 mg produced better accuracy and precision (Slobodkin and Richman 1960). The Phillipson microbomb calorimeter (Phillipson 1964) did the same job faster, and, with improved design (Wiegert and Gentry, U. S. Patent 3,451,267) and a better recorder, could handle samples as small as 1 mg in mass. Recently, this limit has been further reduced to a few μg by the design of Scott (1982).

The trend in instruments for measuring changes in both O_2 and CO_2 has been to miniaturization and portability, and, in the case of O_2, faster response time of probes and greater sensitivity.

Recent developments in O_2 probes include the microprobes of Revsbech and Ward (1983) and Revsbech and Jorgensen (1986), which need no stirring, respond in less than 1 sec., and can differentiate between O_2 concentration in microzones as small as 10 μm in diameter.

Winkler O_2 determinations, 14-C labeling, and CO_2-pH curves have been adequate for measuring productive aquatic systems. However, increased sensitivity is required in probes used to measure photosynthesis in the open ocean, where production is extremely low. The design described by Williams and Jenkinson (1982) has been widely used.

Miniaturization and field portability have been the themes of developments in instrumentation for measurement of CO_2. Infrared CO_2 gas analyzers, once some of the largest and most temperamental of laboratory instruments, can now be obtained in lightweight field-portable forms, thus permitting instantaneous measurement of rates of photosynthesis and respiration.

3.2. Theory

3.2.1. Thermodynamic Basis

The Laws of Thermodynamics set the maximum limits to the rates and efficiencies of ecological energy transfer and transformation. However, the evolutionarily derived requirements of living systems impose additional constraints. In general, the latter comprise the losses due to internal work (respiratory heat loss), egestion of unassimilated food, and losses to mortality. In the following paragraphs, I will compare some of the thermodynamic maxima with the realized ecological rates.

Thermodynamic constraints arise from the degree of irreversibility (= friction) of the energy process in question and the temperature (°K) at which the process takes place, or, in the case of a process operating between two temperatures, the difference between the source temperature and the sink temperature.

Some terms need to be defined: *Potential energy* is literally the energy of position. In biological systems, most interest is centered on *chemical energy* which is the potential energy of elements and compounds. *Kinetic energy* is the energy of movement, measured in mechanical terms as the product of mass and velocity. *Work* is a form of energy, measured in mechanical terms as the product of force and distance. *Heat* is a measure of the kinetic energy of molecular motion. *Power* is the *rate* at which work is delivered. *Free energy* is the energy available to do work. *Entropy* is expressed in the units of heat capacity, joules per (degree × moles).

3.2.2. Ecological Efficiencies

Efficiency is the ratio of some defined output or product to the input or cost. The Second Law tells us that no conversion of energy from one form to another can be 100% efficient (except for conversion into heat), unless the process is completely reversible; i.e., no friction. The final state temperature must also be zero if the conversion is from heat into some other form of energy. A biological process of conversion of free energy to work would have to occur infinitely slowly to meet the criterion of complete reversibility. This is a constraint rather difficult to meet in practice!

Using the thermodynamic laws, Spanner (1963) modeled the plant as a heat engine and showed that the maximum efficiency of photosynthesis is 60–80%, the value depending on the intensity of the light used in the equations. How closely this maximum value is approached during the actual capture and transformation of energy within the chloroplast remains unknown, but the overall efficiency of the process at the level of the plant population is far less, because of the additional ecological constraints (nutrient transport within the plant, seed production and dispersion, leaf replacement, etc.).

Similarly, in other areas of ecological energetics, the efficiency is usually far lower than the thermodynamically permissible maximum. Production of new tissue at the population level, relative to ingestion, seldom exceeds 20%, and in homeotherms is often 1% or even less. The assimilation of the chemical energy of ingested food is usually not more than 75–80%, and may (if much cellulose and lignin is ingested) be less than 10%. During digestion and assimilation, organic materials are degraded in various ways, mainly by splitting complex sugars to simple ones and proteins to amino acids. During the growth phase of metabolism, these compounds are reassembled into new tissues. As noted earlier in the discussion of Battley's work, the maximum efficiency of this transformation cannot exceed 35% in yeast, and very likely is little larger in other organisms. Additional ecological constraints further reducing this growth and ingestion are the sometimes low assimilation abilities and high respiration rates, particularly in homeotherms, necessary for normal activity and survival. Occasionally (for example, in certain sucking insects), ingestion of dilute solutions of organic compounds such as amino acids results in assimilation efficiencies approaching 100% (Wiegert 1964). In organisms with determinate growth, the growth/ingestion efficiency becomes zero when growth stops. In all organisms, growth rates slow with age and growth efficiencies decrease.

3.2.3. Thermodynamic Laws and Ecological Energy Budgets

To analogize between the terms of the ecological energy budget and those of the Laws of Thermodynamics is tempting, but can lead to problems. These laws, as they are stated in the introduction, are valid only for closed systems; i.e., those exchanging only energy, not matter, with the surroundings. Furthermore, the systems are reversible, containing no provision for indicating frictional energy losses. To identify the respiratory energy loss of an organism or population with the entropy containing term, $T\Delta S$, of the Second Law (see Brody 1945; Patten 1959; Wiegert 1964) has been particularly satisfying because each represents an amount of energy that is "unavailable." However, such analogies are superficial and misleading. Organisms and populations are open, highly irreversible thermodynamic systems far from thermodynamic equilibrium. As such, they exchange enthalpy (H), bound heat (TS), and frictionally produced heat (Q) with their surroundings. Thus, the respiratory energy loss that is measured (or computed indirectly from other measurements) represents that part of the heat exchange due to Q, not the bound heat, TS. In living organisms, the former is much larger than the latter.

The direct measurement of entropy changes, ΔS, in living systems is impossible in practice. Computation of such change can only be done with a knowledge of all compounds involved in, or produced by, the life processes. For any ecological system more complicated than the simple microbial systems studied by Battley, the necessary information is unavailable at present. See also Scott's critique of the utility of Second Law considerations in ecological energetics (Scott 1965; and in Wiegert 1976, p. 13).

Because the First Law represents the conservation of energy, analogies between its terms and the ecological energy budget have had more success. By adding terms representing the enthalpy of matter exchanged to both equations, that of the First Law and that of the energy budget, Wiegert (1968) developed the following correspondences (the notation is changed slightly):

$$\text{First Law (open system) } \Delta H_s = H_1 - H_2 - Q_{SR} - W_{SR} \qquad (3)$$

$$\text{Ecological Energy Budget } P = I - E - R - W_{SR} \qquad (4)$$

where:

ΔH_s is the enthalpy change of the system;
H_1 is the enthalpy content of matter imported;
H_2 is the enthalpy content of matter exported;
Q_{SR} is the net heat exchanged with the surroundings;
W_{SR} is the net work exchanged with the surroundings;
P is the enthalpy content of matter "produced";
I is the enthalpy content of matter ingested;
E is the enthalpy content of matter egested;
R is the net respiratory heat loss.

The work term, W_{SR}, is commonly ignored as negligible in ecological energy budgets. Highly irreversible thermodynamic systems such as living organisms perform most of their work against friction. This energy shows up as the respiratory heat loss. Only a small fraction of the work results in an increase in the free energy of the system or its surroundings. This is the quantity represented by the W_{SR} term. For example, the potential energy increase represented by beavers piling up sticks to form a dam is negligible compared to the respiratory energy expenditures of the beavers while they were building. Indeed, if the original potential energy of standing trees were considered, the result could well be a decrease in energy of the surroundings. Thus in *most* ecological energy budgets of the beaver population, W_{SR} could be disregarded. But to the beaver, this small amount of potential energy gain represented by the dam is vital in terms of survival, whatever the overall gain or loss of energy by the environment as a whole.

3.2.4. Work and Power

Odum and Pinkerton (1955) were the first to consider the general problem of optimization of energy utilization, specifically the relation between the speed of a transformation and its efficiency. Drawing on examples from a variety of physical and biological systems, they illustrated the fact that maximum power output depends on a balance between the optimization of each of these attributes. Their conclusions are relevant to understanding the kinds of adaptation that one might find in the various components of any ecosystem, given the thesis that energy is a common limiting factor.

Power is computed as the product of a thermodynamic force and a flux. In electrical systems, for example, the force is voltage (V) and the flux is current (I). Their product, power, is measured in watts (W). In ecological energetics, the forces usually are the useful *energy concentrations* of the materials transferred (joules/g) and the fluxes are the *rates of transfer* or production (j/unit time). Note that in the case of photosynthetic processes, the input force is related to the concentration of photopigments and the input flux is the rate at which usable light energy is absorbed (Odum and Pinkerton 1955). In the class of machines to which living organisms and communities belong, and providing they are assumed to be near equilibrium, Odum and Pinkerton demonstrated the following relationships:

Letting E equal the efficiency measured as the ratio of useful power output to useful power input, and R equal the ratio of output force to input force, (a) as frictional losses (leakage) approach zero, E and R approach identity; and (b) maximum power is always developed when $R = 1/2$, and thus, under ideal conditions, maximum efficiency can never exceed 0.5. This principle says that in an ideal frictionless system, as the ratio of forces changes from near unity (balanced, almost no input or output) toward zero (maximum imbalance, maximum input, zero output), the power out at first increases and then decreases.

Figure 3.1 shows the relationship of output power, efficiency, force ratios, and, by implication, time. When there is no loss to friction (lines labeled a), efficiencies are identical with the ratio of input and output forces R. As R and E approach unity, power input and output, together with velocity of change, approach zero. When R and $E = 1/2$, power is maximized. Finally, as R and E approach zero, so does output power, although power input and velocity are maximized. When friction (leakage or maintenance in a living system) is present, the force ratio R can be chosen for maximizing either output power (at $R = 0.5$) or efficiency (at $R \approx 0.75$ in this case), but efficiency at $R = 0.5$ is always less than 50% (lines labeled b in Fig. 3.1). Note also that although output power is still maximized for $R = 0.5$, the absolute amount of output power is reduced because of losses of energy to friction.

Odum and Pinkerton discussed the "climax" community, noting that when community net production becomes zero, the *plant component* should show an R of 0.5, maximum output, thus maintaining a community of the maximum possible biomass. However, if the R and power values for the *entire community* are used, a different picture emerges: the "climax" community by definition has no useful output; i.e., all input is dissipated as heat, but input power (intercepted sunlight and gross community production are maximum). Such communities would therefore cluster in the lower left of Figure 3.1. The primary successional community has low biomass, and thus a very low input and

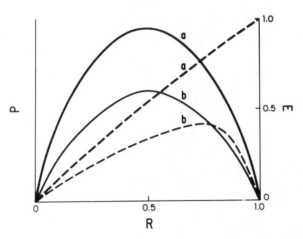

Figure 3.1. The relationship of output/input energy force ratio (R), efficiency of power transformation or transfer (E), and output power levels (P); assuming (1) thermodynamic systems near equilibrium, and (2) a constant input force and flux. Lines (a) show the case for no frictional losses (E = R). Lines (b) show the case for friction > 0. Dotted lines are absolute efficiencies (right-hand scale). Solid lines show relative power values (left-hand scale). Redrawn from Figures 2 and 3 in Odum and Pinkerton (1955).

a low output, but decomposer losses are also minimal, so efficiency is relatively high and such systems would be located to the extreme lower right in Figure 3.1. Intermediate successional communities are close to the maximum in power production (as a community), because productivity is high, yet biomass is intermediate and community respiration is low. Thus, the output power and efficiency maxima of such communities would approximate the peaks of the curves labeled b in Figure 3.1.

3.3. Current and Future Directions

3.3.1. Animate vs. Inanimate Systems

The "maximum power" hypothesis of Odum and Pinkerton can be restated to make an important distinction between energy transformations in living and inanimate systems. Maximum power is equivalent to transforming the most energy per unit time. The forces and fluxes of energy are related by terms (a) specifying the frictional loss or leakage of energy, and (b) specifying the "conductivity" of the system. That is, for a given difference between input and output forces, what will be the associated input and output fluxes? In inanimate systems (electrical, thermal, mechanical, etc.), the answer may be complicated, but, in principle, can be found in a straightforward manner and is *constant for the specified system*. That is, at a given temperature, the current flowing in a copper wire will be the same each time a 10 volt potential difference is applied. In animate systems in which energy flow is of interest, the forces are, to take a predator–prey pathway for example, the unit energy content of prey items and the unit energy content of predator production. Flows (for example, ingestion by the predator and its subsequent production of new biomass) are determined by the biological equivalent of the electrical conductivity, that is, the factors which determine rate of ingestion on the input side and production of new biomass of predators on the output side. These factors are not only vastly more complicated and numerous than those governing an inanimate system, but they cannot, in most cases, be measured directly in their individuality. Only by observing and experimenting with organisms can the relations between forces and fluxes be established. Furthermore, the relationship (the rules of the game) can change, either behaviorally or physiologically in the short run, or genetically in the evolutionary long run. For example, a given force of voltage produces the same current each time in a particular wire, subject only to known physical changes (variable temperature, for example). But, a given energy concentration of prey often produces a very different ingestion or production flux in the same predator population as a result of internalized behavioral and physiological changes by both prey and predator. This phenomenon has the very important result of shifting the emphasis, when considering power production in living systems, from the direct measurement of forces and fluxes, to concern with the "conductivity" analogues; i.e., the factors that both *determine* and *control* the rate of biological energetic processes. This concern manifests itself in different forms in the following areas of ecological research, where, I believe, energetics can make valuable contributions, both currently and in the future.

3.3.2. Resource Foraging

Resource foraging has been one of the most active areas of research in population ecology during the past decade. I have avoided the term "optimal foraging theory," in part because of the different connotations of the word "optimal", and in part because the "theories" are still largely hypotheses. Foraging reflects the fundamental bottom line of the ecological energy budget, the acquisition of free energy. Most of the intellectual interest in this area of ecology has focused on heterotrophs, but I will include autotrophic organisms as well in the following general assertion: Energy limitation is the ultimate control on populations in any natural (or human managed) ecosystem.

As support for this statement, I point out that energy can be used to increase the availability of all other relative limiting factors. This does, of course, require, in many cases, specific evolutionary adaptations. Thus, one cannot simply supply some form of energy and always expect to see an immediate response of the target population. This seems to suggest some justification for separating the "ultimate" control from some more immediate "real time" control. But such a separation here would only cloud the issue I am trying to sharply define. The temporal scale of operation is, in any case, obvious from the discussion of any specific example.

The assertion that energy is the ultimate limiting factor requires a distinction between energy used directly for maintenance and growth and energy used to acquire some other scarce factor, say nitrogenous compounds. We must also inquire as to what circumstances require organisms to choose between acquiring the maximum energy or minimizing mortality. Since these latter two alternatives may often be the two extremes of a gradient, how is the problem solved? To what extent, in specific populations and assemblages, are these choices mediated (or directed) by evolutionarily constrained behavior? The attempt to answer these questions, it seems to me, is the core of future research efforts in resource foraging. If so, then a vital question is: Is energy the most common ultimate limiting factor against which alternative behaviors are selected? The following examples may help illustrate the dimensions of this question.

In terrestrial plants there are three distinct pathways of photosynthesis: C_3, C_4 and Crassulacean acid metabolism (CAM). In C_3, the "normal" pathway, three physiological characteristics may be noted: a tendency (1) to be carbon-limited at modest CO_2 levels; (2) to be light-saturated at less than full sunlight; and (3) to be least tolerant of drought. The C_4 plants, by several physiological and anatomical developments, have become tolerant of high light, lower CO_2 and modest drought. The CAM plants (mostly desert species) are specialists in arid environments.

For the most part, these different adaptations on the part of C_4 and CAM plants have been acquired at the cost of energy, either a decrease in the innate rate of gross photosynthesis under "normal" conditions of light, CO_2 and water, or a sacrifice of the efficiency of net photosynthesis under such conditions; i.e., some of the free energy fixed is used to increase the efficiency of water retention . But why have these modifications been selected? I believe it is because, in the given environment, overall net fixation of energy per unit area/time was increased and mortality decreased.

Plants would be expected to respond to changes in the limiting factor. If the adaptation is a learned behavioral or a labile physiological response, then the response may be rapid. Thus a plant that developed a response to high light by bleaching more exposed chloroplasts and turning on chloroplasts deeper in the plant might respond positively to further increased light simply by deepening the chlorophyll layer. A C_4 plant, however, has a large amount of fixed physiological and anatomical capital to change. Rapid increases in light which overburdened the fixed system could produce (in the short run) no effect, or even a negative effect. This is likely to prove a general problem for most of the secondary limiting factors (N, Fe, moisture, etc.) that commonly vary over the short run, but day length and total light at a given location are much more constant and predictable. Thus, the ability to respond rapidly to short-term changes in the secondary limiting factors often has a selective advantage, while evolutionarily, the plant can maximize its light energy capture mechanisms to a much greater degree.

Many plants fix dinitrogen. Reducing nitrogen to ammonia, whether it is done directly by the organism or by means of symbiotic associations, requires energy. This is, incidentally, one of the areas where many bioengineers have a blind spot. Of scientists interested in producing N-fixing corn, for example, I have encountered none who consider the consequences of the significant decrease in corn yield that would follow the cessation of N-fertilization. The inescapable fact (from thermodynamics) is that the reduction of N_2 requires free energy. Corn is limited by the amount of energy man is willing to subsidize by way of N fertilizers. Soybeans fix dinitrogen, yet farmers are now finding that soybeans often yield more if fertilized with nitrogen. Why? Simply because N-fixation requires energy and the soybean is able to reduce its N-fixation in accord with increased concentrations of available nitrogen in the soil. Are soybeans nitrogen-limited because fertilization produces increases yields? I argue they are in an ultimate sense energy-limited. The nitrogen subsidy in the form of fertilizer simply permits the plant to allocate more energy to production of soybeans. Increasing the light energy would produce the same effect up to the point where the increased light negatively affected some other process, or where a different secondary factor became limiting.

Animals, too, show evolutionary responses to changes in energy. For example, if water is scarce or if embryonic development must take place in a closed environment (an egg), surviving species often are those that produce insoluble uric acid as a waste product containing the N of protein deamination. But uric acid costs more to produce than an N-equivalent amount of ammonia. The relevant question then reduces to an energy cost–benefit analysis. If such evolved mechanisms are behavioral/physiological (labile), then changes in any of the secondary factors (N, water, etc.) may produce a short-term effect. But, as in the plant examples, the ultimate factor to which evolution is keyed is the question of the best allocation of a finite amount or rate of energy supply.

The selection for increased rates of energy acquisition must take place in the context of two qualitatively different kinds of energy losses. First are those which are the essential accompaniments of the physiological activities of the organism, such as respiration, or the intrapopulation costs of social structure, as in defending a territory. Strong selective pressure would exist for minimizing such losses, whenever unrelated to, or only indirectly related to, the second category of losses, predation (in the broad sense, including parasitism). By this, I mean there is a direct and strong selective advantage for minimizing respiratory energy loss as a consequence of a given level of activity, whenever doing so will not increase predatory losses by an unacceptable degree. Increasing energy acquisition, by evolving more energy efficient ways of defending a feeding territory and thus increasing territory size, could also be counterbalanced by increased risk of predation.

Thus, for populations or individuals at steady state, if anything having to do with energy acquisition can be said to be "maximized," it is the rate of free energy intake relative to those exploitative competitors who face essentially the same exposure to predatory losses. Free energy, once taken in and transformed, may no longer be available to competitors. Thus, light, whether intercepted and utilized to reduce CO_2, or reflected into space, or re-radiated as longer wave infrared radiation, is no longer available to competing plants.

Food eaten by heterotrophs and either assimilated or egested is, even in the latter case, seldom available to direct competitors. One could argue further that the first species to evolve the means of using kinds of free energy other than ingested energy to increase the availability of the latter would gain a selective advantage. These additional kinds of energy are work exchanges, either as net work performed by the system on the surroundings (a loss to the system), or as net work performed by the surroundings on the system (a net gain to the system). The first case can be of advantage only if the net work performed on the surroundings (say, by an individual), results in a quantitatively greater return via assimilated food energy. Agriculture is an example.

In its simplest form, "agriculture" is found even in nonhuman populations. For example, it may be seen in the work of elephants converting a forest into savanna or ants cutting leaves and establishing a fungal garden. The work need not directly (or solely) result in increased food acquisition. Beavers, for example, perform work on the environment by cutting and moving sticks and logs into piles, thus creating impoundments. These then serve the dual purpose of reducing energy losses to mortality as well as facilitating the acquisition of food.

Man has carried this process further than any other species, first performing work on the environment by muscle power alone, then harnessing other sources of energy. The domestication of other animals was followed by gathering fuel wood, then using fossil fuels such as coal and oil, and finally harnessing the *strong force* in the atomic nucleus. Thus, for example, agriculture in advanced countries has now become a process of: (1) performing work on the environment to obtain oil, (2) performing work on the oil to obtain lighter fractions (e.g., gasoline); and (3) using these to perform work on the environment resulting in increased rates of solar energy capture by domesticated plants in the form of the free energy of food. Qualitatively, the process is analogous to the beavers harnessing the free energy in falling water to increase their "fitness." Quantitatively, humans have far outstripped all other species in the magnitude of the free energies expended for a return of chemical free energy in the form of food. Note also one other distinction. The beaver (and man when capturing hydropower) uses kinetic energy to increase the availability of potential energy. Most human endeavors, however, transform potential energy into the most unavailable of kinetic energies, heat, released into a large, unilaterally constant temperature sink.

Assimilated energy can be dissipated by work performed on the environment (mostly ending up rapidly as heat), or as heat of respiration, or it can be stored. If the work performed on the surroundings results in potential energy stored, or if there is storage of production, then the system and/or its surroundings are changing; i.e., evolving. To what extent do systems evolve toward a steady state where energy acquisition, and therefore energy dissipated as heat, is maximized? This question has occupied a central position for many decades in the intellectual considerations of physicists, chemists, and biologists interested in the applications of thermodynamics to their respective fields. The answer, however, depends on how the system is defined.

Evolution in living systems, at the population level, driven by natural selection of individuals, produces the maximum possible difference between energy acquired and energy lost as heat, in the context of a given environment. Generally this results in selection for a high standing stock (storage of energy), since, because of the autocatalytic nature of population growth, the highest steady state standing stock will produce the greatest number of descendants. Such a configuration is not the maximum dissipative structure, however, since much of the energy loss from the population will be nondissipative losses of standing stock to natural mortality and predation. Thus, at the population level, the evolutionary tendency will be to develop higher rates of intake and storage (until the steady state is achieved), but lower rates of dissipation (measured as heat lost to the environment).

At the ecosystem level, however, the collection of populations constituting the system will form an integrated whole in the sense that evolution of each population will tend to maximize its net energy intake. Nondissipative energy losses from any one population will form the food of another. At steady state, unprotected storages (detritus, for example) will be utilized at an annual rate just balancing inputs and stored energy, whether small or large, will be constant. Thus, maximization of energy acquisition will have as a consequence maximization of energy dissipation. Such a configuration requires that the autotrophs maximize the fixation of solar radiation, a point made by Odum and Pinkerton (1955). But, note that the above argument concerns only the result of successional evolution of a living system using solar energy. An earth bereft of life would lose all absorbed solar energy as heat.

Thus, in a steady state system evolved to maximize energy input (in the case of the Earth, this is the amount of solar energy fixed), the necessary storages are either protected (biomass of trees) or dissipated at the same rate as produced (detritus). But what if the system manages to accumulate, or already contains, large stores of potential energy? In the case of the planet Earth, part of this energy (fossil fuels) resulted from non-steady state conditions in solar powered systems of the past, with the resulting stored energy protected from immediate degradation by geological processes. The other, much larger, quantity of this energy is the nuclear energy of matter itself. Does the evolutionary tendency of living systems to become more and more dissipative still hold? Certainly the historical record of Earth seems to answer in the affirmative. Once freed from the constraint of dissipating only the free energy acquired by feeding, the human population has become an ever more dissipative thermodynamic structure, first by burning natural and fossil fuels, subsequently by dissipating the energy of the nucleus. Ultimately it seeks to dissipate energy by nuclear fusion while inhabiting a physical structure, the planet Earth, which is far too small to sustain such a dissipative reaction by gravitational forces alone. This dissipation phenomenon has of late been largely one of cultural evolution; i.e., the Industrial Revolution, as opposed to biological evolution. Furthermore, the structures evolved, although aiding the survival and increase of the human population, have been physical, not biological, structures and have themselves increased the rate of energy dissipation.

Thus, Evolution, at the population level of systems organization, proceeds in the direction of increased energy acquisition by feeding and increased transformation, but not maximal dissipation. At the level of the ecosystem, succession proceeds towards increased acquisition, paralleled by increasing rates of dissipation through diversification of trophic structure. Stored (potential) energy of progressively lower degrees of availability will continue to be dissipated as a consequence of the evolution of more and more complex dissipative systems, once biological systems have evolved that can further change, through cultural evolution, to create the physical structures necessary for this dissipation.

3.3.3. Energy Quality

In thermodynamics, heat energy availability is measured in terms of its temperature ($°K$); the higher the temperature, the easier it is to construct a lower temperature sink and thus to convert some of the heat to work. This concept has little meaning for ecological systems, however, because organisms feed on potential energy and exist, for all practical purposes, in a constant temperature. They are not heat engines (with the exception of photosynthetic plants; see Spanner 1963). Several years ago, a different measure of energy quality was introduced by H. T. Odum and called "embodied energy" (for the most recent discussion, see H. T. Odum 1983). Briefly, this concept of energy attributed to the stored potential energy at each trophic level a quality computed according to the energy dissipated at each trophic step leading from solar energy to the trophic level in question. Thus, a joule of energy in the standing stock of a top predator has by definition a higher energy quality than a joule of energy in the standing stock of its prey, etc. This redefinition of energy was useful, but somewhat misleading, because embodied energy was not really a different form of energy at all, but a new quantity. Odum eventually changed the name to reflect this and coined the term "emergy" to replace "embodied energy."

The concept of emergy is extremely useful in certain kinds of energy flow analyses such as those incorporating both energy flow and monetary flow. This is because in most cases the monetary equivalent of a unit of emergy varies much less (to the extent that the energy/emergy equivalent is accurate) than does the monetary equivalent of a unit of energy. Because the concept is so new, emergy analysis in ecology is confined largely to the work of Odum and his group. But, the idea of assigning quality is an exciting one and would seem to be an area where we can expect much work to be done in the future. Major problems needing solution involve reconciling the concept of energy quality based on trophic efficiencies, and that imposed by human economic considerations when the monetary equivalent of emergy does vary. For example, the emergy value of a piece of oyster shell and that of a pearl would be identical if based on trophic energy efficiencies. But from the standpoint of human economics, the value of the pearl is much greater than that of the mussel shell. Different emergy standards will obviously be needed for different kinds of analyses.

3.3.4. Flows and Cycles

Currently, it is again fashionable to argue the merits of energy cycling instead of "flowing," and material flowing instead of "cycling." Perhaps we are doomed to repeat this sterile argument every 10 years or so. To my mind, the matter was conclusively settled more than a decade ago. I quote from Rigler (1975a): "Traditionally we say that energy flows unidirectionally, whereas materials cycle. This is only a crude approximation of the real difference, because energy and materials both cycle and both flow out of ecosystems irreversibly." Much more relevant to the advancement of ecological theory is the recognition that *usually* inorganic nutrients cycle very efficiently, whereas energy cycles relatively inefficiently. The point is made explicitly by Figure 3.3 (using the energy symbols shown in Fig. 3.2). The pathways of N and energy through the producers, consumers, and decomposers of a simple diagrammatic model are shown

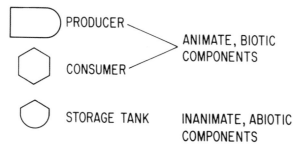

Figure 3.2. H. T. Odum energy symbols for biotic and abiotic components.

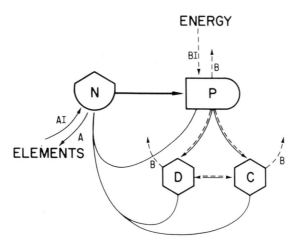

Figure 3.3. Generalized static model of energy/element flows and cycles in ecosystems. Solid Lines—Elements (A). Dotted Lines—Energy (B). N—Nutrient Pool. P—Primary Producers. C—Consumers. D—Decomposers.

as solid and dotted lines, respectively. Neglecting immigration and emigration (in which all of both the energy and the element are lost), the pathways of input and output for the two are different. Energy enters such systems as fixed solar energy and leaves as heat. The heat losses for living compartments are always a large fraction of the net input (assimilation), usually more than 50%, and in the case of homeotherms, more than 90% (Wiegert 1976). Thus, energy is typically very inefficiently cycled, from the standpoint of the total flux. There is no theoretical reason why the input–output fluxes of nutrients should be small relative to the flows within the system, and, if sedimentation is included in the loss pathway, the cycle can indeed be very inefficient. Most populations must nevertheless evolve conservation physiologies for essential elements, because they are scarce and/or costly (in terms of energy) to obtain. Thus, most ecosystems show highly efficient cycles for such elements (for some examples, see Pomeroy 1975).

Although the argument about cycling vs. flow as outlined above may be partly semantic, the "cycling" of the chemical potential energy of organic compounds has not received the attention it merits in ecological energetics. That lack is currently being addressed and the analysis of cycles in ecosystems and their models is receiving attention (Patten 1985). However, we must, in trying to encompass both material and energy cycling in the same theoretical framework, not lose sight of what is and is not a cycle.

There are some ecologically important differences between inorganic and organic nutrient cycles. Organic cycling is concerned with the available energy content as the "essential" part, whereas inorganic cycling is usually concerned with some essential element or compound without which energy cannot be made available to, or used by, the organism (there are exceptions, as in the case of vitamins, which are organic materials needed not for their energy but because of their chemical structure). Patten (1985) noted that elemental cycling is understood by most ecologists to involve movement of material out of a biotic pool into an abiotic pool from which re-assimilation ensues after "degradation" of organic to inorganic forms. From this, he poses a problem in defining energy cycling, since energy is degraded to heat, and thus cannot be reused by organisms. Using a similar type of definition of elemental cycling needs careful wording, because cycling is occurring even if the compound returned to the abiotic pool is not the same as the one initially taken up. For example, nitrate taken up by a plant may be returned as ammonia, which is also available, and thus in one sense completes the nitrogen "cycle." Sometimes the distinction even between organic and inorganic cycles is blurred, as when essential elements are taken up (cycled) as organic compounds. Furthermore, defining a cycle as transfer between the biotic (living) and abiotic (nonliving) states is also ambiguous. First, it is ambiguous because in a predator–prey transfer the transition between the living–nonliving–living state is so brief, with no storage, that the nonliving state is usually not represented in diagrams of the system. Secondly, such a definition ignores the very real trophic distinction between noneaten material or material egested and that which is assimilated by the consumer. I shall consider this distinction in more detail shortly. The most useful definition of a step in the cycling of matter or energy thus remains simply whether it is chemically transformed and/or incorporated into the protoplasm of the producer or consumer. This definition is, I believe, the one implied, if not explicitly stated, in almost all ecological writings on the topic. It is also the definition of a "trophic transfer."

3.3.5. Trophic Levels

The concept (not a theory) of trophic levels is another area addressed by Rigler (1975a) and others. The concept has been recognized for decades as being difficult to apply, unless one proceeds from a detailed knowledge of the energy flow through a species food web. Such information is now becoming more available for a variety of ecosystems, however. When coupled with detailed simulation models, both those based on empirical data and those constructed on theoretical grounds, useful theoretical constructs about the trophic dynamics of ecosystems can now be postulated. I will consider some general results from theoretical trophic model computer simulations in the final section of this chapter. Here, I want to discuss specifically the effects of the number of trophic transfers on both the total energy ingestion within the system and the total energy dissipation.

Consider two systems with identical food energy input and assimilation efficiencies. For the sake of clarity, I consider unassimilated matter/energy to leave the system, although this constraint in no way affects the qualitative validity of the conclusions. Figure 3.4 shows two trophic configurations. In the first, only one trophic transfer is explicitly modeled, with all subsequent transfers in the chain represented as N-1 self-ingestion loops. In the second, N separate trophic transfers are explicitly modeled. The simplification in the first instance is typical of the process of constructing ecological models wherein we wish to represent only the detail necessary to incorporate "significant" transfers. Each trophic level must have an input, an ingestion of energy (I). The models are used to examine the effects of two different modifications: (1) What does adding successive trophic transfers do to the total ingestion of energy by all levels, assuming a constant proportion of ingestion by level i is available as ingestion by level $i + 1$. (2) What proportion of the energy ingested by the first level (I_1) is the total energy ingestion by all higher levels, ($^n\Sigma_{i=2} I_i$)? This proportion is one measure of the degree to which energy is recycled as the potential energy of matter passed along a trophic chain. Figure 3.4 makes clear the following two conclusions. First, the total energy ingested by all levels rises quickly for increasing N. For trophic levels greater than 4, the value is virtually indistinguishable from the maximum reached when N equals infinity. Second, as the proportion of ingestion available to the next level decreases below 0.5, the ingestion of all levels above trophic level one (expressed as a proportion of ingestion by level one) rapidly decreases and becomes relatively insensitive to the total number of trophic transfers. Because this transfer efficiency cannot, on theoretical grounds, often exceed 20%, and seldom reaches 15% in real populations (Slobodkin 1959; Wiegert 1964), energy cycles are very

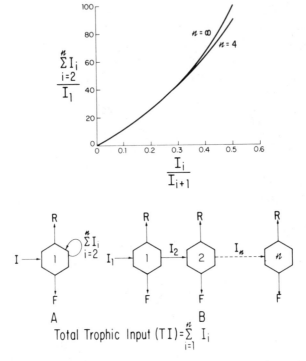

Figure 3.4. Effects of increased number of trophic levels and efficiencies of transfer on summed ingestion of energy by consumers, expressed as a proportion of ingestion by the first trophic level.

inefficient compared to most nutrient cycles. Also, the inclusion of more trophic steps in models will have a vanishingly small effect on the summed ingestion of energy by all consumers.

Static population energetics models of the kinds shown in Figures 3.5, 3.7, and 3.9 can be used to further illustrate the points just made. They show the energy flow through the system in the revised Lindeman-type trophic scheme offered by Wiegert and Owen (1971) and shown in Figures 3.6 and 3.8. The essential new features of this revised trophic scheme were: (1) to represent two distinct lineages or pathways of energy transfer through trophic steps wherein the material ingested was living (biophagy), shown on the left in Figures 3.6 and 3.8, or dead (saprophagy), shown on the right in Figures 3.6, 3.8; (2) to provide a lateral pathway of energy flow for ingested but unassimilated material from a biophage to a saprophage on the *same* trophic level; and (3) to permit crossing between the vertical pathways, as when a biophage eats a living saprophage, or a saprophage utilizes a dead biophage or its nonliving protoplasmic products.

In Figures 3.5 and 3.6, four trophic transfer (assimilative) steps are possible (components 1, 2, 3, 6). Component 6, a carrion feeder, transfers unassimilated

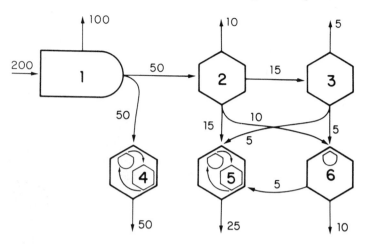

Figure 3.5. The energy flows in a system with four trophic levels. Ingestion is shown as the total for each trophic step and as a ratio to the ingestion of Level 1.

	Trophic Level			
Flux	1	2	3	4
F01	200	—	—	—
F12	—	50	—	—
F23	—	—	15	—
F14	—	50	—	—
F26	—	—	10	—
F36	—	—	—	5
Total	200	100	25	5
Ratio to F01	1	0.5	0.125	0.005

material to the decomposer group, component 5. The movement of un-ingested material is not a trophic transfer, because the material is neither altered chemically nor incorporated in the protoplasm of the consumer. Indeed, Wiegert and Owen (1971) made the point that such material is, in a topological sense, never inside the organism. Arranged as trophic transfers, the flows are shown in Figure 3.6. Unassimilated material passes from a biophagy level to the saprophage side without changing trophic level. The self-loops indicate the relative apportionment (trophically) of the net production of the top consumer group. For example, from Figure 3.5 we see that the top consumer, the carrion feeder in compartment 6, ingests 10 units via a pathway of three trophic steps, and 5 units via a pathway of four steps. The 5 units of uningested material passed to compartment 5 then represent 3.3 units remaining at trophic level 3 and 1.7 units at level 4 (the self-loops of Fig. 3.6). In this manner all of the flows to *compartments* in Figure 3.5 can be reassigned to *levels* as in Figure 3.6. Biophagy has a qualitatively different effect on interactions in the system, but is trophically (in the sense of the efficiency of transformation) similar to

R. G. Wiegert

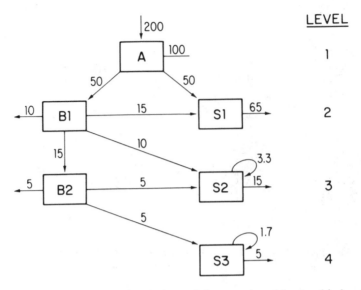

Figure 3.6. Energy flows of Figure 3.5, apportioned by trophic level.

saprophagy, shown on the right. Natural mortality; i.e., death and subsequent utilization by a saprophage, is shown in the transfer from B1 to S2.

Adding predation on saprophages (Figs. 3.7, 3.8) creates an additional trophic transfer pathway of five steps in length, but has no effect on the flow of energy over paths of length 3 and 4 (Figs. 3.7 and 3.8). Finally, Figure 3.9 shows the effect of adding paths of length 6 and 7. The flows along paths of length 1–5 are unaffected to the 3rd decimal place, and flows through paths of length 6 and 7 are vanishingly small.

These results emerge from a model which has been deliberately biased toward an efficient energetic system; i.e., the respiratory losses, as a fraction of the flow into each compartment, are lower than those in most natural ecosystems, particularly in the case of the primary consumers and saprophages, the components through which the largest energy flows occur. Yet these conclusions are in opposition to the position of Patten (1985), who shows large amounts of energy flowing through very long paths. In part, this difference is due to Patten's definition of a trophic transfer as any transfer whatever, including ingestion and fecal production. I think the case for not including the latter as a trophic transfer has been made sufficiently clear by Wiegert and Owen (1971) and the brief discussion above. But the major reason the conclusions above differ from the energy cycling development of Patten (1985) has to do with the way that storage is considered.

3.3.6. Energy Storage

Patten (1985) regards matter or energy as "cycling" when it simply sits in storage from one time to the next. But a trophic transfer must, if it is to be

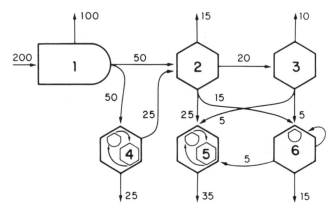

Figure 3.7. Energy flows in a system with 5 trophic levels.

	Trophic Level				
Flux	1	2	3	4	5
F01	200	—	—	—	—
F12	—	50	—	—	—
F42	—	—	25	—	—
F23	—	—	13.4	6.6	—
F14	—	50	—	—	—
F26	—	—	10	5	—
F36	—	—	—	3.3	1.7
Total	200	100	48.4	15	1.7
Ratio to F01	1	0.5	0.24	0.075	.009

ecologically meaningful, involve some transformation to which an energy tax or a leakage term (for elements) can be related. A "transfer" from storage to storage cannot be assigned a meaningful time dependent rate. The path length when such a definition of "transfer" is used can be extended without limit, by simply choosing a smaller time step or extending the total time, since storage by definition has no loss. Indeed, there is no necessary positive correlation between storage and total energy flow, unless the stored biomass is living and thus has a metabolic rate. Much, usually most, of the stored energy in ecosystems is nonliving. Examples of high storage–low flow can be cited, as can examples of low storage–high flow and intermediate systems. In fact, both storage and energy flow are determined by the net specific rate of transfer through the component, a rate whose dimensions are inverse time. Thus, differing net specific rates of energy flow commonly produce situations where energy flow is relatively constant in the face of greatly different storages. The point is clarified by Figure 3.10, with examples of energy flow and storage in a physical and a biological system. The physical system is a copper bar surrounded by a cylindrical adiabatic cover. One open end of each of the bars projects into an infinitely large heat source at a temperature of 100°C, and the

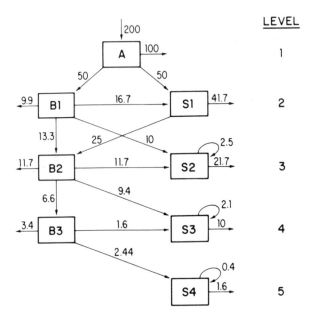

Figure 3.8. Energy flows of Figure 3.7, apportioned by trophic level.

other into a heat sink at 0°C. The area of metal open to the sources and sinks is the same in each case. At steady state, the larger mass will have approximately 10 times the stored heat energy, but the flow of heat through each will be the same. Similar examples could be made with electrical flow and capacitors.

The biological example uses two simple plant detritus/decomposer systems, in each of which the surface aerobic decomposition of detritus is the same. But in the second, a large, low specific rate anaerobic decomposition creates a large storage with no change in energy flow per unit area/time. The same result would be found in the case of two aerobic systems with different net specific rates of degradation of detritus.

3.3.7. Energy Flow Modeling

Models of energy flow are beginning to be used in a number of ways in ecological research. Energy is a conserved unit, at least under the conditions wherein it is transformed in living organisms. Thus, models that use energy as the "bookkeeping" unit are much easier to design, construct, and analyze than are those which employ nonconserved units such as biomass or individuals. The same is true of those models using conserved elements, and there is a useful future for models simultaneously tracking energy and elements such as nitrogen, phosphorus, etc.

Models are being used more and more to suggest and even, in combination with field and laboratory experimentation, to test the conclusions from theory

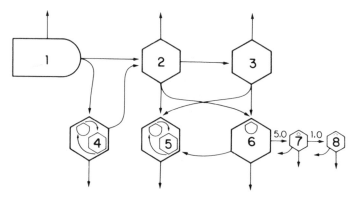

Figure 3.9. Energy flows of Figure 3.7, with two additional trophic transfers added to prey on compartments 6 and 7. In the example, net production going to predation is estimated at the high end (20%). Comparison of trophic ingestions with ingestion at the first level shows a rapidly diminishing increase in percentage as trophic levels increase, as follows.

Input, as Ratio to 1	Trophic Level						
	1	2	3	4	5	6	7
Example 1	1	.5	.125	.03			
Example 2	1	.5	.24	.075	.009		
Example 3	1	.5	.24	.088	.023	.003	.0001

(Fontaine 1981; Wiegert 1986). Static species food web models of the kind discussed in the previous section can easily be converted to dynamic mechanistic simulation models and used to examine the possible ecosystem consequences of adding trophic levels, changing the maximum specific rates of energy transfer, etc. My own work and that of my students with simple yet mechanistically realistic energy flow models with three and four trophic levels suggests some general conclusions that could be tested in the laboratory and/ or field (Elliott et al. 1983). For example, we found entirely deterministic models with ecologically realistic feedback controls of the biotic compartments oscillating in time depending both on the rate at which potentially limiting resources were supplied to the first trophic level; i.e., level of enrichment, and on the relationship between the growth potential and factors limiting the top level consumers (predators) and their resources (prey).

Some additional general attributes of such ecosystems are predicted if models are constructed to mimic the general characteristics of some extremes of trophic arrangements in the field. For example, in open water aquatic systems, one often finds small autotrophs with high growth potential occupying the first trophic level, and large, relatively slow-growing consumers occupying the progressively higher levels. I call these the High–Low systems. In forests, the reverse is generally true: trees have low growth rates and are large, whereas their consumers and their predators are small arthropods with high growth

Physical Example

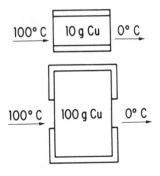

Biological Example

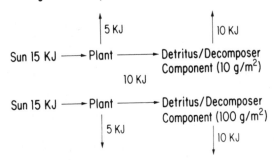

Figure 3.10. Relationship of energy storage and energy flow in a physical and a biological system.

potential. I call these the Low–High systems. Grasslands are an interesting intermediate case. Also, the pelagic zones of aquatic systems very commonly have four well-developed trophic levels (phytoplankton, zooplankton, small planktivorous fishes, predatory fishes), while the aboveground zones of terrestrial forests often have but three (trees, insect consumers, predatory birds). By well-developed, I mean a number of species or life history classes, accounting for a significant energy flow, that can be clearly associated with but a single trophic level. At present, this is certainly a subjective judgement when applied to any specific example, but it is useful in terms of suggesting some of the theoretical consequences of systems with few vs. those with several trophic levels (Smith 1969).

The behavior of such models of the extremes of trophic organization suggests the following generalities: (1) timedependent oscillations in the standing stocks of energy are common in the Low–High simulators; and (2) response to the relaxation of control of the density of higher level consumers also differed markedly in the two types of system. The High–Low systems produced large oscillations in response to such perturbations, but were remarkably stable in the sense that all trophic groups maintained themselves. The Low–High, however, usually responded with small but increasing oscillations that often resulted in the extinction of all but the autotrophic and primary consumer groups and a smaller but more heavily grazed producer standing stock. This is suggestive of a grassland system in which the third trophic group is dependent on, rather than in control of, the large ungulate grazers. The point I wish to make with this brief discussion is simply that such energy flow models can integrate theoretical insight and synthesize conclusions that could be arrived at in no other way.

A further area of ecological energetics that I foresee growing in the future also has to do with energy flow models, but the focus is on the functional controls that are built into these useful mechanistic models, not the time dependent simulations. Remember the comment in an earlier section about the focus of interest shifting from an emphasis on input/output per se to the factors controlling the biological equivalent of energy "conductivity." Most functional controls used in mechanistic models represent some kind of information transfer. Some examples are: (1) if the standing stock of some compartment rises or falls below some threshold values, then some new rules come into play; (2) if a given amount of a particular group occupies a given part of a spatially (or temporally) heterogenous model, then one rule applies, or else some other rule holds; and (3) if the temperature, current velocity, odor, etc. of the system changes, then the rate(s) of energy transfer change.

All of these examples are controls keyed on some kind of perception. There is indeed a transfer of energy accompanying the signal and an energy cost of the perception. Yet these are small, and, to my knowledge, they have never been quantified. For example, what is the energy cost of the pheromones produced by animals? The situation is somewhat analogous to the fact noted earlier that the energy cost of work performed on the environment or on the population is acknowledged, but then ignored as trivial compared to the other components of the energy budget. Is the energy equivalent of the informational controls in energy flow model controls measurable? If so, is this measure worthwhile in the sense that the energy equivalent has meaning for the quantity and quality of control of energy flow in ecosystems?

Certainly a knowledge of the controls themselves is vital to the explanation of how energy is transferred and transformed in ecological systems. This field has come a long way from the earliest descriptive studies. Current technological development, including the capability for dynamic modeling, when coupled with the spirit of experimentation that is rampant in current ecology, may be signaling the beginning of an exciting and productive period of advance for ecological energetics.

4. Freshwater Ecosystems: A Perspective

James E. Schindler

Mathematics may be compared to a mill of exquisite workmanship, which grinds you stuff of any degree of fineness; but, nevertheless what you get out depends upon what you put in; and as the grandest mill in the world will not extract wheat flour from peascod, so pages of formulae will not get a definite result out of loose data.

T. H. Huxley (1897)

Present day ecosystem ecology is founded on the premise that behaviors of homogeneous landscape units (ecosystems) can be adequately characterized by a few "emergent properties" that are generated by biotic energy flows and material cycles (E. P. Odum 1983). This approach also assumes that behaviors of local ecological systems converge into distinct regional ecosystem patterns that are deterministically generated by interactions of biota with their environment. This popular "black box" approach to the study of ecosystems apparently evolved from the uniformitarianist views of James Hutton (1795) (see Simpson 1970). These views, advanced by Playfair (1802), and fully expressed by Lyell (1830–1833) contain both a method (research technique) and a system (an all-embracing theory) of research (Hallam 1983). Lyell and Darwin advanced uniformity as a methodological postulate that reinforced the notion that scientists should work with small-scale events since they can be seen and investigated. The assumption that natural laws are spatially and temporally invariant is also a part of the uniformitarian method. For example, Darwin used artificial selection by animal breeders and small differences in geographic variation of species' races as observable examples of evolution (Gould 1986). In the same way, paleolimnologists compare nutrient and pH requirements of plankton in modern lakes with stratigraphic analyses of lake sediment cores to generate a composite picture of lake ontogeny (Rymer 1978).

As a system, uniformitarianism was used in geology to replace the commonly accepted preternatural causation with the notion of landscape evolution that depends on slow, cumulative change, produced by natural processes operating at relatively constant rates. Ultimately, natural steady state processes were also invoked in biology and natural history to replace the Creator as the source of unity and harmony in nature (Egerton 1973; Kingsland 1985). The assertions of modern ecosystem ecology were anticipated by Playfair's (1802) notion that time collects and integrates the parts of a natural progression into a whole that is more than the sum of its parts; Huxley's 1869 (in Hallam 1983) concept that living forms are adapted to the time-frame of geology provided a model for evolutionary research. Thus, various aspects of the uniformitarian system, most of them with origins in geology, laid the foundations for other sciences (Simpson 1970). Landscape–organism typologies, superorganismic concepts, succession, climax, eutrophication models, and steady state perspectives became 20th century ecological–limnological concepts that were derived ultimately from the 18th and 19th century uniformitarian system of Hutton and Lyell.

By the 20th century, ecologists metaphorically imposed harmony on nature by borrowing laws of order and process from organismic (homeostasis), social (hierarchy), engineering (electricity, machine systems), and economic (production-optimization) models (Kingsland 1985). Mathematicians, trained in laws and principles of physical sciences, invoked equilibrium ($dx/dt = 0$) for analytical solutions to mathematical expressions. Biologists, who were trained to see heterogeneity and individuality, accepted mathematical equilibrium and the uniformitarianism system (things operate this way—always) because it seemed to be an acceptable way to integrate the complexity of natural associations and processes. The observed slow changes of terrestrial processes in temperate regions supported acceptance of the uniformitarian system and its balance of nature assumptions.

Although ecologists tacitly accepted uniformitarian views, the natural history foundations of ecology constrained the science to the notion that species populations should be conspicuous, tangible objects of ecological inquiry. Descriptions of species and kinds of species were combined with 19th century geologists' landscape–evolution concepts, which, in turn, led to the assumption that natural ecosystems could be defined in terms of lists of constituent species. Groups of organismic associations were classified by static climate factors to produce climax communities and biomes, and subunits were isolated into physically bounded landscape units (watersheds and lakes) for study of ecosystem processes. While ecosystem and community classifications, based on dominance criteria, keystone and indicator species, and functional characteristics of organisms subsumed species lists as criteria for examining ecological relationships, concepts of balance, order, and steady state remained as vestiges of the uniformitarianist system in ecosystem ecology. These historical perspectives became an essential feature of modern "holological" (holistic) ecology (Rigler 1975 b; Odum 1983). This approach assumes that ecosystems operate as large homogeneous landscape units that can be characterized by measurement of a few "emergent properties" over long (> annual) time scales. This perspective is now part of the research tradition of modern ecosystem ecology.

According to historicist philosophers such as Kuhn (1962), Laudan (1977, 1981), and Lakatos (1970), a research tradition is a broadly-based foundation of many theories and an accepted way of viewing fundamental phenomena. Historicist philosophers understand science as a process of human thought and behavior exhibited by practicing scientists, where the basic unit of progress is the solved empirical or conceptual problem. With this, the function of any theory within a research tradition is to explain scientific observations, solve problems, reduce ambiguity, and show that what happens is intelligible and predictable. From this point of view the issue of whether or not a theory is true is irrelevant, since the progress of a theory is defined by the degree to which it solves more scientific problems than its rival theories (Jacox 1981).

While the holological approach is supported as the dominant perspective in modern ecosystem research, a different sense of organization emerged from the ecological survey work conducted by Forel on Lac Leman (Forel 1895). Elton (1966) noted that Forel's work was the first survey of a "great ecological system" that presented a broad account of how the features of a lake are "integrated into a whole working system in which the flora and fauna form part of the network of channels for matter and energy." Elton and Miller (1954) echoed Forelian perspectives in their approach to the survey of ecological systems by noting that "an ecological survey is not just a catalogue of biological properties of individual species or a list of species, or a series of censuses. It is, or is headed towards, a synthesis that will describe not only the parts of a complex system but the interaction and balance between them, and the dynamic properties of the system as a whole." Elton (1966) viewed the pattern of communities in nature as a product of environmental dynamics wherein the parts of a complex system are "arranged in a hierarchy of recognizable and repeated patterns that are produced, maintained or changed by potentially violent dynamic forces that do not normally operate at full power." Ecologists use the "recognizable and repeated patterns" of nature (i.e., populations and communities) for classification (Gilmour 1951). And, classified units are then used for observations and inferences about biological interactions. However, if the very patterns themselves are products of "dynamic forces," then interpretations of the biological interactions must be ultimately related to the dynamic properties of the "system as a whole."

The notion that ecosystem properties can be assembled from information about specific component processes has been termed the "merological approach" (reductionist) to ecosystem research (Odum 1983). However, merological research preserves the influence of the uniformitarian tradition of ecosystem ecology by assuming that the whole system is more than the sum of the parts (Playfair 1802), and by adopting the perspective that ecosystems change gradually over time.

The research tradition of modern ecosystem ecology has limited investigations of the processes that contribute to the unique features of local aquatic environments. The assumption that lakes can be characterized by a few emergent properties of homogeneous landscape units ignores what is known about lacustrine dynamics. However, aquatic scientists are now beginning to realize that short-term (< annual) dynamic processes play an important role in the development of ecological and evolutionary patterns in nature (Legendre and Demers 1984). Hydrodynamics influence the physical and chemical characteristics of environment that define habitats of organisms and delimit the ranges of species. Southwood (1976, 1977) recognized the importance of habitat dynamics as a template for population responses, and placed emphasis on the relationship between habitat duration and generation times of organisms of different sizes. Habitat dynamics should also provide a template for ecosystem interaction and biogeochemical processes. If aquatic ecosystems are viewed as hierarchical ensembles of species' habitats that define the natural patterns of biological interaction, then natural order and function should emerge as products of the real behavior (physical, chemical, and biological) of ecosystems in time and space. These local dynamics should converge with regional ecosystem patterns if the assumptions of modern ecosystem ecology are accurate. However, such assumptions should not be accepted without consideration of the influence of temporal and spatial dynamics on the behavior of lacustrine ecosystems.

A basic premise of this chapter is that lacustrine ecosystem structure and function is dependent on the relative scales of spatial and temporal dynamics of the local environment. A corollary of this premise is that perceptions of the scales of environmental dynamics affect interpretations of the dominant processes that control ecosystem structure and function. This point was emphasized by Harris (1980 a) for temporal and (horizontal) spatial scales by illustrating the "scales of interest" of physiologists, ecologists, and climatologists. Figure 4.1, while not exhaustive, indicates that scientific interest in processes leads to distinct perceptions of the nature of the environment. Ecosystem ecologists must be capable of deriving an independent perception of the scales of environmental change and ecosystem behavior. This relational approach should not be constrained by any observer dependent perception of order or structure (Wicken 1985), and may be illustrated with examples from aquatic systems wherein biological processes are related to pelagic habitats that are defined by scales of hydrodynamic interaction.

4.1. Aquatic Ecosystems

4.1.1. Habitat Dynamics

Research on hydrodynamics has advanced far beyond original classification systems based on climate or complete, annual mixing cycles (Hutchinson and

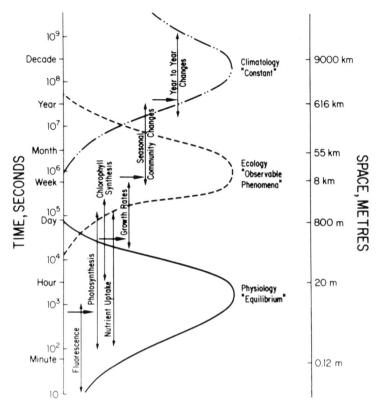

Figure 4.1. The hierarchy of algal responses to the spectrum of environmental fluctuations. The three sigmoid curves roughly define scales of interest to physiologists, ecologists, and climatologists. The horizontal arrows indicate that higher frequency (lower level) processes collapse into higher level responses. From Harris (1980a), by permission of Can. J. Fish. Aquat. Sci.

Loffler 1956; Hutchinson 1957; Walker and Likens 1975), and more refined estimates of temporal and spatial scales of mixing are now accessible to investigation (Imberger et al. 1978; Spigel and Imberger 1980; Imberger and Patterson 1981; Patterson et al. 1984; Imberger and Hamblin 1985; Schindler et al. 1986). As a consequence, much of the observed structure of plankton communities is now considered to be a function of hydrodynamic processes in aquatic environments (Harris 1980a, 1980b; Legendre and Demers 1984). Missimer (1986) identified four major meteorological cycles or periodicities that determine temporal and spatial (vertical) scales of mixing in a temperate lake (Fig. 4.2). At one end of the scale, solar radiation contributes to the annual cycle of heating and cooling that is the dominant period for most holomictic (complete annual mixing) classifications. At the opposite end, diel cycles of heating and cooling determine daily depths of mixing. The two intermediate cycles, operating with periodicities of days to months, are generated by major

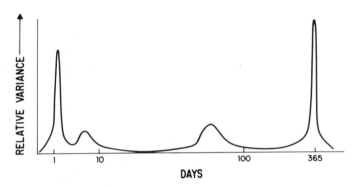

Figure 4.2. Estimates of the relative variance of vertical mixing from four major meteorological cycles. From Missimer (1986), with permission of the author.

weather patterns and the passage of warm and cold fronts (Mitchell 1976). Figure 4.2 is an idealized representation of the amount of variance that is contributed by the four major weather cycles to the total variation in the depth of mixing. Annual and diel cycles are shown as sharp spikes because of their astronomically determined periodicities and high contribution to the total variability of mixing depth. The intermediate cycles of variability are shown with broad maxima because of their quasiperiodic nature and lower contribution to the total variability of mixing depth.

While temporal scales of mixing are determined by regional characteristics of meteorological processes, the overall response characteristics of a lake are system-specific and influenced by the local characteristics of the ecological system. Morphometry, orientation of the lake, wind fetch, and dissolved and suspended materials in the water (optical clarity) are among the system-specific characteristics that influence local meteorological–hydrodynamic interactions. For example, Ford and Stephan (1980) demonstrated that differences in morphometry affect depth and timing of mixing in lakes under identical meteorological influences. Imberger and Parker (1985) also noted that separate basins in large systems may have their own characteristic responses to meteorological variables. Organisms also influence interactions, since radiation penetration, heating depths, and intermediate mixing cycles are altered by amount, size, and composition of suspended matter in the water column (Stavn 1982,1983). For example, proliferation of microplankton decreases heating depths and increases frequency of partial mixing, which, in turn, contributes to the segregation of intermediate water masses. This can be accompanied by changes in temperature, density, oxygen concentration, and nutrient regeneration (Waldron 1985). In any case, the system-specific signature of the temporal and spatial scales of meteorological and hydrodynamic interaction provides an important habitat feature of local aquatic environments. Many organisms that occupy lacustrine ecosystems have life cycles that are comparable to these time scales of environmental change. This can be illustrated by examining the relationship between habitat dynamics and generation times or turnover times of organisms.

4.1.2. Biological Dynamics

According to Southwood, the generation time of organisms, t, is an important concept, for the reasons listed below:

1. It provides an estimate of the time it takes a population to return to equilibrium following a disturbance.
2. Generation time is correlated to organism size, probably due to the inverse relationship between longevity and total metabolic activity per unit weight. Size also affects interspecific competition, predation, and effects of spatial and temporal variation in habitat.
3. Per capita rate of increase, r, depends on net reproduction, R_0, and generation times:

$$r = (\ln R_0)/t.$$

Turnover time, T, might be considered a surrogate of generation time. Leopold (1975) considered the turnover time of biological units to be a central theme in biology, since it provides the basis for metabolism, homeostasis, evolutionary change, and adaptation. Both Southwood and Leopold note that generation times or turnover times are directly related to size and longevity of organisms. An estimate of the range of organismic (trophic level) turnover times for lacustrine ecosystems can be derived from the reciprocal of the ratio of production to biomass (P/B = turnover rate). Brylinsky (1980) listed a table of the ranges and average values of turnover rates for phytoplankton, zooplankton, and benthos from IBP study sites; Morgan et al. (1980) listed a range of values for fish turnover rates. These values, combined with estimates of longevity, provide the limits for the trophic levels illustrated in Figure 4.3. A comparison of the estimated food web turnover times (T) with habitat (mixing) periodicities (H) provides a simplified relational perspective for comparing physical and biological properties of lacustrine ecosystems. Figure 4.3 illustrates how relational time scales of physical and biological processes might influence the various approaches (physiological, behavioral, population, community structure) that are used for ecological studies. It also follows, however, that ecologists must have an understanding of the relational characteristics of the ecosystem subsets *before* invoking a particular theory to account for the biological interactions. This point is emphasized because modern ecosystem theory is based on the untested assumption that any emergent property can be used for comparisons of ecological systems. However, if lacustrine habitat characteristics are determined by local conditions, then specific trophic level and biological responses (physiological, population, behavioral) to environmental change may also be expected to be determined by local conditions. This outcome would then invalidate the (holological approach) assumption that lacustrine ecosystems can be studied by using properties such as annual

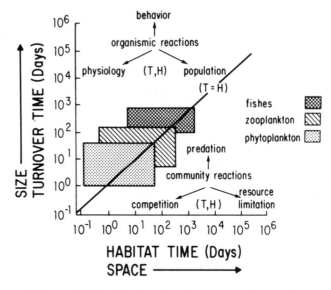

Figure 4.3. Idealized scales of ecosystem interaction.

chlorophyll or energy concentrations of plankton in a euphotic zone that is assumed to be isotropic. This conclusion would also apply to assumptions that community characteristics are determined entirely by immediate biological responses such as predation, competition and resource limitation.

4.2. The Relational Ecosystem

4.2.1. Organisms and Populations

Graphically combining population responses with habitat mixing periodicity provides a visual characterization of the relative habitat–organism scales that might be anticipated in aquatic ecosystems (Fig. 4.3). The figure is divided by a line that represents the steady state condition ($T = H$), where organism turnover time matches the periodicity of habitat. This is the only condition in which the theories of population ecology (e.g., May 1976) provide an adequate means of resolving the effects of habitat dynamics. Increasing organism size (and turnover time) should decrease sensitivity to spatial-temporal variations of the habitat subsets of the ecosystem. Hence, larger organisms should integrate short-term dynamics of environment in their population responses. In lacustrine ecosystems, fish populations should provide an adequate means of assessing long-term behavior of aquatic ecosystems. This is essentially confirmed by the work of Ryder et al. (1974) and Roff (1986), as well as by recent research by Schindler et al. (1985) on the effects of acid rain on Canadian lakes. Of course, the converse is also true, since population responses of smaller short-lived organisms should be more sensitive to short-term habitat dynamics.

This illustrates the point that the assessment of population dynamics of organisms in nature must be defined by referring to the time scales that are appropriate to the organism and the habitat that influences them.

In a time series analysis, assessment of population dynamics would require sampling at a frequency of at least $1/2\ T$ to avoid aliasing (Stearns 1981). Figure 4.4 (Legendre and Demers 1984) illustrates the effect of sampling system properties at incorrect time scales. The 60 day cycle illustrated by the broken line is an artifact of sampling at incorrect intervals. Populations of small-sized organisms are usually not sampled adequately, since most sampling schedules are not based on environmental processes but on time scales that are convenient for the observer (weeks and months). By failing to sample at appropriate time scales, an observer may witness one of several possible conditions that will influence their interpretations of the processes that occur in the aquatic environment. These conditions are represented in Figure 4.3 by the equality, $T = H$, and the inequalities, $T > H$ and $T < H$.

At time scales $T > H$, the turnover times of organisms are much greater than habitat duration times, and several processes are affected by these conditions. According to McFarland and Houston (1981), homeostatic responses might be expected if the action of environmental factors contributes to physiological imbalances. Organisms detect environmental imbalances and correct them by physiological and behavioral mechanisms that may act separately or in concert. If short-term physiological corrections cannot restore homeostatic balances in the immediate environment, behavioral mechanisms are activated (gulping, panting, sinking, etc.) that result in the intake of the required commodities to facilitate the restoration of physiological conditions. If adequate conditions are not found in the immediate environment, other motivational systems are activated. Large, long-lived nektonic organisms move to preferred zones based on food resources and temperature or oxygen concentrations (Carr

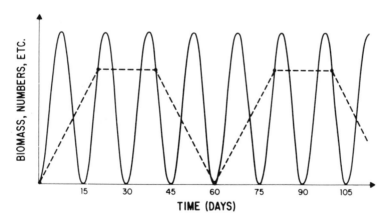

Figure 4.4. Aliasing of a 15 day cycle to spuriously suggest a 60 day cycle as the result of sampling every 20 days. From Legendre and Demers (1984), by permission of Can. J. Fish. Aquat. Sci.

1986). Small organisms such as phytoplankton are entrained to the dynamics of the physical environment and generate heterogeneous distributions (Denman 1977; Lewis et al. 1983; Legendre and Demers 1984; Abbott et al. 1984; Imberger 1985). Hence, environmental dynamics affect physiology, behavior, and the spatial distributions of organisms in aquatic ecosystems. Allogenic forces may be responsible for changes in the distribution of organisms as well as changes in abundance due to reproduction and population growth (Reynolds 1984). Reynolds (Fig. 4.5) noted that phytoplankton communities may exhibit periodic progression from one dominant assemblage to another. Autogenic successional changes may be traced to nutrient concentrations and the ratio of nutrients. Increased mixing (perturbation) at any time may shift the system to new conditions where a new succession may be initiated, or, as the system becomes less mixed, it may revert to a previous dominant community. Reynolds' argument illustrates the point that without knowledge of the antecedent habitat conditions, observers may incorrectly interpret the processes that are influencing organismic responses in the ecosystem. In any case, assessment methods based on organismic responses that are assumed to occur in stable isotropic environments will give unreliable interpretations of ecosystem function and structure (Legendre 1981; Harris 1980a, 1980b; Legendre and Demers 1984).

4.2.2. Community Processes

At time scales such that $T < H$, habitat dynamics are relatively constant compared with life history phenomena, a condition that is required for classic autogenic succession. Hence, interspecific community responses should dominate biological organization. Although it is often assumed that resource competition structures communities in stable environments (Chapter 3), there are other possible contingencies that can influence this organization. For example, Grime (1977) considered stress and disturbance to be two major external factors that limit phytoplankton biomass in any habitat. Stress includes conditions

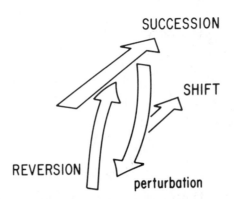

Figure 4.5. Three directions of periodic progression from one dominant assemblage to another. From Reynolds (1984), with permission of Blackwell Scientific Publications.

that restrict production, such as shortages of light or mineral nutrients, and temperature extremes. Disturbance includes conditions that lead to partial or total destruction of plant biomass, including the activities of herbivores and pathogens, chemical toxicity, and physical destruction. A consideration of four permutations of these two external factors leaves three possible contingencies [deleterious effects (−); neutral, or positive effects (+)] that must lead to the development of distinct biological strategies to cope with environmental conditions. These combinations are:

		Stress	
		High	Low
	High	−/−	−/+
Disturbance			
	Low	+/−	+/+

It may be assumed that if organisms cannot escape, habitat conditions of high stress and disturbance preclude survival. However, the three remaining contingencies permit accommodation if populations follow adaptive strategies that include:

Low stress–low disturbance = competitive organisms

High stress–low disturbance = stress-tolerant organisms

Low stress–high disturbance = ruderal (garbage) organisms

4.2.3. Competition

Although Grime presented the contingencies as extremes, they fit a number of theoretical community models that are used to account for the structure of biological communities in constant environments. For example, low stress, low disturbance "competition" can be equated with classical Lotka–Volterra models that do not explicitly consider mechanisms of organismic interaction. Competition between two species requires an estimate of values of six constants $(\partial, \beta, r_1, r_2, K_1, K_2)$. Interactions between species are described in terms of the competition coefficients. Coefficient ∂ measures how much each individual of species 2 depresses the potential carrying capacity, K_1, of species 1. Similarly, β is a measure of how much each individual of species 1 reduces the potential carrying capacity, K_2, of species 2. The maximal per capita rates of increase for species 1 and 2 are r_1 and r_2, respectively. With these terms, Lotka–Volterra equations for competition between two species are written as:

$$\partial N_1/dt = r_1 N_1 \ (K_1 - N_1 - \partial N_2)/K_1$$

$$\partial N_2/dt = r_2 N_2 \ (K_2 - N_2 - \beta N_1)/K_2$$

The theoretical power of classical competition models is empirically limited by their equilibrium requirements. These requirements state that, given a constant environment ($T < H$, but unspecified in terms of resource characteristics), the populations within the community will not change (Levins 1976; Leigh 1976). Tilman (1982) warned that application of Lotka–Volterra models to environments with a specific resource gradient may be dangerous, because it is difficult to know which mechanisms of competition are reflected in the competition coefficients (∂,β) and carrying capacities (K_1,K_2). However, while the models do not appear to be applicable to planktonic assemblages (Hutchinson 1961; Richerson et al. 1970; Petersen 1975; Smayda 1980), the relational perspective indicates that they may be applicable to problems that involve long-lived species (i.e., fish, benthos) that are not resource limited and have a turnover that is long relative to the habitat dynamics of the lacustrine ecosystem.

4.2.4. Stress Tolerance and Resource Limitation

High stress, low disturbance conditions may also affect perceptions of community organization, since limitation of nutrients or light can lead to competition for those resources. Tilman (1982) considered a resource to be any substance or factor that is consumed by an organism and leads to a direct relationship between growth rates and supply. Competition for those resources may be another factor that determines the relative abundances of organisms within a habitat (Tilman and Kilham 1976; Tilman 1977; Kilham and Kilham 1980). The advantage of resource models over classical Lotka–Volterra competition models is that explicit mechanisms are invoked for competition. For a given resource and species, physiological abilities are characterized with a Monod-type (rectangular hyperbola) equation that relates resource-specific growth rates to supplies or concentrations of that resource. For example, competition for resources has been used to account for phenomena such as diatom succession and alterations of community structure in nitrogen-limited environments. Nitrogen resource models of algal succession are based on the premise that the environment is relatively undisturbed ($T < H$), and that cyanobacteria are free from destruction by herbivores and pathogens. Under these conditions, cyanobacteria are capable of adjusting N/P ratios by nitrogen fixation (Schindler 1977). Kilham and Kilham (1984) noted that when nutrient loading ratios of nitrogen to phosphorus are 31 or greater, green algae such as *Scenedesmus* dominate the phytoplankton associations. When the ratio is dropped to 11:1, lakes become dominated by nitrogen-fixing cyanobacteria. Presumably, the competitive advantage conferred to cyanobacteria by their ability to fix nitrogen, their limited destruction by herbivores, and their ability to float in a low turbulence environment favors their succession and proliferation, and ultimately, the formation of undesirable algal blooms.

However, strict resource ratio hypotheses do not apply to all environments. Although several reports indicate that nitrogen fixation is enhanced by increasing nutrient conditions (Horne and Fogg 1970; Stewart et al. 1971; Vanderhoef et al. 1972, 1974), there is good evidence that it can be inhibited by point and nonpoint source effluents (Horne and Goldman 1974; Pearson and Taylor 1978). Horne and Galat (1985) note that nitrogen fixation by cyanobacteria is not easily related to aquatic system fertility and that levels of nitrate or ammonium in the environment do not give a reliable guide to the need for nitrogen fixation.

Although resource models have been used at all trophic levels, extreme care must be exercised in their general application since multiple factors influence responses of organisms to changing conditions. Resource models often assume that total nutrient concentrations adequately portray environmental supplies available for plant nutrition. However, Liebig (1855) was the first to point out that there is a difference between the amount of a nutrient measured in soil and the amount available for plant nutrition. Availability was perceived to be a function of chemical factors and physical quality of soil. It is now known that physiological and chemical reactions complicate interpretations that depend on simple growth rate and resource supply relationships. Insoluble or complex chemical formations limit biological availability (Giammatteo 1986), and changes in physiological and chemical states alter patterns of resource availability, utilization, storage, and growth rates of consumer populations (Speziale 1985; Giammatteo et al. 1983). Chemical conditions such as high pH favor formation of solid chemical complexes so that precipitation of solids in a physically stable environment effectively separates organisms from their resources. The nutrient conditions also affect macromolecular composition of phytoplankton populations (Giammatteo 1986), which, in turn, affect behavioral and reproductive responses of herbivore populations. In extreme cases of stress (temperature, low light, salinity, pH, or anoxic conditions), slower growing, stress adapted species may succeed in the absence of direct competition.

Brock (1986) recently reviewed the special biochemical and physiological requirements for thermally adapted species. Spatial separation of aquatic communities into euphotic and aphotic regions also creates habitats which support organisms with specialized adaptations that are normally ignored in ecological literature (Fee 1976, 1978; Richerson et al. 1978; Cullen 1982; Moll and Stormer 1982; Pick et al. 1984). For example, Bird and Kalff (1986) recently demonstrated that organisms common to the metalimnetic association, such as *Dinobryon sertularia*, are capable of deriving at least 50% of their nutritional requirements from the ingestion of bacteria. Anoxygenic photoautotrophy, chemolithotrophy, buoyancy regulation, and shade adaptation are just some of the other processes that accompany life in stable deep layer habitats (Sorokin 1978; Parkin and Brock 1981). Assumptions of organismic equilibrium or interpretations of trophic conditions and biogeochemical interactions based on slow gradual changes of environment obscure the chances for observing these biogeochemically complex processes in nature.

4.2.5. Predation

In addition to classic competition and resource competition, other interorganismic relationships, such as predation, affect community structure (Taylor

1984). Predation modifies the primary role of interspecific competition, and predator structuring of aquatic communities is a well-known phenomenon (Zaret 1980). Many aspects of a species' niche (body size, feeding station and method, etc.) that are considered to influence classic competition also affect relationships between predator and prey. Jeffries and Lawton (1984) argue that an organism's niche is influenced by many variables including physical environment, nature and rate of supply of food resources, competition for limiting resources or space, and natural enemies. However, the consequences of sharing a common natural enemy are identical to conventional forms of interspecific competition for limiting resources, and there may be competition between "victim" (prey) species for "enemy-free space." Jeffries and Lawton defined enemy-free space as "ways of living that reduce or eliminate a species's vulnerability to one or more species of natural enemies." According to this idea, the invasion and maintenance of a "victim" population in a community requires its rate of increase to be higher than the rate that polyphagous natural enemies find and destroy the population. For successful invasion and establishment:

$$r_j > a_j P$$

where r_j = instantaneous rate of increase of invading victim species j; a_j = attack rate (or area of discovery) of an established natural enemy on species j; and P = density of established polyphagous enemies.

Hence, characteristics of j that reduce a_j (size, color, and feeding) favor the establishment of j in a community. Enemies are less likely to recognize j as prey if it differs from other victim populations in the community. While an enemy-free space continuum from total vulnerability to complete indifference to presence or absence of particular predators could exist, Jeffries and Lawton speculate that communities constrained by natural enemies will not converge unless the enemies are the same or have similar properties. On the other hand, standard competition for limiting resources should lead to convergence in community structures under similar habitat conditions. Given that only a limited number of prey types are able to coexist with a particular predator, competition for enemy-free space should lead to a broadly constant ratio of prey to predators in food webs.

Classic ecological theory views predators as limited by resources, and the diversity of predators, in particular, as being limited by the diversity of prey (Pimm 1982). However, Briand and Cohen (1984) argue that in both constant and fluctuating environments the number of predators is causally more important in controlling the numbers of prey than the reverse—a conclusion that is consistent with the hypothetical constructs of Jeffries and Lawton. Briand and Cohen also argue that on the average, the fraction of top, intermediate, and basal species in a food web are independent of the total number of species. While Briand's (1983; 1985) analyses of food webs support the broad claim that the physical nature of environments is the major determinant of food web structures, the mode of operation is uncertain.

4.2.6. Ruderal Growth

Ruderal species that rapidly proliferate in low stress–high disturbance environments should have many of the specialized features that are considered in both resource competition and enemy-free space models of interspecific interaction. Ruderal species are capable of rapid accumulation of available nutrients and rapid growth responses. These characteristics favor successful colonization by these species in the presence of predators. Modifications of the enemy-free space would also insure success in the presence of predators. For example, Cladocera and rotifers are ruderal species that are capable of rapid proliferation, even in the presence of predators. Parthenogenetic reproduction assists the rapid expansion of populations in warm, food-rich waters, and large broods and rapid developmental rates ensure the colonization of populations under favorable food conditions. Cladocerans and rotifers also show morphological changes (cyclomorphosis, size) that can be interpreted as alterations of the enemy-free space.

Food web nutrient recycling is an example of a synergistic interaction that may confuse interpretations that community structuring results from resource limitation or predator interaction. Sterner (1986) recently demonstrated that herbivores can cause an increase in algal reproductive rates by regeneration of nitrogen. As Sterner notes, however, it is not just the response of a particular taxon to nitrogen fertilization that affects this response, but a combination of the effects of nutrient uptake, growth response, and differential mortality due to the herbivore feeding preferences. Differential effects of herbivore defecation can also influence nutrient recycling. Cladocera, such as *Daphnia pulex*, disperse their feces into the environment, whereas other herbivores such as copepods form membrane bounded fecal pellets. Nutrient regeneration will be affected by fecal production if the "packaged" fecal units are large enough to sink out of the mixed layer before they are degraded by microorganisms. Since mixed layer depth and persistence is influenced by interactions of meteorological processes with local system characteristics, even the effects of defecation can be influenced by specific system properties.

Finally, predator reproduction and foraging rates are affected by the macromolecular composition of the food. Eidson (pers. comm.) observed that *Ceriodaphnia reticulata* fed on high protein diets reproduce more rapidly and are more active than genetically identical organisms that are fed on high carbohydrate diets. If this observation holds true, then fat storage, reproduction, and foraging behaviors may all be intimately linked through complex food web relationships.

4.3. Discussion

In the past decade, aquatic scientists have become increasingly aware of the effect of physical hydrodynamic processes on biological components of pelagic environments. Interest in effects of hydrodynamic processes increased during the 1970s with the development of new analytical techniques such as in vivo fluorescence assays for continuous measurements of chlorophyll a (Lorenzen 1966), and spectral analysis procedures permitted development of new models of phytoplankton variability (Denman 1977). The steady increase in literature on the effects of hydrodynamics on biological processes led Legendre and Demers (1984) to claim that a new discipline of dynamical biological oceanography and limnology was emerging from the scientific community.

Aquatic sciences are fortunate in that habitat dynamics and biological interactions are observable on relatively short time scales. But studies of physical processes have advanced far beyond the biological theories and methods of data collection (Wang and Harleman 1982). Although recent literature has featured a number of very exciting research results for interpreting physical processes, most of the biological literature still emphasizes long-term ($>$ annual) behaviors of lakes that are assumed to be homogeneous and regionally similar. Annual heat budgets and mixing behaviors are generally utilized as primary descriptors of lake–climate interaction (Hutchinson and Loffler 1956, Hutchinson 1957; Walker and Likens 1975; Ragotzkie 1978). However, some exceptions to this generalization have appeared in the literature of tropical limnology. W. M. Lewis (1973) noted that Lake Lanao in the Philippines had an increased sensitivity to changing meteorological conditions due to low density differences of water layers and a reduction in deep Eckman mixing because of reduced Coriolis Force effects at low latitudes. He identified three classes of weather phenomena (breezes, squalls, and storms) that were capable of mixing the lake to progressively deeper depths. Isolation of adjacent water layers between these partial mixing events was considered to be a major process contributing to nutrient regeneration. This partial and frequent mixing due to meteorological events was termed ateleomixes.

Observations on partial mixing in surface waters of temperate lakes have also demonstrated that other scales of meteorological interaction are active in all lakes (Zimmerman et al. 1979; Frempong 1981). Although Legendre and Demers (1984) point to a growing body of literature that considers the effect of hydrodynamic forces on biological processes, most ecologists ignore environmental dynamics as an important feature of ecosystem structure and process. Typically, sampling schedules do not adequately reflect the nature of the dynamics and processes. If habitat mixing dynamics are considered to be a suitable template for bionomic strategies, inadequate sampling can lead to incorrect interpretations of biotic interactions. For example, antecedent mixing processes in lakes ($T > H$) leads to redistribution of organisms with the potential for reversion of species succession (Reynolds 1984), and ruderal species would be favored in turbulent environments in spite of predator activities ($r_j > a_j P$). However, since long-lived predators integrate short-term system dynamics into their population responses, an observer might conclude that community structure is ultimately affected by predator interactions. This conclusion would be inescapable since other levels of community or organismic interaction such as competition and resource limitation would occur on time scales that are shorter than the observer defined sampling of the ecosystem.

On the other hand, long periods of habitat stability will increase the chances that an observer will witness direct biological interactions within the environment. Resource competition and autogenic succession may then appear to be the dominant processes that structure biotic interactions. In either case, the observer cannot interpret how the biota are responding without some prior knowledge of the system dynamics. The problem of data interpretation is compounded by considerations of environmental interaction at the spatial scales considered. Nonseasonal, partial mixing dynamics are not adequately characterized by thermal profiles or seasonal thermoclines. Furthermore, the assumption that the epilimnion is *always* turbulent to the level of the thermocline is incorrect. Mixed layer dynamics depend on the *immediate and the antecedent* meteorological conditions (Patterson et al. 1984; Imberger 1985); the assumption that the epilimnion is isotropic is patently wrong and misleading.

The tradition of modern ecosystem research assumes that natural systems can be described and classified without looking at their individual parts. This perspective is supposed to represent one of the highest levels of integration of natural phenomena (Bartholomew 1986). However, ecosystem research should not be restricted to approaches that depend on untested assumptions of the constancy and uniformity of processes in nature. A scientific understanding of how ecosystems operate can only be advanced by a careful analysis of the natural interactions that occur between levels of system organization. At the same time, subsystems must be considered in their own particular physical, chemical, and biological time frames, if interconnections are to be found.

It is apparent that the choice of scales for observing natural interactions is a very important feature of ecosystem research. This point was emphasized by Hutchinson (1964) when he noted that:

> It is desirable to think for a moment about certain scale effects characterizing the lacustrine microcosm when viewed by a human observer. If we suppose that an organism reproduces about once every week for the warmer half of the year and on an average about once every month in the cooler half, it will have about thirty generations a year. This corresponds in time to about a millennium of human generations, and considerably longer for those of forest trees. In the case of the latter, we should expect in thirty generations some secular climatic change to be apparent. We should not expect in a tree the seeds or resting stages to remain viable while thirty generations passed, and in the larger animals no such stage exists. The year of a cladoceran or a chrysomonad, in both of which groups rapid reproduction may alternate with the formation of resting stages, is thus in some ways comparable to a large segment of post-glacial time, though in other ways the comparison either to several millennia, or to a year in the life of a human being or tree, is definitely misleading. Another peculiar scale effect is that, in passing from the surface to the bottom of a stratified lake in summer, we can easily traverse in 10–20 m. a range of physical and chemical conditions as great or greater than would be encountered in climbing up a hundred times that vertical range of a mountain.

While ecologists typically utilize human time scales (weeks and months) for making observations and drawing inferences from nature, new methods are now available for defining the physical nature and dynamics of aquatic habitats. If habitat dynamics are used as a template for the study of biological inter-actions and adequate spatial and temporal scales are considered, then existing theoretical ecological models may provide some insights into the processes that contribute to ecosystem organization in nature. Once the significance of the dynamic interactions is assessed, biological processes can be analyzed. The recognition of habitat dynamics in biologically sensitive time frames minimizes the potential for rejecting hypotheses or failing to formulate hypotheses because natural phenomena lie unseen (Allen and Starr 1982). In the final analysis, recognition of the relational features of aquatic ecosystems provides an avenue for eliminating the traditional dichotomy between organismic (merological) and black box (holological) approaches to ecosystems research.

5. Achievements and Challenges in Forest Energetics

William A. Reiners

5.1. Historical Review

This historical review begins at a benchmark year and occasion—the refresher course on energy flow and ecological systems organized by Frederick Turner for the AIBS meetings in College Station, Texas in 1967 and published in the American Zoologist in 1968. The leading presentation was by Eugene P. Odum and was titled "Energy flow in ecosystems: a historical review." In that paper, Odum succinctly sketched the development of our thinking on ecological energetics. Beginning with Forbes's classical ideas on "The lake as a microcosm," Odum proceeded with the contributions of Thienemann and Elton on the concepts of niche and pyramids. He showed the beginnings of thermodynamics in the writings of Lotka and the first conceptions of energy budgets and primary production among limnologists such as Birge and Juday in the 1930s. Odum noted that it was limnologists who were the first to employ these ideas, perhaps because of the convenience of aquatic systems for measurement. The trophic-dynamic concepts of Lindeman, with contributions by Hutchinson, Clarke, and MacFadyen, were keystone writings of the 1940s. Odum also introduced the energy flow diagram and measures of community metabolism which he and his brother, H. T. Odum, conceived of in the 1950s. He cited a number of studies on secondary production and energy flow in populations, and the use of laboratory studies. Finally, he pointed to the growing attention paid to energetics in the areas of succession and more formalized systems ecology.

Of the 47 benchmark papers cited in Odum's paper, 14 might be described as general treatises like his own books, 20 involved aquatic or marine studies, and only 13 involved terrestrial studies. Most interesting is that only one of the terrestrial studies was of a forest. This distribution probably did not misrepresent "mainstream" ecological thought. At that time, there had been much less study of forests as ecosystems than of aquatic or marine systems. Most terrestrial work to that date had involved old fields or grasslands. This poor representation of forest ecosystem work is somewhat ironic since forests occupy 11% of global area and 38% of land area, their primary production represents 46% of the world's total, and their biomass constitutes 90% of total world biomass (Whittaker and Likens 1975).

In fact, work on forest energetics was underway by 1967, even if it had not been absorbed by mainstream literature at that time. The second paper published from Fred Turner's course was by George M. Woodwell and Robert H. Whittaker, who described the confluence of their respective works—dimensional analysis and gas analysis—which led to measures of gross primary production (GPP), autotrophic respiration (Ra), net primary production (NPP), heterotroph respiration (Rh), net ecosystem production (NEP), and the patterns of change in NEP with succession in the Brookhaven forest. As far as I know, this was the first direct estimate of these ecosystem parameters at that time.

Woodwell and Whittaker noted that aboveground litter and root contributions to the forest floor soil complex were about equal, and that, collectively, about 58% of NPP went to this detritus pool. On the other hand, they estimated that only about 3% of NPP went to the grazing pathway of energy flow from aboveground sources. Significantly, they did acknowledge that they could not account for grazing belowground. With these observations, they established two critical points about forest energetics: (1) that most of the energy not stored in tree biomass flows via the detritus energy flow pathway–herbivory is normally trivial; and (2) that root turnover could be an important source of energy flow to either detritus or grazing consumers. A third point, made without much emphasis, was that forests possess a very large capacity for increase in plant size and organic matter storage. Woodwell and Whittaker understood that through this capacity, ecosystem parameters such as NPP and biomass would undergo extensive change with successional status. The magnitudes of such changes occurring over successional time, the relatively long time scales over which succession can occur, and the predominance of energy flow via detritus in all stages of succession, are the three distinguishing characteristics of forest ecosystems.

While these images were coming into focus for Woodwell, Whittaker, and some of their colleagues, such ideas seem not to have been widely adopted. Apparently, they were relatively new for the discipline in 1967. Completing this AIBS short course and the subsequent series of papers in the 1968 American Zoologist were presentations by Charles Goldman on aquatic productivity, Larry Slobodkin on predation, Frank Golley on secondary productivity, and Manfred Engelmann on soil arthropods. While the Golley and Engelmann papers dealt exclusively with terrestrial heterotrophs, only one example involving a forest animal was cited in either of those papers. That example was from MacFadyen's work on British soil arthropods. Evidently, the study of forest energetics was largely confined to autotrophs at that time, except for treating the detritus pool (forest floor soil complex) as a black box partially composed of heterotrophs, especially the dominating microbes.

This was 1967. In his concluding remarks, E. P. Odum pointed out that the International Biological Program (IBP) was being planned worldwide around the theme, "the biological basis for productivity and human welfare." He foresaw intensive, multidisciplinary studies of landscapes (including forests), important processes, and key problems. The year 1967 was a pivotal point in disciplinary history, because IBP did become the largest organization in ecosystem level research of the following decade. A number of biome studies were initiated in the U.S., and two of these focused on forests: the eastern deciduous forest and the Pacific Northwest coniferous forest. These, together with forest-centered IBP projects in other countries, and the Hubbard Brook Forest Ecosystem study, became major arenas (although not the only ones) of advances in forest energetics.

At least four other lines of activity were taking place during this time, and they came to make important contributions to forest energetics. One was the professional forestry tradition of the U. S. and Europe (particularly Germany and the Scandinavian countries). Extensive syntheses of stand development had accumulated by this time (Loucks et al. 1981), as manifested in data-rich treatments like that of Ernst Assmann (1970). His volume contained detailed descriptions of the allometric growth of trees, dimensional analysis, and the change in basal area, height, volume, and volume increment with time. All of this material was fundamental to the clear understanding of tree and stand growth and provided valuable material for estimation of biomass and productivity in ecological terms.

The second source of influence and impetus in forest energetics was the work of U. K. ecologists and foresters, particularly Amyan MacFadyen and Derek Ovington. Ovington translated forestry concepts to the woodland ecosystem concept, vigorously and effectively providing some of the first interpretations of forest productivity and biomass increment in the literature of the early 1960s (Ovington 1962, 1965). Ovington also worked with American ecologists in developing these techniques, leading a team in the comparative estimation of productivity and biomass of prairie, savanna, and woodland ecosystems in the vicinity of historic Cedar Creek Bog, made famous 20 years earlier by Raymond Lindeman (Ovington et al. 1963).

A third influential source was a group of ingenious Japanese ecologists—H. Ogawa, K. Yoda, T. Kira and others—who presciently asked interesting questions about forest ecosystem metabolism and developed clever and efficient methods for measuring energy fixation and respiration in ecosystems of both Japan and the tropics (Ogawa et al. 1965; Satoo 1970). Their methods were incorporated into American thinking to various extents, and their results formed an important part of the array of data that eventually was used for estimating global productivity and biomass.

The fourth major influence came from the USSR in a compendium of international (though mainly from the USSR) data on major fluxes and accumulations of matter in terrestrial ecosystems (Rodin and Bazilevich 1967). This book contained an enormous compilation of data on comparative biomass, productivity, litterfall, and detritus in both mass and chemical terms. It still stands as a valuable source of comparisons across a wide range of ecosystem types from both biogeochemical and energetics viewpoints.

With the exception of this USSR contribution, the international nature of the study of forest energetics was captured in one of the earliest IBP publications (Reichle 1970). This volume represented the state of the art at "kickoff time" for the IBP. It captured not only the energy aspects of forest ecosystems, but the nutrient cycling and hydrologic considerations as well. Through the decade of the 1970s and into the 1980s, the lines of approach described above proliferated, hybridized, and evolved to give the very complicated contemporary scene of research and concepts in forest energetics in the mid-1980s. Forest energetics is now much more sophisticated in detail, but less coherent as a single subject area. We have learned much about ecosystem functioning, but it is more difficult to relate interacting factors in an integrated model and to generalize about forests as a whole.

5.2. General Patterns of Energy Flow in Forest Ecosystems

Few detailed studies of energy flow have been done in forests. Three prominent examples are the study of a Puerto Rican rain forest (H. T. Odum 1970), a study of an Amazonian rain forest (Fittkau and Klinge 1973), and the Hubbard Brook study (Gosz et al. 1978). I will review some salient points from the more recent Hubbard Brook study that seem to be generally characteristic of forests. The data provided in that study represent the four month growing season only, but most biological activity takes place during that time.

Gosz et al. depicted energy flow in the style originated by H. T. Odum (Fig. 5.1). The system is organized into three principal structural blocks: plant biomass, detritus with its attendant consumer biomass, and the members of the grazing energy flow pathway. Whereas the first block represents the first trophic level in the Odum style, the other blocks are actually conglomerations of second and higher trophic levels. Most of the energy consumed by heterotrophs is respired in the second trophic level within these aggregate blocks. Only trivial amounts of energy go into higher trophic levels, although these levels may be extremely important in terms of regulatory functions, as discussed later. Plant biomass is high compared with other ecosystem types, although the Hubbard Brook forest is not particularly high among forests (Whittaker and Likens 1975). More interestingly, detritus mass is even higher than plant biomass—1.7 times greater in this case. This is not uncommon for temperate and cold region forests (Reiners 1973; Ajtay et al. 1979), but the balance shifts toward higher plant biomass in the wet tropics (Odum 1970; Fittkau and Klinge 1973; Ajtay et al. 1979).

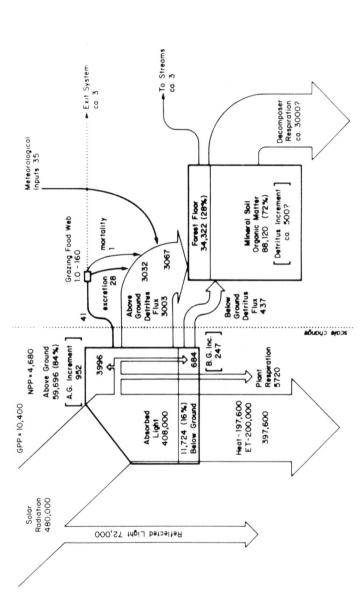

Figure 5.1. Energy flow through the Hubbard Brook Forest over the four month summer period. Pool sizes are in kilocalories per square meter; energy flows are in kilocalories per square meter per four months for physical fluxes and per year for biological fluxes. Except for the scale break between autotrophs and heterotrophs, pools and fluxes are proportional to one another. Redrawn from *The Flow of Energy in a Forest Ecosystem*, by J. R. Gosz, R. T. Holmes, G. E. Likens, and F. H. Bormann. © 1978 by Scientific American, Inc. All rights reserved.

Turning from standing states to energy flow, while gross primary production (GPP) is relatively high compared with other ecosystems, much of it is consumed in the maintenance of the large plant biomass (Fig. 5.1). This "maintenance" term actually includes the energy lost in synthesis of organic compounds plus catabolic processes. In the Hubbard Brook case, that maintenance load is calculated at 55%, which is a reasonable approximation for mature forests. Thus, while forest GPP is relatively high among ecosystems of the world, NPP is not particularly high compared with other ecosystem types in fertile environments (Edwards et al. 1981; Long 1982).

A further conclusion from Figure 5.1 is that most of the flow of NPP is primarily into two sinks: accumulation in plant biomass itself (26% here) and to the detritus compartment of the system (74% in this case). Normally, energy flow through grazers is minor. Especially after the earliest stages of forest growth, the detritus pathway is typically the most important pathway of energy flow. Flow via the grazing pathway, the pathway most often represented in general diagrams of ecosystem energetics, is usually minor ($<$ 1% at Hubbard Brook most of the time). The grazing pathway is important, however, to the functioning of the forest, and is subject to explosive increases in insect and pathogen outbreaks. Unlike the detritus pathway, the grazing pathway is not directly donor controlled. Thus, it is subject to interesting dynamic changes that may emanate from ecological interplay between grazing or pathogenic organisms, their enemies, the weather, and the physiological state of the plants themselves (Waring and Schlesinger 1985). The ecological ramifications of variation in the grazing intensity of forests have been reviewed by Matson and Addy (1975) and Seastedt and Crossley (1984).

Most NPP goes into two sinks that may represent long-term storage. Organic matter and energy are aggrading through most of the history of a given forest. In this Hubbard Brook case, the accumulation of phytomass alone is about 26% of NPP. Assuming that detritus increment is about 500 kcal·m^{-2} (this author's estimate), then accumulation of all biomass is about 36% of NPP. There are apparently exceptions of near steady state forests of large scale (Grier and Logan 1977; Gosz 1980), but, for the most part, forests go through long periods of biomass and detritus accumulation, both of which may be abruptly decreased by the action of some disturbance such as fires, logging, or blowdowns. In environments in which disturbances are rare or lacking, the same phenomenon is driven by gap formation so that the upgrade–downgrade cycles occur in a more or less balanced gap mosaic (Bormann and Likens 1979a; Peet 1981; Pickett and White 1985). In such cases, whether or not the forest is aggrading depends on the scale of one's view.

While there is initially a clear demarcation between functional groups of heterotrophs in the grazing and detritus pathways, this distinction becomes blurred at the third trophic level and above. In the Hubbard Brook forest, most of the grazing is done by insect larvae, particularly lepidopterans, while most detritus consumption is probably done by fungi and bacteria (although it has not been confirmed in this case; see Swift et al. (1979) for other examples). At the third and higher trophic levels at Hubbard Brook, birds, chipmunks, and mice derive part of their diets from grazing (foliage, fruits, and seeds) and part from detritus-based fungi and arthropods. In fact, during the breeding season and in nonoutbreak years for insect grazers, songbirds derive 81% of their diets from detritus-based insects. Salamanders and shrews, usually considered part of the grazing food chain, derive nearly all their energy from detritus-based insects.

These examples come from a particular ecosystem. Some basic questions arise from this single example:

1. To what extent are the points made above generally true for forest ecosystems?
2. How do these patterns change with forest development?
3. What factors set rates and magnitudes of change?
4. How do we account for episodic changes in patterns caused by herbivore and pathogen outbreaks?

Explaining and predicting these dynamic patterns constitutes some of the interesting challenges in forest ecosystems. There are at least three approaches to answering these questions. The first is the comparison of ecosystem attributes along geographic or environmental gradients; the second is through system or subsystem level experiments, such as application of toxins to eliminate grazers or decomposers; the third is observation of energetic patterns with disturbance and successional development of forests.

5.3. Changes in Energy Flow with Forest Succession

The term succession usually refers to species turnover as ecosystems change following disturbance, or with exposure of a new surface. However, in many forest successions there is very little species turnover. Rather, there are merely changes in the age class structure of the initial dominants, and in the system level characteristics before harvest or natural disturbance reoccurs (Fig. 5.2). Succession in the sense of change in ecosystem structure and function will be used in this review even though species turnover may not occur. This decision results from the admonition of McIntosh (1985) that the term development itself has organismic implications worth avoiding.

The issue of limited species turnover deserves further consideration. Most temperate forests are first generation forests recovering from some kind of disturbance (Delcourt et al. 1983; Sprugel 1985, p. 336). For other forest types which do not seem to undergo large scale turnovers very often, turnovers of canopy dominants apparently occur at a high frequency but a small spatial scale (Lieberman et al. 1985; Runkle 1985). Thus, forest energetics can be evaluated either in terms of the individual units of the resulting patch mosaic, or as an integration of the entire mosaic. The proper procedure depends on the scale desired.

In this review, mass must serve as a proxy for energy flow, because very few forest energetics studies have actually converted organic mass to energetic

Figure 5.2. Forest strips in various stages of succession separated by avalanches. Grand Teton National Park, Wyoming. W. A. Reiners photo.

values. When this conversion is made, the range of variation of caloric content for different forest tissues is so limited that caloric values essentially parallel mass, whether expressed as organic matter or carbon, within the limits of precision of most methods used in estimation of forest energetics (Reiners and Reiners 1970; Reiners 1972). I will seek general patterns in changes in mass and energy flow in forests, but will also highlight the variability that has been recognized. Variability in itself suggests questions we can ask to lead to better explanations for natural patterns. Ideally, extensions of this analysis would lead to a diagnostic framework by which ecologists could ask a series of questions that would lead to a reasonable prediction of how energy flow would vary with time in any kind of forest.

5.3.1. Patterns of Primary Production with Forest Succession

Fortunately, this subject has recently been reviewed by Peet (1981), Reiners (1983), Sprugel (1985), Waring and Schlesinger (1985), and to a lesser degree by O'Neill and DeAngelis (1981). Patterns of productivity and other ecosystem characteristics were summarized by Bormann and Likens (1979) for northern hardwood forests, and by Long (1982) for western conifer forests. Most of these reviews were based on numerous case histories, providing a convenient body of summarizing literature from which to draw generalities.

Gross Primary Production (GPP)

In their recent review of forest dynamics, Waring and Schlesinger (1985) present a series of curves for changes in energy allocation with stand succession (Fig. 5.3). They generalize GPP as sharply increasing, peaking, then declining slightly. It should be noted that their time axis covers a period of only 60 years. Other representations attempt to portray behavior further into succession to include old age stands in which the decline from the peak may be steeper. In any case, the general pattern is the same as portrayed by Moller et al. (1954) and E. P. Odum (1969).

Waring and Schlesinger (and others) point out that GPP is never directly measured, because of physiological and technical difficulties. GPP estimates are sums of direct measurements of NPP plus estimates of plant (autotroph) respiration (*Ra*). *Ra* is very difficult to measure because of the variability in rates with respect to surface area, temperature effects, seasonal changes in plant metabolism, the lack of access to undisturbed roots, and especially photores-

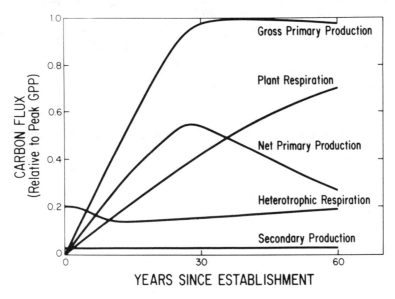

Figure 5.3. Generalized relationships showing how components of ecosystem metabolism might change over the course of development of a forest from establishment to maturity. Gross primary production is shown to peak after about 30 yr., corresponding to canopy closure. Plant respiration continues to increase as living tissue accumulates. Net primary production peaks with canopy closure, and then decreases as a result of continuing increases in plant respiration. Heterotrophic respiration of microbes and other nonphotosynthetic organisms is initially high, following removal of a previous forest. With canopy closure, respiration decreases slightly. Later, as gaps in the canopy appear, respiration of heterotrophs again may increase. Secondary production of nonphotosynthetic organisms is always a small component of GPP. Reprinted by permission of Academic Press, Inc., from Waring and Schlesinger (1985).

piration of leaves. Large-scale measures of night respiration estimated from microclimatic data (Woodwell and Whittaker 1968) or large chambers (H. T. Odum 1970) do not measure plant respiration alone, even in the dark, but measure whole system respiration (Ra plus Rh (heterotroph respiration)). Hence, GPP, while of theoretical interest, can only be approximated by current methods.

Given this extreme difficulty in measuring GPP, we are left speculating on how it varies with time. It is reasonable to assume that GPP is roughly proportional to the leaf area index (LAI) or foliar biomass of a stand (Long 1982). Waring and Schlesinger (1985, p. 53) demonstrate a relationship between GPP and the product of LAI and number of growing season months for seven broad-leafed forests. While that approach would not provide a real measure of GPP, it would at least represent the time trend, the main question of interest here. There are abundant examples of changes in LAI or foliar biomass with successional time (e.g., Cooper 1981; Bormann and Likens 1979a; Tadaki et al. 1977; Rauner 1976; Long 1982; Johnson et al. 1982). Typically, these increase asymptotically, or else reach a peak and decline to a lower level late in succession. This latter pattern confirms the trend for GPP suggested by E. P. Odum (1969) and Waring and Schlesinger (1985).

Under what circumstances might we expect an asymptote, or when might we expect a decline in GPP? An asymptote fits our expectation if we think of a site as having a finite carrying capacity, but how is a leveling off actually brought about? An asymptote might occur when the LAI and distribution of leaf area with respect to the extinction curve for radiation remain constant, and when the species contributing to LAI maintain the same photosynthetic and respiration rates for given light conditions. These conditions could be met early in most successional sequences, but probably change later in a successional sequence. Conditions would change if there were marked changes in the extinction curve through self-shading by branches (Swank and Schreuder 1974), or if there were changes in species composition, or if there were a decrease in LAI such as would result if the canopy broke up into a gap mosaic having a lower overall LAI. Given these considerations, an asymptotic pattern in GPP would be most probable in simple successional situations with short regrowth-disturbance cycle times such as occur in many western and boreal conifer forests where mean fire return interval is short and a succession of species composition is limited (Kilgore 1981; Heinselman 1981).

GPP may be expected to peak and then decline to lower levels in systems in which there is species turnover, longer times between disturbance events, and a breakdown in canopy structure through gap formation. Such circumstances seem to prevail in moist tropical forests (Lieberman et al. 1985) and temperate deciduous forests (Runkle 1985). Of course, GPP would have to be averaged over the entire gap mosaic in such conditions. Otherwise, each gap unit would presumably show the same patterns of a rapid rise culminating in an asymptote or its own particular decline pattern.

Net Primary Production (NPP)

Three general questions are addressed with regard to net primary production (NPP) and succession: (a) what is the shape of the NPP curve with time, (b) what sets the upper limit of the curve; and (c) how is NPP allocated among the plant parts? NPP is, of course, the difference between GPP and Ra and is the easiest of these three values to actually measure. As Waring and Schlesinger (1985) have generalized over the relatively short time span represented by Figure 5.3, GPP is nearly asymptotic, and Ra increases almost linearly, so that NPP has a maximum rate at 30 yr. The humped pattern for NPP reflects the vast majority of documented cases (Rodin and Bazilevich 1967; Loucks et al. 1981; Sprugel 1985; Long 1982; Pearson et al. 1987; Albrektson 1980). The amplitude of this hump can be severalfold. For example, Rodin and Bazilevich (1967) show spruce forest NPP rising from 3,500 to 20,500 kg/ha between 20–40 yr, then descending to 6000 by the 100th yr of succession. Steeper and better defined humps might be associated with short, simple successions as with fire-prone coniferous forests. Temperate deciduous or tropical forests may have broader humps, depending on the spatial scale being assessed. Counterexamples for this idea are balsam fir wave successions described by Sprugel (1984), and succession in the Douglas fir region which was, and to some extent still is, burned over very large areas (Franklin and Hemstrom 1981). These cases do not show steep maxima in NPP during midsuccession. Sprugel (1985) and others attribute the decline in NPP after midsuccession to either increasing Ra relative to a steady or slightly declining GPP, or to a decrease in nutrient supplies caused either by accumulation in biomass or decrease in mineralization rate (Van Cleve et al. 1983). In this latter circumstance, a slight reduction in LAI and increased allocation belowground can contribute to the decline in NPP.

Peet (1981, p. 330) has examined alternative time courses for NPP in some detail. The initial ascent to the maximum or asymptote depends on site quality and establishment rates. He says that stands with dense initial stocking will plateau in only a few years, whereas on poor sites with low establishment rates, production will only slowly increase to an asymptote. According to Peet, there is a theory and a small amount of evidence that, following attainment of this peak, NPP can maintain an asymptotic level. This might occur on poor sites, stressed environments, or in primary succession. He shows, however, that there is more evidence that NPP drops after midsuccession. He presents three scenarios for such a decline. The first requires that GPP remain constant while Ra increases through time because of changes in species composition, tree geometry, and stand stocking. This leads to a gentle decline in NPP (Fig. 5.4). The second decline scenario is one in which GPP decreases for reasons suggested above, so that with increasing Ra the decline in NPP is more precipitous. Normally, there would be a return to higher levels after turnover of stand dominants (Fig. 5.4). In the third scenario, retrogression occurs in which the very process of succession leads to a deterioration of site conditions. These

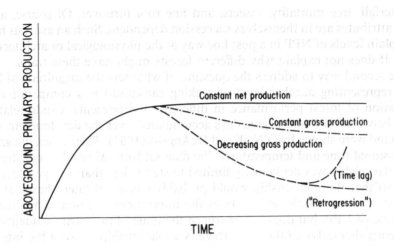

Figure 5.4. Four scenarios for changes in NPP during succession. See text for details. From Peet (1981).

scenarios provide a demographic and site character basis for predicting both the rate at which a hump shaped NPP curve would be attained and the relative magnitude of the decline on the right side of the hump.

An obvious consequence of the oftentimes steep hump in the NPP curve is that characterizations of NPP for forest types are very time dependent. Unfortunately, this is rarely taken into account when tables are compiled to provide average rates for different forest types, and, subsequently, for global totals. It would be interesting to see how such averages would be altered if the component NPP estimates for each category (e.g., boreal forests) were reexamined for successional state and somehow prorated over age distributions of the entire forest type. Another consequence of the time varying NPP curve is that expressing NPP as a proportion of GPP requires specification with regard to successional status. Simple ratios for all forests are not appropriate.

Among measures of NPP for various forests (Jarvis and Leverenz 1983), NPP varies about fortyfold. Some of this variation is dependent on the successional stage in which the estimates in this range were taken, but some of it is dependent on characteristics of the ecosystem. What factors control the absolute values of NPP at comparable stages of succession? This question can be addressed in at least three ways, all of which represent different levels of causation. The first way is a detailed, physiological–structural approach, as represented by a thorough analysis of trees themselves. In such an analysis, Jarvis and Leverenz (1983) broke down productivity variables into: (a) radiation interception as set by amount of leaf and branch area, vertical distribution of foliage, grouping or clumping of foliage, and absorptive characteristics of foliage; (b) photosynthetic efficiency as dependent on leaf properties; (c) respiratory losses to maintenance and growth respiration; and (d) mortality losses to litterfall, tree mortality, insects, and fine root turnover. Of course, all of these attributes are in themselves succession dependent. Such an analysis helps to explain levels of NPP in a post hoc way at the physiological or architectural level. It does not explain why different forests might have these traits.

The second way to address the question of what sets the magnitude of NPP (and representing another level of seeking causation) is a comparative examination of forest performance in different environments. Crude relationships between NPP and degree days accumulated over the development span of a stand were shown by O'Neill and DeAngelis (1981). Their index integrated successional time and temperature. The data set from which they constructed this relationship was deliberately limited to stands less than 130 yr. old, however, so that the relationship would probably change at later phases of succession. This approach demonstrates the importance of resource availability and time to NPP, but does not approach the issue of the intrinsic qualities of the plants themselves. Of course, there is a relationship between the intrinsic nature of plants and the environment they have survived in through migrational history, or have adapted to through selection. Circular causal loops (Hutchinson 1948) exist between plant qualities and site qualities ("reaction" by the earlier jargon of Clements, 1916).

Another comparative approach is to examine the pattern of productivity with changing LAI, which itself is a function of successional development. Waring and Schlesinger (1985) show that NPP rises, then falls with increasing LAI, and that the magnitude of the rise and the optimum LAI depend on the availability of resources (Fig. 5.5). Both this and the first comparative approaches confirm that the magnitude of productivity is a complex function of both resource availability and stand structure. The amount of organic energy that can be allocated to maintaining LAI, and the amount of energy produced in foliage that can be allocated to aboveground growth both seem to be related to belowground allocation needs for water and nutrient resources. Irrigation or fertilization would be expected to decrease belowground energy allocations and permit a higher LAI and growth per unit LAI. Environmental characteristics will largely set the potential amplitude of NPP, while stand structural characteristics, both at the species level and at the physical level, will set the actual amplitude within the environmental limits available.

Still, the evolutionary and biogeographic explanations for why plants available for forest succession in a given region have the structural and physiological attributes they do is not addressed by this type of analysis. Such explanations can be found in the reasons trees have the foliage type (e.g., evergreen needles vs. deciduous broad leaves), the amount of foliage, the patterns of array in the canopy, the physiological rates, and the duration of extended growth that they do. These questions have not, in my opinion, been dealt with sufficiently in the analyses of forest productivity.

An exception to this is the extremely thoughtful and stimulating analysis of the evergreen coniferous forest of Pacific Northwestern North America by Waring and Franklin (1979). In that paper, the authors explored the adaptive nature of evergreenness, growth rates, extended periods of individual tree

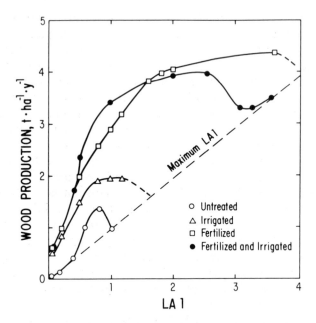

Figure 5.5. Stem wood production is maximized at optimal LAIs, and LAIs vary with nutrient and water resources in Scots pine plantations. From unpublished data by B. Axelsson, A. Aronsson, and R. H. Waring, Swedish University of Agricultural Sciences, Uppsala, Sweden. Reprinted by permission of Academic Press, Inc., from Waring and Schlesinger (1985).

growth, and longevity, in terms of the historical heritage of the flora, the present climate, soil qualities, and the occurrence of disturbances through wind storm and fire. While that was an informal analysis, it enhanced understanding of how plant traits evolved as a long-term function of the total environment.

Another example of broad evolutionary considerations for tree adaptation was presented by Jordan (1971) and Jordan and Murphy (1978). They postulated mechanisms to account for the allocation of fixed energy in trees across global environmental gradients. In this case, less definitive statements could be made. Understanding and predicting limits to NPP for all forest regions could benefit from this kind of evolutionary-based consideration. In so doing, we should not assume that native plants are always the best adapted species for a particular environment in terms of NPP. The commercial success of North American trees in New Zealand, and direct comparisons of physiological performance of exotic and native species in a common environment (Benecke and Nordmeyer 1982; McEwen 1983) sometimes show that native species may not be the best adapted in terms of photosynthesis or water use efficiency.

The third question raised under the topic of NPP is that of allocation within phytomass. Understanding the trends in allocation allows one to predict how energy flow patterns from phytomass to storage or consumers might change with succession. The investment of photosynthate into new biomass can range from bole wood, which represents a long term increase in phytomass, to fine roots, which probably represent the most ephemeral investment. In most productivity studies, the component of phytomass in mobile reserves is not usually distinguished. These reserves, which typically are included in the gross weight of twigs, boles and roots, may be substantial, and are vital to producing new tissues, sometimes in the face of successive losses (McLaughlin et al. 1980; Webb 1981). Few data sets of this type are available for review. While there are numerous data sets on biomass allocation with time, that is quite different from the allocation of NPP itself. For NPP, allocation must be inferred from incremental changes in biomass of long-lived tissues, litter contributions both belowground and aboveground, and from herbivory rates. In most cases, the degree of investment into underground parts is relatively poorly known and probably underestimated so that NPP itself is underestimated (Harris et al. 1980; Ulrich et al. 1981).

Three examples of NPP allocation are shown in Figure 5.6. The first is a generalized case for a Scots pine (*Pinus sylvestris* L.) succession involving one generation of one tree species and without consideration of fine roots (Fig. 5.6a). Allocation priorities in early phases are needles > branches > stems > large roots. One might guess that allocation to fine roots would parallel that to needles, if establishment of canopy and fine root systems are of highest priority in tree establishment. From midsuccession to late succession, allocation to needles peaks and then drops to an asymptote; this is paralleled in branches. Allocation to large roots shows a broad hump and sustained decline. Stems become the main sinks for NPP allocation in all but the earliest phases of succession, and the broad hump is largely responsible for the hump shaped pattern for total NPP described earlier. Except for the absence of fine root data, Figure 5.6a may be considered an ideal case.

The case in Figure 5.6b is from a Japanese fir (*Abies*) chronosequence found in subalpine "fir waves". Fir waves are synchronized waves of growth, maturation, and death of more or less even aged fir trees. This is another simple case of stand development without much species turnover and without sufficient time for the stand to acquire old age characteristics. In this example, allocation priorities at the earliest measured age (12 yr.) are stems > needles > roots = branches > cones. It is possible that allocation to needles exceeded that to stems before the earliest age of estimate (12 yr.). "Roots" only include large, coarse roots as in Figure 5.6a. Allocation to branches lags behind allocation to needles in time and quantity, and allocation to reproductive parts—the cones—is the last in the sequence. The allocation to the various sinks at peak NPP differs from Figure 5.6a in that a relatively smaller proportion goes to stems and that allocation to roots equals that to branches.

The example in Figure 5.6c differs markedly from the ideal pattern in Figure 5.6a. It represents changes in productivity over the first 60 yr. of a balsam fir (*Abies balsamea* (L.) Mill.), wave complex very much like the situation described for Figure 5.6b. In this case, there are no data for roots at all. Initial allocations are very similar to the 5.6b case, but the trajectories of change with

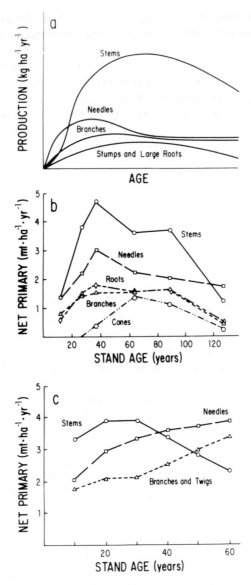

Figure 5.6. Three examples of NPP allocation to phytomass. A represents generalized stand development of Scots pine in Sweden, from Albrektson (1980). B represents stand development in a Japanese fir wave (redrawn from Tadaki et al. 1977), and C was redrawn from Sprugel 1984, for a northeastern US fir wave chronosequence.

time are unique, with allocations to foliage increasing asymptotically, allocations to branch components increasing almost linearly, and bole wood and bark peaking early and declining rapidly to a ranking well below that of the canopy components. The peculiar transition at 40 yr., in which productivity of bole and bark declines while canopy components accelerate, may be caused by the combination of death of old trees and concomitant growth of young seedlings and saplings. The productivity of twigs and needles may also be enhanced by a regular "harvest" of current twigs and needles by abrasion and rime ice breakage during the winter (Foster 1984). This is an interesting case in which overlapping generations of trees, growing and dying in the same phases of the wave complex, may integrate NPP to more or less constant rates for the entire area (Sprugel 1984). This case also illustrates how enhanced "harvest" of current twigs and foliage with increasing age can alter NPP allocation.

These three examples show similarities and differences in allocation of NPP over successional time. From these and other examples, the following tentative generalizations about NPP allocation with succession are offered.

Ground Layer

The biomass of the ground layer (including tree seedlings), and presumably its annual productivity, normally rises to a high level early in succession before a taller canopy restricts sunlight at the ground level. The ground layer descends to a nadir when canopy closure occurs, and rises again to an intermediate level as thinning of the dominant stratum provides more sunlight penetration to the forest floor (Reiners et al. 1971; Tadaki et al. 1977; Bormann and Likens 1979 a; Long 1982; Romme et al. 1986). The rate at which these changes occur depends on the nature of the disturbance, site quality, life history characteristics of ground layer species, and overstory restocking rate.

Foliage

NPP allocation to foliage rises very steeply, faster than for other tree components, but then declines into a lower asymptote as the stand goes into a gap mosaic or other old age pattern (Fig. 5.6a). The energy allocated to foliage will go partly to herbivory but mostly to the detrital energy flow pathway. With species changes, herbivory rates could change, altering this flow pattern.

Twigs and Branches

NPP allocation to twigs and branches will lag behind that to foliage, but will precede the major peak in allocation to boles (Fig. 5.6a). This energy may be stored for a long time as phytomass, or will be routed to detritus. Little will be routed to herbivory.

Boles

NPP goes last to boles, but this is the largest allocation (except possibly to roots) and that of longest duration (Fig. 5.6a). Most of this energy will go eventually into detritus.

Reproductive Parts

Reproduction is the last in the sequence to receive energy allocation and the amount recieved is the least (Fig. 5.6b). This energy represents a very important food for herbivores, although most will be captured by detritivores.

Roots and Mycorrhizae

The situation here is very complicated and very poorly understood. Strong and La Roi (1983) found that rooting depth increased through successional time on well-drained soils, but not on poorly drained soils. On the other hand, the number of roots per square meter was stable throughout several successional sequences, possibly even decreasing through time in some cases. Santantonio et al. (1977) found a consistent allometric relationship between stem size and coarse root biomass. This suggests that coarse root biomass increases as long as bole biomass does. We can only presume that NPP of these coarse roots parallels their biomass and that turnover rate of these large roots is low.

The situation for fine roots, with their much higher turnover rate and energy demand, is quite different. According to Santantonio et al. (1977):

> One might infer that complete occupation of the forest site by fine roots occurs early in stand development, peaks, and levels off as physiological and ecological factors limit fine root biomass per hectare at some upper level, independent of large root and aboveground biomass.

Vogt et al. (1983) found that fine root and mycorrhizal biomass increased in parallel with crown closure, then decreased on a good site, while remaining constant on a poor site. This suggests a relationship not only with successional development, but with site quality as well. Vogt et al. also differ from Santantonio in suggesting an actual decline in fine root biomass, rather than a leveling to an asymptote.

Particularly with fine roots, the relationship between biomass and productivity is not clear. Vogt et al. (1986a) found fine root turnover to be in the range of fine root biomass, but this relationship probably is not constant within successional sequences (Persson 1979). Turnover rate can be a function of site quality, the carbohydrate status of the tree (Glerum and Balantinecz 1980), and other regulatory factors. Marshall and Waring (1985) showed that the longevity (and by implication, the turnover rate) of fine roots was a function of the original carbohydrate content of the individual fine roots and their respiration rates as set by soil temperature. Persson (1983) found fine root turnover to be faster in a young stand than in a mature stand. Keyes and Grier (1981) showed that root turnover was partly a function of site fertility. The present evidence suggests that coarse root NPP increases asymptotically, and then parallels bole biomass trends. Fine roots, on the other hand, probably have a hump shaped NPP curve, higher than that of coarse roots, and with different declining asymptotes for sites of different quality. The evidence is still very limited, however, and these trends should be viewed as highly tentative.

The example in Figure 5.6a best represents a general pattern for NPP allocation. It does not include fine roots, however, and only represents growth of a single generation of trees. Our ability to generalize about NPP allocation through succession is very limited. Clarifying these trends and explaining them in terms of physiological, demographic, and environmental mechanisms is one of the challenges facing us.

5.3.2. Major Pathways of Energy Flow

Forest NPP faces three short-term fates: storage in phytomass (including storage carbohydrates (McLaughlin et al. 1977, 1980)), flow into detritus with consequent utilization by a very complex network of decomposers (or transport out of the system), or immediate utilization through herbivory. A generalized graph of how energy might be apportioned between these three sinks, and how the total balance of energy fixed and energy consumed (net ecosystem production = NEP) changes with time is given in Figure 5.7.

Phytomass Storage

Phytomass is often represented as increasing asymptotically with succession (e.g., E. P. Odum 1969), although such representations do not usually go beyond the initial phase of first generation growth. Sprugel (1985) provides an excellent review of general trends in the accumulation of phytomass components. Briefly, foliage increases rapidly, then levels off at a resource-determined level. Foliage mass may decline as stands go into internal regeneration patterns when the first generation of trees becomes senescent. Changes in species composition, particularly from conifer to deciduous (or the reverse) will cause large changes in foliar mass. Bole mass increases more slowly at first, but continues its increase well past the leveling point for foliage. Branch biomass is intermediate between foliage and bole mass. According to Ulrich et al. (1981), root biomass increases slowly through the period of biomass accumulation in stands, but represents a decreasing proportion of total biomass. Some data suggest that root/shoot ratios decline in a slightly exponential manner, reaching an equilibrium level in time, the exact ratio being determined by the forest type (Ulrich et al. 1981). There are no data, to my knowledge, on changes in root mass with fluctuations in aboveground mass caused by small-scale turnover (e.g., tree gaps).

Of more direct interest is how patterns of biomass accumulation might vary and what factors underlie these changes. Peet (1981) presents four alternative patterns for total biomass accumulation (Fig. 5.8). The first is the classical asymptotic increase proposed by E. P. Odum (1969) and represented in several data sets either for relatively short-term sequences or for sites with slow growth rates. The second pattern is one in which biomass rises to a peak, then drops to a somewhat lower steady state level (Fig. 5.8). This pattern was proposed for northern hardwoods by Bormann and Likens (1979b). The decline after the maximum was ascribed to a transition from an even aged stand of aging

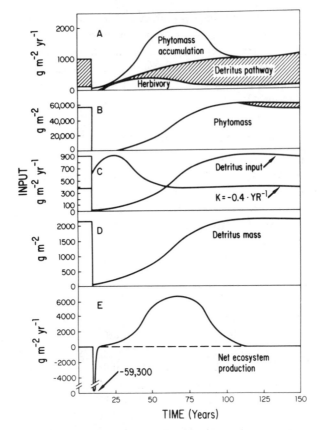

Figure 5.7. A. Change in NPP and the apportionment of energy fixed following disturbance. The scaling is for a forest, but the allocation to herbivory is graphed unrealistically high, and the period of positive accumulation is made unusually short, for illustrative purposes.
B. The accumulation of phytomass based on the integration of the phytomass accumulation area in A. A shallow, hump shaped maximum is suggested by the shaded area, representing a possible shunting of accumulated biomass to detritus.
C. Detritus input calculated from the integral of the detritus area in A over the disturbance recovery time course. Also shown is a suggested change in the decay coefficient (k), with changing conditions following disturbance.
D. Change in detritus mass calculated from input and decay variables in C.
E. Net ecosystem production calculated from the changes in phytomass and detritus mass above. The sudden consumption or export of all phytomass and detritus at the time of disturbance is considered to be net negative production. Animal biomass is too small to appear at this scale. From Reiners (1983).

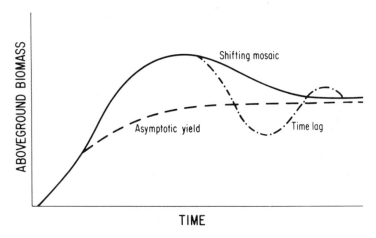

Figure 5.8. Biomass accumulation over the course of three successional scenarios. See text for details. From Peet (1981).

trees to a shifting mosaic of gaps caused by treefalls and subsequent regeneration.

Peet's third pattern is an increase, followed by a decline, and then a series of oscillations (Fig. 5.8). He suggests that the steeper decline and continuing oscillations of this pattern are caused by a longer time lag between the formation of gaps, and (I would interpret) larger gaps than those represented by the shifting mosaic model. Collectively, these three patterns, or models, of biomass accumulation can be integrated into a general model in which the first case (simple logistic) represents no lag in regeneration, the second case (shifting mosaic) represents a moderate lag, and the third case represents long lags (and possibly larger gaps) in regeneration.

Peet's fourth pattern or model (not shown in Fig. 5.8) is one in which biomass peaks, then declines to very low levels, which is the result of nutrient immobilization in accumulated detritus in very cold or infertile sites (e.g., Van Cleve et al. 1983). These alternative models are based largely on the nature of stand maintenance following growth of the initial generation of trees. They explicitly address the nature of stand turnover and allow us to speculate on the mechanisms that can produce different results in different situations. By evaluating these dynamics, we can predict the probable trends in NPP allocation, the distribution and chemical nature of biomass, and potential trends in energy flow beyond the autotrophs.

Energy Flow Through Detritus Pathways

That most energy flowing from the autotrophic level in terrestrial ecosystems goes through the detritus pathway has become conventional wisdom (Coleman et al. 1976), although this wisdom is threatened by a growing appreciation for grazing potential in the rhizosphere of grass and croplands. Nevertheless, this energy flow pathway probably remains the dominant one and is of vital importance to forest ecosystems from many points of view. These viewpoints include provision of habitat as well as nutrition for small animals, facilitation or limitation of plant reproduction, cycling of nutrients, storage and routing of water, maintenance of soil structure, and influence on fire probabilities. Normally, energy flow from autotrophs into the detritus pathway is strictly donor controlled (although this can be changed by pathogen-caused mortality) and highly stable from year to year through most of the life of a forest ecosystem. In this pathway, energy flows from phytomass to a detritus pool—the accumulated dead organic matter in the canopy (Carroll 1980), forest floor, and the soil. The consumption rate of this material by microorganisms and animals of all sizes (Fittkau and Klinge 1973; Gosz et al. 1978) is partially dependent on environmental factors and partly on the amount and kind of detritus that has accumulated.

There is an enormous literature on detritus flux from aboveground sources and on consumption (decomposition) of this aerially derived material, but extremely little literature on the contribution and fate of material from belowground sources. As is quite obvious to all, research on belowground processes remains a critical need for all terrestrial systems.

Detritus Flux from Phytomass to Detritus Pools

Aerial litter can be divided into three categories: foliar, fine woody, and coarse woody material (e.g., large branches and boles). The phasing of fluxes for each of these categories will be different and predictable. According to Meentemeyer et al. (1982), 70% of fine litter (the sum of foliar and fine woody litter) is foliar material, but this percentage undoubtedly changes through successional time. Johnson et al. (1982) note that in Douglas fir stands, the per cent of litter that is woody is 25 at 22 yr., 50 at 95 yr., and 90 at 450 yr. It would seem reasonable that foliar litter fall would parallel changes in foliar biomass. As discussed earlier, foliage biomass will generally increase rapidly to an asymptote as crown closure is reached. Thereafter, foliar biomass, and thus foliar litter, will vary according to stand dynamics following maturation of the first generation of trees.

Similarly, it seems reasonable that equilibration of fine woody litter would lag behind that of foliar litter until first generation turnover dynamics accelerated all woody litter deposition. Finally, coarse woody litter inputs would depend on the fate of the previous generation of boles (Harmon et al. 1986). If this last input is excluded, woody litter input would increase with succession, with greatest variability occurring when stand turnover occurs in either small patches or large-scale areas. Chronosequence data for a number of first generation stands largely support this scenario (Cooper 1981), although the examples from high latitude stands of the USSR show hump shaped curves that are subdued parallels with NPP (Rodin and Bazilevich 1967, pp. 50, 123).

Changes in belowground litter production with successional development are much less clear than those in aboveground production. If coarse root biomass parallels aboveground biomass, as suggested earlier, then the turnover of that coarse root biomass should follow a pattern like that of boles outlined above. In any case, litter production by coarse roots would seem to be a small proportion of this subterranean detritus flux, as all evidence points to very high turnover rates of fine roots. For example, Cromack (1981) showed that 71% of belowground detritus was derived from fine roots in an old growth *Pseudotsuga menziesii* (Douglas fir) stand. Grier et al. (1981) showed that the turnover rate of fine roots and associated mycorrhizae was 5.6 times that of the coarse root increment in a 23 yr. old Pacific silver fir (*Abies amabilis* (Dougl. Forbes)) stand. In the same study, fine root turnover was 16.5 times the biomass increment of coarse roots in a 180 yr. old stand of silver fir. It is true that coarse root increment is not the same as coarse root loss to detritus, but it is unlikely that the loss to detritus would have been greater than the increment in these stands. Also, fine root losses to herbivory could not be discriminated from losses to saprophages. If fine root biomass were to parallel that of foliage as suggested in the discussion of NPP above, and if turnover rates were to remain constant, then detritus inputs from fine roots might approach an asymptote in parallel with foliar biomass, and then decline as succession proceeded into second-generation dynamics. That scenario was not observed by Grier et al., however. They found that fine root contributions to detritus were higher in a 180 yr. old stand than in the 23 yr. old stand, possibly because the accumulated detritus created a higher energy requirement for nutrient acquisition by roots in the deeper forest floor. Thus, the change in fine root contributions might rise rapidly in forest succession and approach an asymptote, or increase slightly through late succession in response to an effective deterioration of site quality in terms of nutrient availability.

How do aboveground and belowground contributions to detritus compare? Most of the few data available show a much higher importance of belowground contributions than formerly suspected. Vogt et al. (1986a) indicated that fine root inputs ranged from 20%–77% of total inputs to the forest floor. Harris et al. (1980) calculated that belowground transfers in a *Liriodendron tulipfera* (yellow poplar) stand were 2.3 times the aboveground transfers. Persson et al. (1980) calculated that root inputs were 0.97 times aerial inputs in a 120 yr. old Scots pine stand. Cromack (1981) estimated root and mycorrhizal inputs to be 5.4 times aerial inputs in the old age Douglas fir stand described above, while Grier et al. (1981) estimated root/shoot ratios for detritus inputs at 4.7 and 5.2 for the 23 yr. old and 180 yr. old silver fir stands, respectively.

Clearly, time trends for underground transfers to detritus will dominate the time trends for total detritus transfer in most cases. These time trends have been carefully considered by Aber et al. (1978) in developing a simulation model of detritus dynamics for a northern hardwoods forest following logging (Fig. 5.9). Successional patterns that they observed are consistent with data that have been produced since 1978 and probably still represent the best approximation of these dynamics. Figure 5.9 does not go beyond first generation succession, however, so that changes in detritus fluxes for forests undergoing patch dynamics at all scales remain to be determined. In general, these proc-

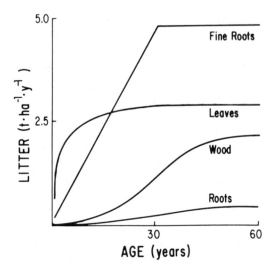

Figure 5.9. Litter input from four sources as a function of time since cutting. From Aber et al. (1978), Canadian J. Forest Res.

esses would probably follow the same kinds of patterns as shown in Figure 5.9, except for the incorporation of coarse woody debris missing in a postlogging scenario. The exact patterns would depend on the scale of patches occurring in that environment and on the scale of forest stand represented.

Detritus Consumption Rates

Energy flow via forest detritus is accomplished through the metabolic actions of an extremely complex web of diverse, small organisms (Birch and Clark 1953; MacFadyen 1963; Swift et al. 1979; Persson et al. 1980). While animals (mostly invertebrates) are important in implementing this energy flow, most of the detrital matter is mineralized to carbon dioxide and other inorganic compounds through the action of bacteria and fungi. These microorganisms have high specific metabolic rates and a rich suite of enzymatic capabilities, but are limited by their relative immobility in the substrate, by nutrient supply, and by other environmental factors, especially temperature. For details of detritus consumption, see Swift et al. (1979). The net process of detritus consumption, or decay, will be discussed here.

The most widely used expression of decay rate is the familiar exponential decay model of Olson (1963). The critical parameter is the intrinsic decay rate, k, which integrates the influences of physical variables, such as temperature and moisture, with chemical qualities of the litter, such as cellulose, lignin, nitrogen, and base content (Swift et al. 1979). Given changes in forest microclimate and the composition of litter inputs with forest succession, then k would be expected to change as well (see example in Fig. 5.7).

Aber et al. (1978) and Sprugel (1985) considered early postdisturbance changes in k caused by changes in microclimatic and litter quality. They postulated that k would at first increase steeply, then decrease slowly to levels occurring in the predisturbance forest. Berg and Staff (1980) measured decay rates of needle litter in Scots pine stands ranging in age from 5–120 yr., finding no significant difference over that time span. Birk and Simpson (1980) and Cromack (1981) reviewed evidence that changing litter quality over successional time would decrease k. Reviewing data from various sources in the Douglas fir region, Johnson et al. (1982) provided a more detailed sequence of events. Needle decomposition rates increased until canopy closure, then slowly decreased, at least through 97 yr. Meanwhile, the woody contribution to litter input increased, exaggerating the decrease in decay rate of the forest floor in total.

The general picture that emerges from these results is that following disturbance, decay rates are usually relatively high. After canopy closure, microclimatic effects (lower temperatures and moisture levels) will decrease k (see exceptions below). In addition, fine litter is likely to become increasingly recalcitrant and the proportion of woody litter should increase. Wood not only decays more slowly than do leaves, but its high carbon to nutrient ratios make it act as a nutrient sink, removing available nutrients and thereby inhibiting the decomposition of fine litter surrounding the wood (Melillo and Aber 1984). Successional variables of interest for particular cases would include the microclimates of recently disturbed sites compared with closed canopy situations. It is conceivable that the temperature–moisture regime could be improved following canopy closure in summer dry climates. Another variable would be the size and chemical nature of coarse, woody debris. Graham and Cromack (1982) point out that the very large size of boles falling to the floors of Pacific Northwest forests give them an influence on forest floor dynamics that is greater than in forests with trees of lesser bulk. All of these variables are extensively discussed in Harmon et al. (1986).

Standing State of Detritus

The amount of mass (and thus energy) that accumulates in detritus (all dead organic matter) is now recognized to be very large, in fact, often larger than the energy stored in phytomass itself (Schlesinger 1977). For temperate and cool climate forests of high elevations and latitudes, the balance between detrital inputs and decompositional outputs is not reached until forest floor and soil organic matter is quite massive. Generally, this organic matter provides good hydrologic and nutritional qualities for forest ecosystems, but sometimes it depresses reproduction and requires more root growth, so that apparent NPP decreases (Sprugel 1985).

The pattern of detritus accumulation is classically portrayed through the model of Olson (1963). With assumptions of steady state detritus inputs and decay rates, that model expresses an asymptotic accumulation of detritus over time (Fig. 5.7). Although this model has problems (e.g., Minderman 1968; Birk and Simpson 1980; Wieder and Lang 1982), it works reasonably well for early phases of many secondary successional sequences and for some examples of long-term, primary successional chronosequences (Crocker and Major 1955; Crocker and Dickson 1957). In other cases, it is not so successful (Dickson and Crocker 1953). It is unrealistic to expect this model to represent behavior over long periods of time when the steady state assumptions are not upheld. Departure from asymptotic behavior can be seen, for example, in systems undergoing periodic disturbance through fire (Birk and Simpson 1980). In the secondary succession presented by Long (1982) for Douglas fir, fine litter of the forest floor does follow an asymptotic trajectory, but large woody material continues to increase linearly through 450 yr. Apparently, in Douglas fir forests wood decay never does come into balance with inputs. Instead, wood accumulates until fire terminates the succession and consumes the wood. This seems to be a common situation in Rocky Mountain lodgepole pine forests as well (Fahey 1983; Pearson et al. 1987). Woody debris accumulation does not seem to follow asymptotic behavior in situations where logs are very large (Graham and Cromack 1982), or where the growing season is short, cool, and dry (Fahey 1983).

Consideration of how coarse woody detritus might vary with different disturbance situations has been graphically portrayed by Lang (Fig. 5.10) and by Harmon et al. (1986). Lang (1985) illustrates how dead wood is a very dynamic component of some forests, and even, given a constant decay rate, how the standing state of dead wood may vary enormously with different histories of disturbance. The importance of woody debris to forest ecosystems is probably greater in terms of nutrient cycling, fire probability, geomorphic processes, and biotic activities (Franklin et al. 1981) than it is in terms of energetics.

Obviously, the degree of change in forest floor mass in response to disturbance–regrowth sequences depends on the nature of the disturbance, the rate

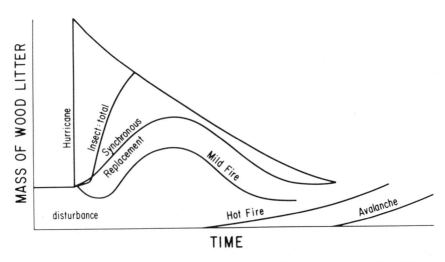

Figure 5.10. Predicted changes in mass of bole wood litter on the forest floor following several hypothetical disturbances in a subalpine balsam fir forest. From Lang (1985), Canadian J. Forest Res.

of regrowth and productivity of the phytomass, and the decay rate of the various forms of detritus in the system (Aber et al. 1978). Forest floor material derived from aerial inputs can accumulate or decay relatively quickly, or can burn relatively easily, and thus, frequently. Organic matter accumulated in the mineral soil, which often exceeds that of the forest floor (Schlesinger 1977), is mainly derived from belowground detritus production and leached dissolved organics from the forest floor. It is thought to accumulate more slowly, decay more slowly following a change in input rates, and to be much less affected by fire. Thus, organic matter in mineral horizons is less dynamic than the forest floor and is an important buffer through disturbance–succession sequences. Dead wood can have violent oscillations in input rates, but, except for combustion by fire, has a relatively slow decay rate. As a result, phytomass, forest floor mass, soil organic mass, and dead woody material will have independent responses to different disturbances and recovery regimes.

The steady state model for litter dynamics of Olson (1963) is a very useful model for ecologists, providing a valuable "first approximation" for our thinking about the behavior of litter. In this way it is analogous to the Lotka–Volterra competition model for population ecology, or the Hardy–Weinberg Law for population genetics. When considering how detritus dynamics might change over a range of successional circumstances, however, the parameters of the model change, causing concomitant changes in the resulting pattern of detritus accumulation. To predict the behavior of detritus, we must take into account the nature of the disturbance, the amount of inherited components of detritus (dead wood, soil organic matter, forest floor litter), the influence of microclimate on decay rates, the changing contribution of dead organic matter from various sources of regrowing vegetation (and thus its chemical composition), and other episodic events that overlie these other factors as succession proceeds.

Energy Flow Through Grazing Pathways

Normally, energy flow through grazing pathways of forest ecosystems is minor compared to that flowing through the detritus pathway (Fig. 5.7). Nevertheless, the fraction that does flow through direct herbivory and consequent carnivory is extremely important in the functioning of the ecosystem. Matson and Addy (1975), and more recently, Seastedt and Crossley (1984) and Romme et al. (1986) have shown how direct effects of herbivores on forest plants can drastically influence NPP, production efficiency, microclimate, decomposition rates, fire probability, plant reproduction, and subsequent forest composition.

Whereas energy flow through detritus is largely donor controlled, it is intuitive and traditional to assume that energy flow via grazing is acceptor controlled. In fact, as we shall see, this is probably not the case. It now seems that the likelihood of plant material being consumed will, in large part, be determined by the chemical nature of that material—an inherent, biological characteristic of the plant, influenced by its current state of vigor and possibly by its recent encounters with damaging organisms (Schultz and Baldwin 1982;

Baldwin and Schultz 1983; Bryant et al. 1983). Thus, energy flow via grazing is actually controlled by interactions between the donor (plant) and the acceptor (consumer) organisms. The major actors in forest grazing pathways are arthropods, mainly insects, especially aboveground (Franklin 1970; Rafes 1970; Seastedt and Crossley 1984). These heterotrophs can consume seeds, buds, twigs, foliage, sap, phloem through bark boring, wood, and roots (Franklin 1970). Little is known about belowground activities. The larger vertebrates receive much attention as grazers but, in fact, the amount of energy flow directed through vertebrates is minor compared with that directed through invertebrates (Gosz et al. 1978).

Belowground Grazing

Given the evidence for very high turnover rates for fine roots (discussed above in terms of detritus), we must ask whether a large fraction of fine root biomass change could not be caused by root grazers rather than decomposers. In fact, there is very little evidence for it at this time. Harris et al. (1980) believed that less than 10% of root turnover in forests was attributable to herbivory. Likewise, Persson et al. (1980) calculated root herbivory to be less than 1% of energy flow through roots in a 120 yr. old Scots pine stand. The paucity of data is expressed by Ulrich et al. (1981), who stated, "There are no known studies which consider total herbivory on roots." The present consensus seems to be that herbivory is of minor importance belowground in forests, although root pathogens can be significant (Wargo 1972). The question will be left with this consensus, but it deserves further investigation and confirmation. The remainder of the section will address aboveground herbivory only.

Chronic Herbivory Levels

The literature on herbivory in forests can be divided into two categories; i.e., that regarding chronic, usually low, levels of herbivory, and that regarding very high levels associated with episodic outbreaks of herbivores. When outbreaks occur, herbivory can have enormous impact on the ecosystem, not only by changing the pathways of energy allocation, but by altering total NPP and the distribution of productivity among plant populations. If severe enough for long enough periods of time, outbreaks may reset succession.

There seems to be general agreement that in non-outbreak situations, the amount of herbivory in forests is less than 10% of NPP (Golley 1972; Wiegert and Owen 1971; Matson and Addy 1975; and Crawley 1983). Still, the nature of herbivory (Franklin 1970), its rate, and its proportion of NPP probably change with successional stages (Sprugel 1985; Reiners 1983; see Fig. 5.7). In general, early stages of succession should be more frequently subjected to herbivory because of greater accessibility of foliage, lesser biochemical protection, and richer resources levels, at least during secondary succession. Herbivory may be lower in middle stages when the first generation of trees reduce light levels at forest floors and are growing rapidly (Sprugel's stem exclusion phase, 1985). However, with breakup of the first generation of regrowth into patches of various sizes, resources should again become available to plants, providing better forage for herbivores (not shown in Figure 5.7a). Thus, herbivory might follow a shallow trough-shaped curve in a typical successional sequence, with the bottom of the trough corresponding to the period of peak biomass increment.

Coley et al. (1985) presented an appealing hypothesis for differential herbivory of various plants that would partially explain such a pattern. They suggested that some plants have inherently higher rates of growth than others in the same general environment; and that these kinds of plants are typically found on habitats richer in resources, whether light, water or nutrients. Fast-growing plants, or plants in these conditions, typically devote a small proportion of their energy resources to biochemical defenses and suffer a relatively high rate of herbivory. In contrast, inherently slow-growing plants, often associated with poorer sites or later stages of succession, typically devote a larger proportion of their resources to biochemical defenses and have relatively low rates of herbivory. Figure 5.11, from Coley et al. (1985), illustrates the relationship between realized growth rates of plants and the levels of defense investment characterizing these plants. If a successional sequence involved species replacement from fast growing plants to inherently slower growing plants, then there would be a descent through the family of curves towards low growth rates, high chemical defense outlays, and low grazing rates. Of course, many successional sequences may be different. For example, primary succession or some sorts of severely disturbed secondary successions may have such low nutrient resources that the first occupants are slow growing, high defense plants that will be succeeded by faster growing, low defense plants after the site has been modified and resource levels raised. Also, the very common fire-dependent conifer forests of mountain regions and boreal zones typically have very little species turnover, so that changes of this type would have to take place within a limited number of species (often only one). The Coley et al. hypothesis is extremely useful for relating biochemical defenses to many ecological situations, but its application to successional sequences must be used with care.

Herbivore Outbreaks

While the Coley et al. hypothesis is useful for interpreting and predicting chronic levels of herbivory, it poses problems for outbreaks in which defoliation can exceed 100% of the initial standing crop of leaves (Matson and Addy 1975). We must turn from a comparative analysis of species behavior, which is useful for understanding site differences or successional sequences in the long- to mid-term, to an analysis of changes in plant–herbivore interaction with single species and in the short-term. The bulk of evidence for occurrence of outbreaks is that they are associated with low growth rates of plants, either because they occur on poor sites (Matson and Addy 1975) or because they are undergoing severe competition for limiting resources such as light, water, and nutrients (Waring and Schlesinger 1985). The origin and spread of outbreaks

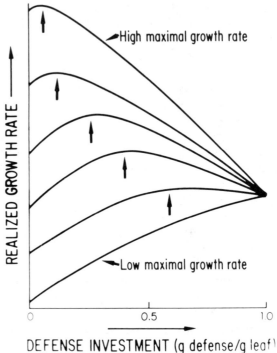

Figure 5.11. Effect of defense investment on realized growth. Each curve represents a plant species with a different maximum inherent growth rate. Levels of defense that maximize realized growth are indicated by an arrow. Realized growth (dC/dt) is calculated as $dC/dt = G*C*(1 - kD^a) - (H - mD^b)$, where $G (g\,g^{-1}\,d^{-1})$ is the maximum inherent growth rate permitted by the environment (without herbivores), $C\,(g)$ is the plant biomass at time zero, $D\,(g\,g^{-1})$ is the defense investment, and $k\,(g\,d^{-1})$ and a are constants that relate an investment in defense to a reduction in growth. The entire term $(1 - kD^a)$ is the percentage of reduction in growth due to investment in defenses. The term $H\,(g\,d^{-1})$ is the potential herbivore pressure in the habitat (assuming no defense). Potential herbivory is reduced by a function of defense investment, (mD^b), where $m\,(g\,d^{-1})$ and b are constants that determine the shape of the defense effectiveness curve. The entire negative term $(H - mD^b)$ is the reduction in realized growth $(g\,d^{-1})$ due to herbivory. Since it is subtracted from growth, this assumes herbivores consume fixed amounts of leaf tissue and not fixed percentages of plant productivity. The model's results depend on the extent to which this assumption is true. To further conform to biological reality, the herbivory term $(H - mD^b)$ cannot be less than zero, regardless of the value of D. From P. Coley et al. 1985, Science, Vol. 230, No. 4728, pp. 895–899. © 1985 by A.A.A.S.

are associated with the vigor of trees, which in itself is a function of site quality, stocking levels, and successional stage. Thus, the very conditions under which Coley et al. would postulate high defense levels and low herbivory levels are the very conditions in which sometimes devastating outbreaks are most likely to occur. This seeming contradiction is reconciled by realizing that the patterns interpreted by Coley et al. are mainly for interplant comparisons, not for intercondition comparisons within species.

Waring and Pitman (1985) described results of an experiment in which a 120 yr. old lodgepole pine stand was treated with additions of sawdust and sugar, nitrogen fertilizer, and thinning plus fertilizer. Through baiting, these experimenters insured uniform dispersal of mountain pine beetle (*Dendroctonus ponderosae*) throughout the treated blocks of the stand. Waring and Pitman found that tree mortality was intimately related to tree vigor, which was assayed as growth efficiency—wood growth per unit of leaf area. Upon thinning by whatever means, light intercepted per unit leaf increased and raised the level of vigor of surviving trees so that insect attacks were reduced and trees survived. From this experiment, it seems that susceptibility to attack and vulnerability to mortality from attack are related to the condition of trees which, in turn, is related to stocking levels and successional status.

In another experiment, shrubs of lowland tropical rain forest were examined for levels of defensive compounds in canopy openings and under deep shade of intact canopies. The defensive levels were clearly higher in shrubs intercepting more light and having higher resource levels to produce defensive compounds (J. Schultz, pers. comm.). Tannin production has been experimentally shown to increase at high light levels, regardless of nutrient status, by Waring et al. (1985). To what extent this effect is related to UV photoinduction (McClure 1979) or higher fixed carbon levels is unclear. Of course, for plants whose chemical defenses are based on nutrients such as nitrogen, rather than carbon, nitrogen supply, not light, is the critical, limiting resource (see example below). Thus, within species, higher resource levels lend better protection against herbivory, not less, as we might find in interspecies comparisons (Coley et al. 1985).

Insect herbivory is not the only kind of consumption of trees that may be associated with low tree vigor. Fungal infections, while usually termed parasitism, have the same trophic status as insect herbivory, and can also be linked with tree vigor. For example, Matson and Waring (1984) showed how laminated root rot in mountain hemlock (*Tsuga mertensiana* (Bong.) Carr.) is related to the vigor of the hemlock trees. In old age stands, nitrogen becomes very limiting. That, together with decreased light per unit of leaf area, reduces the vigor and defense levels of hemlock. Laminated root rot infections develop to the extent that most of the trees die, leaving a wavelike zone of dead trees that are eventually replaced by young trees which enjoy the benefits of newly mineralized nitrogen and abundant light. These more vigorous young trees are then able to prevent root rot infections.

At a higher level of generalization, herbivore or pathogen outbreaks can be viewed as a special form of disturbance (Waring and Schlesinger 1985). There seems to be a positive relationship between time since the last disturbance, whether it be wind damage, fire, or herbivory, and the severity of damage done when the disturbance occurs. This suggests that there may be an optimum minor disturbance rate that will maintain tree vigor so that major disruptions are minimized. What are the consequences of insect or pathogen outbreaks in forest ecosystems? Clearly, a minimum consequence will be a larger fraction of energy flow through the primary consumer pathway at the expense of the detritus pathway. At an intermediate level of attack, whether in terms of intensity or extended time, outbreaks can change the light distribution pattern within the tree canopy or to the forest floor. This can lead to reallocation of NPP to larger amounts of foliage, and to shrubs and herbs of the understory. Increased light on the forest floor can change decomposition rates and the availability of some nutrients. Wood increment may remain unchanged and NPP may even increase under these circumstances (Rafes 1970; Matson and Addy 1975; Seastedt and Crossley 1984; Romme et al. 1986) because of compensatory effects of reducing the overstory canopy (Oren et al. 1985). With even higher intensity of herbivory leading to tree death, the accelerated growth of advanced regeneration or understory species will reset succession. This kind of reset will not mean a drastic drop in NPP, however. Because of the overlap in generations of overstory and understory plants, NPP may actually vary little or even increase during this process. There will be a large variation, however, in the amount and proportion of NPP that will flow through grazing pathways.

Net Ecosystem Production (NEP)

The difference between NPP and energy consumption rates by both grazing and detritus pathways is the net change in accumulated organic matter (net ecosystem production, or NEP). NEP can be negative during destructive episodes, or positive during the aggrading stages of succession. NEP is presumed to reach a maximum positive value around midsuccession, when NPP is very high and before decomposition rates equal detritus input rates. Figure 5.7 shows a sharp negative spike for NEP during a destructive event like logging or fire that resets succession, and then shows a "hump shaped curve" beginning and ending with zero. This probably is a good generalization for simple, ideal successional sequences and a good point of departure for predicting trends for particular cases. Vitousek and Reiners (1975) and Gorham et al. (1979) used this general model to advantage in hypothesizing how succession would influence chemical balances in ecosystems.

What kinds of departures from this idealized case in Figure 5.7 might we find in real successional sequences? Such data are rare. Through a clever process of reconstructing stand histories, Pearson et al. (1987) provide data for four lodgepole pine stands in southeastern Wyoming (Fig. 5.12). Data include plant, forest floor, and dead wood mass—the main components of NEP—through succession. Evidence suggests that these forests are often even-aged and that successional development is usually terminated by fire—not an unusual history for many forests of the world. In the French Creek example, however, regeneration was slow and staggered through time. In three of the four cases, phytomass peaked (although at different times) and was in a declining phase at the time of reconstruction. In the Rock Creek case the decline had not yet occurred.

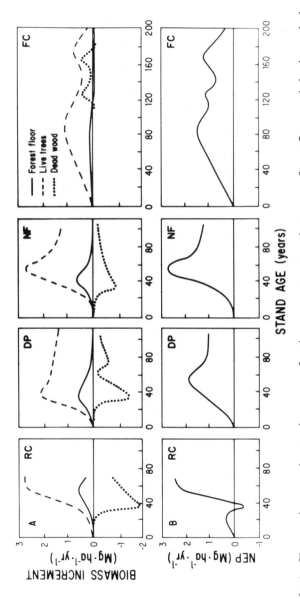

Figure 5.12. A. Changes in annual mass increment for three components: phytomass, forest floor, and dead wood, during stand development of four lodgepole pine stands (*Pinus contorta* (Engelm *ex.* Watts.) Critchfield) in the Medicine Bow Mountains of southeastern Wyoming. B. Net changes in NEP resulting from increments in the three components in part A. The stands are identified as follows: RC = Rock Creek, DP = Dry Park, NF = Nash Fork, and FC = French Creek. From Pearson et al. (1987).

In all four cases, the forest floor increment peaked, then approached zero in classic asymptotic fashion. In three of the four cases, the net change in dead wood inherited from the initiating fires was negative for 100 yr. or so. As the new generation of trees began dying through self-thinning or other causes, inputs to the wood pool compensated for losses through decomposition and the net wood increment approached zero. In the fourth case, French Creek, there was no inherited dead wood, so the dead wood pool did not begin to accumulate until older trees in the slowly regenerating forest began dying and contributing to that pool. Therefore, dead wood increment in that stand was always positive.

The sums of these three increments compose an estimate of NEP. A wide range of variation was observed among these four monospecific stands located relatively close to one another. In general, NEP did follow a "hump shaped curve" but dips, rates of ascent, and maximum NEP values were important sources of variability. The French Creek case was particularly interesting because it represented a slow succession on a site assumed to have been previously occupied by meadow. It comes closest to Peet's (1981) case of asymptotic increase on a poor site, but within 200 yr., turnovers in tree populations and increases and decreases in the dead wood pool created quite an irregular pattern. This study is unusually valuable because it provides a long-term history based on the tree ring record for single stands rather from a chronosequence of stands. As Pearson et al. point out, this approach was possible because of the simple forest structure which undergoes pulse–regrowth successional dynamics.

5.3.3. Energetics and Alternative Courses of Succession

The concept of succession has presented many difficulties in the last 15 yr. (McIntosh 1981), perhaps because we have tried to overgeneralize the process, or perhaps because we have become confused with different scales in time and space (White and Pickett 1985; Delcourt et al. 1983). In fact, there are many alternative scenarios for forest succession, and prediction of changes in energy flow will depend on these scenarios.

In my opinion, the relationship of energetics to forest succession is reasonably well understood. The principal problem now is to find formulae that will allow us to predict energy flow and other ecosystem phenomena for the very large range of alternative courses that succession can take. I do not see any way of proceeding beyond very general cases of succession until we define classes of succession in more specific terms. Classes of successional circumstances might be defined in a manner analogous to a medical doctor's diagnosis of a behavioral syndrome in a patient. For example, a set of dichotomous questions might be asked in order to provide increasing specification of the particular case. The diagnostic process is itself an analysis derived from a holistic understanding of the patient, or the ecosystem. This analogy between ecosystem science and medicine is drawn by Rapport et al. (1981), except that they put diagnosis in the context of unusual stress rather than normal development. An example of such an approach might be as follows.

1. Is the case under study an example of primary succession with extensive site modification (resource enhancement), or secondary succession in which there will be little modification (facilitation sensu Connell and Slatyer 1977)?
2. Is the site resource-rich or resource-poor relative to the surrounding environment?
3. Is there any limitation to stocking rate (e.g., seedling establishment)?
4. Is this region one in which succession involves a single generation of one or a few species and which normally is, or historically has been, terminated by large-scale disturbances such as fire or hurricanes?
   a. If yes, what is the nature of the disturbance?
      i. Given the disturbance type, what is the fate of destroyed phytomass?
      ii. What effect does the disturbance have on site fertility?
5. Do successions in this region or zone involve species turnover by the dominant plants?
   a. If yes, do the succeeding species have evergreen or deciduous habits, do they have faster or slower growth rates, are they more or less shade tolerant, and do they have higher or lower levels of quantitative vs. qualitative biochemical defenses?
   b. If yes to 5a, will detritus quality change?
   c. If yes to 5b, will forest floor microclimate change?
6. What is the typical fate of first generation successional trees if they are not replaced by other species or are not destroyed by large-scale disturbances? Do they turn over in patches?
   a. If they turn over in patches, what is the average patch size, shape, turnover rate?

Questions such as these can be elaborated further. If organized into a dichotomous key for classes of situations, then the alternative outcomes for energetics described under NPP, etc., could be tested against these situations. It would be interesting to see this approach attempted in a comparative way, as it could lead to a basic improvement in our ability to scale our generalizations to the appropriate scale of circumstances, and in our understanding of forest energetics.

5.4. New Approaches and Challenges

Throughout this review, new ways of looking at forest energetics and new questions have been highlighted. In this last section, I wish to list some of the areas of endeavor that have led to these new perspectives, and perhaps point to areas in need of special attention.

5.4.1. Evaluating Stand Architecture and Tree Physiology

Since the late 1960s, scientists from several disciplines have developed a sophisticated capability for evaluating stand architecture and tree physiology. Jarvis and Leverenz (1983) illustrate the ability of foresters, tree physiologists, and ecologists to measure, model, and evaluate the distribution of foliage in the canopy with respect to the distribution of sunlight. With such data, these workers can estimate photosynthetic and respiration rates, calculate allocations of photosynthate to various sinks (albeit in a crude way), and calculate overall NPP. To be sure, these capabilities are more or less limited to single species, first generation stands, but the basic approaches are established for application to more complex and typical situations.

One of the most interesting aspects of this work is the relationship between NPP of a single tree, or an entire stand, and leaf area per unit ground area (LAI). It appears that LAI is limited by climatic and site resources (Gholz 1982), which sets limits on NPP for the local site irrespective of successional state. Further, on a given site, LAI seems to accumulate beyond an optimum for maximum individual tree growth and perhaps even stand NPP. What are the ecological consequences for accumulation beyond this optimum? How does it influence plant demography, competition, reproduction, and vigor as related to defenses against insect herbivores and microbial pests?

This physiological approach to forest energetics is already being applied to understanding declines as well as growth phases of forest dynamics (Waring and Schlesinger 1985, p. 225). Further applications would seem appropriate to other well-known examples of synchronized mortality such as spruce budworm (Royama 1984), red spruce decline (Johnson and Siccama 1983), and *Metrosideros* decline (Mueller-Dombois et al. 1983). This physiological approach ought to provide us with the tools to understand and interpret forest dynamics across the enormous spectrum of ecological circumstances, and to diagnose the general health and degree of vulnerability to pests and diseases of managed forests (Waring 1985).

5.4.2. Root Dynamics

Possibly the most important pioneering work on forest energetics in the last two decades has been on estimating root and mycorrhizal biomass and energetics. This problem area has not benefited from technological breakthroughs as much as from hard work. Tedious as it has been, this effort has given us a very different perspective on forest energetics, particularly with respect to raising our former estimates of productivity by inclusion of surprisingly high rates of root turnover, and by demonstrating how variable root–mycorrhizal biomass and turnover rates can be. Much work remains; the data are still limited (Vogt et al. 1986a). It is especially important to develop guidelines on how fine root biomass and turnover rate vary with site quality and stand age (e.g., Grier et al. 1981; Keyes and Grier 1981; Nadelhoffer et al. 1985).

5.4.3. Detritus Dynamics

Enormous advances have been made in understanding the details of litter decomposition and mineralization in forests in the last 20 yr. These advances have resulted from increased experimental manipulations and attention to organic as well as inorganic chemistry of substrates. Unfortunately, most of our attention has been focused on aboveground litter and not enough on belowground processes. The relative degrees of knowledge can be assessed by comparing textbook space on aboveground and belowground processes in Swift et al. (1979) and Waring and Schlesinger (1985). We can now predict reasonably well which general types of organisms will be involved in surface litter consumption, and the rates at which this consumption (decomposition) will be accomplished in a wide range of environments. We still have little knowledge of detritus inputs below the surface, a problem connected with the root dynamics issue discussed above. Most of our estimates of belowground rates are based on changes in mass over discrete intervals ranging from weeks to months. Most investigators realize that this approach, the only one available at present, surely must underestimate turnover rates and energy flux belowground because it misses short-term inputs and outputs (Vogt et al. 1986b; Lauenroth et al. 1986).

5.4.4. Interplay Between Energy and Nutrients

Models and concepts that use the interplay between energy and nutrients have become more evident (Chapin et al. 1986). The importance of site quality in terms of fertility has been long established in certain extreme cases (Van Cleve et al. 1983) and is now incorporated in some large-scale models of forest growth (Kimmins et al. 1981; Aber and Melillo 1982). The role of fertility in maintaining tree vigor and defenses against consumers has been exemplified by the experimental work of Matson and Waring (1984). The relationship between site fertility and the proportion of NPP that must be allocated to fine root production has been demonstrated by Grier et al. (1981) and more generally developed by Nadelhoffer et al. (1985).

Much remains to be done. There still is a tendency to rate the fertility of a site, or its resource qualities in general, in terms of some ecosystem performance such as tree growth rates or nutrient cycling rate. This would seem to indicate that we do not have good a priori methods for evaluating site quality before these performances occur and can be measured. More experiments are needed for separating the role of inorganic nutrition from other factors involved in growth rates, species composition, episodic pest attacks, and forest declines.

5.4.5. Controls on Rates of Forest Grazing or Infection

The last decade has seen an interesting convergence in the activities of forest pathologists, concerned with "diseases and pests" of forests, with those of tree physiologists, entomologists, and ecologists interested in the evolutionary interactions of plants and their grazers. All of this work helps ecologists interested in energy flow through grazing pathways to understand the factors that set low level, chronic rates of grazing, and how these factors may be changed, leading to explosive oscillations in consumer populations.

The separation of extrinsic factors such as pest dispersal patterns and weather from intrinsic factors such as stand condition has not yet been fully accomplished. Also, the variable kinetics of outbreaks need explanation. Why, in some forest types, can the impact of grazers or disease organisms lead to a gradual stem thinning and resumption of sufficient defense levels to halt the destructive process, while in other forest types, or in other areas, the outbreak proceeds so fast that entire populations are wiped out and succession must begin anew?

Why are roots not grazed much more than is presently perceived? Fine roots would seem to be excellent, succulent material for underground grazers such as nematodes, particularly considering the amount of belowground grazing suffered by some crop plants. Are forest roots better defended? If so, by what mechanisms? How reactive or dynamic are these defense mechanisms?

5.4.6. Modeling Forest Stand Dynamics

With the advent of computers in the last 20 yr., modeling and simulation of complex systems such as forest stands has become possible. Silviculturists and ecologists have taken full advantage of this opportunity, so that a remarkable array of forest growth models is now available. Shugart and West (1980) and Loucks et al. (1981) reviewed the kinds of models available at the beginning of this decade. These models were stimulated by different questions, so they stress different processes, interactions, and scales of time and space. However, all have played an important role in forcing investigators to ask different kinds of questions than had been formerly addressed, and in allowing new questions to be asked of the simulation models. Examples of these can be seen in Shugart and West (1981), Loucks et al. (1981), several chapters in West et al. (1981), and Shugart and Seagle (1985). These models more directly benefit our understanding of forest succession and decline dynamics than energetics per se, but they also allow us to explore energy flow patterns through the application of energy parameters to growth, accumulation, and energy flow parameters.

5.4.7. Energetics on a Landscape Scale

Success in modeling succession and ecosystem processes on a stand level has expanded the vision of ecologists so they may study and interpret ecosystem processes at landscape, regional, and global scales. For most of the short history of ecosystem ecology, we have conceived the world as composed of contiguous, nonoverlapping, largely autonomous ecosystems. I believe that this view, while necessary in preliminary assessments of ecosystem dynamics, is generally unrealistic, leading us to underestimate the intersystem interactions resulting from transfers across landscapes. These transfers may involve materials and genetic information as well as energy (Reiners 1986). With new modeling capabilities, it is no longer necessary to restrict our view and our predictions of the dynamics of single stands.

The opportunity now exists for ecosystem ecologists to take up the nascent ideas of landscape ecology, as introduced by Naveh and Lieberman (1984) and Forman and Godron (1986), and reshape them in ways that are meaningful for ecosystem phenomena. For example, new modes of spatial modeling are needed so that we can map areas of relative homogeneity on a landscape unit, and designate source areas and receptor areas on that landscape with regard to energy, material, and information. If we were to map pathways of flux, we could better interpret why some parts of a landscape remain relatively immature in successional terms, while other parts are relatively senescent. This would allow us to deal with transfers from the atmosphere (Lovett and Reiners 1987), down slopes (Orndorff and Lang 1981; Welbourn et al. 1981; Swanson et al. 1982a), and into aquatic systems (Swanson et al. 1982b). The distribution of critical organisms across the landscape can have far-reaching effects on the forested landscape mosaic. Knight (1987) has suggested that trees damaged by lightning, fire, or wind may serve as epicenters for the spread of insect or fungal consumers which, in turn, may enhance flammability. Thus, factors controlling the probability of lightning, fire or wind damage can interact with the predisposition of consumer organisms to set a probabilistic pattern to the landscape mosaic.

The landscape approach raises other research challenges. For example, how do we systematically manipulate data from landscapes so that phenomena occurring at different spatial and temporal scales are brought into proper focus? How do we integrate such phenomena when we extrapolate to very large scales (e.g., Cropper and Ewel 1984; Meentemeyer 1984; Meentemeyer et al. 1982, 1985; Berg et al. 1984; Weinstein and Shugart 1983)? What are the consequences of landscape grain size and shape? Franklin and Forman (1987) have raised the question of what critical balance between a matrix of old age forests and clearcuts will lead to a breakdown of the matrix forests through windfall and other edge related disturbances. Such conceptual and technical problems will probably be best solved with diligent attention to scaling techniques and awareness of hierarchical relationships (O'Neill et al. 1986; Urban et al. 1987).

That the Earth's surface is composed of complex landscapes and seascapes is recognized by environmental scientists besides ecologists. This is apparent in a recent National Research Council document which discusses planning for an International Geosphere–Biosphere Program—a program that could be the logical successor to the IBP. In that document the authors say:

> The Earth's topography is complex at all scales, providing a mosaic of habitat types that differ in suitability for different organisms. Adding to this complexity is the fact that any habitat presents an assemblage of organisms whose makeup is a function of the time since the last ecological disturbance. Communities of organisms generally progress through a series of distinct stages. The activities of man have greatly increased the complexity of the landscape units by a variety of manipulations. Accounting for the detailed interactions of these landscape units with the atmosphere and geosphere is obviously a complex task, yet it is essential that we do so in order to assess fully the consequences of the increasing tempo of landscape modification.
>
> (U.S. Committee for an IGBP 1986, p. 51.)

McIntosh (1985) noted that when "big ecology" began with the IBP in the 1960s, the ecological community was "preadapted," in terms of conceptual and technical capabilities, from several decades of research and scholarship.

In most respects, ecosystem ecology is now preadapted to participate as a keystone discipline in the partnership of disciplines needed to understand the functioning of the biosphere. Of prime importance now is the development of methods for dealing with variable scales, and for the appropriate modeling of ecosystems at all scales.

Acknowledgements

I extend my sincere appreciation to colleagues I. C. Burke and D. H. Knight, The University of Wyoming; to F.S. Chapin III, University of Alaska; to Robert P. McIntosh, Notre Dame University; and to R. H. Waring, Oregon State University, for their many constructive suggestions for this review. This work was supported by NASA contract No. WAG 2-355.

6. Abiotic Controls on Primary Productivity and Nutrient Cycles in North American Grasslands

Paul G. Risser

At simple and general levels, the occurrence and physiognomy of grasslands and savannas are easily related to the spatial and temporal patterns of such abiotic factors as growing season, precipitation, and temperature (Borchert 1950; Curtis 1959; Lauenroth 1979; Singh et al. 1983; Transeau 1935). Across the Great Plains from the North American Rocky Mountains to the tallgrass prairie and the savannas of the eastern prairie–forest border, precipitation increases and the frequency of droughts decreases (Risser et al. 1981). From the southern plains of central United States to south central Canada, the growing season becomes shorter, average temperatures decrease, and greater proportions of annual precipitation fall as snow. Across these broad gradients from west to east, shortgrass steppes are replaced by mixed grass and tallgrass prairies and, eventually, by savannas. Similarly, species composition changes from south to north, especially increasing the proportion of cool season species. These gradients, however, are confounded by more immediate weather and climate conditions, by topography and soil characteristics, and by management practices (Schimel et al. 1985a; Sims et al. 1978; Weaver 1954; Weaver and Albertson 1956; Weaver and Bruner 1954).

This chapter considers more specific relationships between abiotic factors and structural and functional characteristics of grassland and savanna ecosystems. This exploration will examine the correlation of climate and nutrient factors with various measures of ecosystem energetics and nutrient cycles. A synthetic discussion of our current understanding of these relationships and suggestions for potentially fruitful future investigations follow.

6.1. Abiotic Factors and Ecosystem Energetics

During the past two decades, numerous attempts have been made to relate various climatic measurements to properties of ecosystem energetics (Lauenroth 1979; Sims et al. 1978). This relationship is of inherent interest since it suggests which abiotic factors act as controls on ecosystem processes. In addition, when close relationships are detected, measurements of climatic factors can be used to predict rates of energy capture and transfer among components within grassland ecosystems.

The grasslands of North America probably originated in the Pliocene, and native grasslands of today are similar to those found after the Pleistocene (Singh et al. 1983). Except for successional and human induced grasslands, the grasslands of the Central Plains are characterized by at least one annual season when soil moisture is inadequate for forest growth, but with sufficient available water to support grasses (Gramineae) as a major component of the vegetation (Fig. 6.1). Worldwide grasslands occur in areas receiving 250–1000 mm of annual precipitation and having mean annual temperatures between 0–26°C (Lauenroth 1979; Lieth 1975). North American grasslands nearly encompass this range of amounts of annual precipitation, but the range of temperatures

Figure 6.1. Shortgrass steppe. P. Risser photo.

is only about 5.0–20.0°C. Perhaps as a result of the lower average temperatures and the infrequent occurrence of grasslands in the upper ranges of annual precipitation, the maximum rates of productivity from North American grasslands are not as high as those of some grasslands elsewhere.

A more detailed examination of these relationships in North America can be observed from the analysis of Sims et al. (1978), who compared the relationships between abiotic variables and vegetation characteristics at several North American grasslands (Fig. 6.2). Among eight sites, the length of the growing season varied from 316 days at a New Mexico desert grassland site to 168 days on a mixed grass prairie in North Dakota (Table 6.1). Precipitation ranged from about 100 mm at a northwest bunchgrass site in Washington to over 700 mm at a tallgrass prairie site in Oklahoma. Potential evapotranspiration during the growing season was 550 mm at a Colorado shortgrass steppe and at a mixed prairie site in North Dakota, and greater than 800 mm at a tallgrass prairie site in Oklahoma. Multiple regression models were used to compare various combinations of abiotic variables with vegetation characteristics from these sites. Selected combinations of these variables from ungrazed grasslands are presented in Table 6.2, and the numerical values indicate the percentage of variance accounted for in the dependent biotic variable by one or several independent abiotic variables. In general, variability in aboveground components of grasslands was best accounted for by precipitation or vegetation

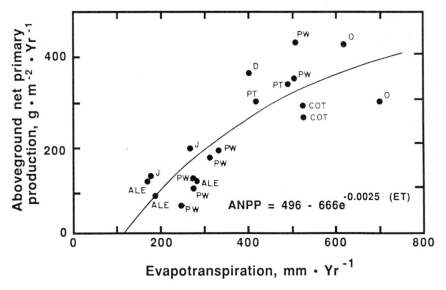

Figure 6.2. Aboveground annual net primary production and actual evapotranspiration from several grassland sites in the United States, as described by Sims et al. 1978. Site abbreviations are: ALE, bunchgrass, Washington; COT, mixed grass, South Dakota; D, mixed grass, South Dakota; J, desert grassland, New Mexico; O, tallgrass, Oklahoma; PW, shortgrass, Colorado; and PT, shortgrass, Texas. From Webb et al. (1978), Ecology.

Table 6.1. Growing Season and Related Factors of Eight North American Grasslands[a]

Grassland Type	Growing Season (Days)	Air Temperature (°C)	Potential Precipitation (mm)	Evapo-transpiration (mm)
Tallgrass prairie				
Oklahoma	272	19.2	710	815
Mixed prairie				
North Dakota	168	15.8	380	550
South Dakota	200	16.0	370	610
Kansas	226	18.4	420	740
Shortgrass steppe				
Colorado	227	15.0	220	550
Texas	269	17.5	380	760
Desert				
New Mexico	316	16.8	220	790
Northwest bunchgrass				
Washington	257	16.3	100	750

[a] From Sims et al. (1978) Journal of Ecology, by permission of Blackwell Scientific Publications, Ltd.

water use variables; solar radiation or temperature factors were more predictive of belowground components. Combinations of from 1–3 abiotic variables accounted for 66, 80, and 87% of the biomass dynamics of the live, total standing crop and root components, respectively. Only 54% of the crown and 60% of the litter biomass were accounted for by abiotic variables. The amount of crown biomass material at the leaf base of perennial bunch grasses demonstrated considerable fluctuation, depending in part on the growth and senescence dynamics of the plant during the growing season and in part on the translocation pattern of carbohydrates stored in the crown and elsewhere in the plant. Litter may be disarticulated from plant growth because of longer turnover times and relatively slow dynamics of biomass weight changes during decomposition.

Although the data are not presented in Table 6.2, Sims and his colleagues considered both grazed and ungrazed grasslands (Sims et al. 1978), and found that these combinations of from 1–3 abiotic variables accounted for variances on grazed grassland even more accurately than for ungrazed grasslands. The authors conjectured that large herbivore grazing might be more natural, so that, under these conditions, one might expect a closer linkage between abiotic factors and biotic components. That is, grassland ecosystem components have evolved under, and therefore are adapted to, the simultaneous combination of grazing and environmental variability. Grazed grasslands, therefore, would be expected to correlate more closely with abiotic factors than ungrazed grasslands. Alternatively, Sims and his associates suggested that grazing shifts the system from one that is biotically-controlled to one that is controlled to a greater extent by abiotic factors, with the latter accounting for a closer relationship between dependent biotic and independent abiotic variables.

As a further elaboration on the tracking of abiotic condition and grassland ecosystem behavior, Parton and Risser (1980) used simulation models to show

Table 6.2. Selected Relationships Between Abiotic Variables and Percent of Variance Explained in Biotic Variables on Ungrazed Grasslands[a]

Dependent Biotic Variable	Independent Abiotic Variables	r^2
Live biomass	Long-term mean annual precipitation	0.54
	Long-term mean annual temperature plus Annual precipitation plus Growing season actual evapotranspiration	0.66
Total standing crop	Growing season precipitation	0.68
	Growing season precipitation plus Annual usable incident solar radiation plus Growing season actual evapotranspiration	0.80
Litter biomass	Growing season actual evapotranspiration	0.27
	Average growing season temperature plus Growing season precipitation plus Annual potential evapotranspiration	0.60
Crown biomass	Growing season daily usable incident solar radiation	0.13
	Long-term mean annual temperature plus Annual usable incident solar radiation plus Growing season daily usable incident solar radiation	0.54
Root biomass	Growing season usable incident solar radiation	0.83
	Mean annual temperature plus Growing season usable incident solar radiation plus Growing season actual evapotranspiration	0.87

Continued

Table 6.2. (*Continued*)

Dependent Biotic Variable	Independent Abiotic Variables	r^2
% Cool Season plants	Long-term mean annual temperature	0.51
	Growing season precipitation plus Annual usable incident solar radiation plus Annual actual evapotranspiration	0.81
% Warm Season plants	Long term mean annual temperature	0.42
	Growing season precipitation plus Annual usable incident solar radiation plus Annual actual evapotranspiration	0.70

[a] From Sims et al. (1978) Journal of Ecology, by permission of Blackwell Scientific Publications, Ltd.

that grazing increased the rate of energy flow through the tallgrass prairie ecosystem. More rapid turnover rates, therefore, might make the system more responsive to short-term changes in abiotic factors. As a result, the more rapid tracking and the consequently closer correlation between abiotic variables and vegetation parameters in the grazing condition may be attributable to the longer turnover times and the relatively slower response times in ungrazed systems.

Lauenroth and Sims (1976) indicated that for most years, actual water loss during the growing season on the shortgrass steppe was approximately equal to precipitation during the growing season. The ratio of annual evapo-transpiration (AET) to potential evapo-transpiration (PET),

$$AET/PET = 0.10 \text{ to } 0.30,$$

existed throughout the year, and, during the May through September growing season, the ratio was 0.25–0.30 (Parton et al. 1981), as shown in Figure 6.3. An examination of daily moisture dynamics makes clear that, in eastern Colorado, most of the annual precipitation comes from infrequent, but relatively large, precipitation events (> 16 mm), and that most of the water loss from the system (52%) occurs as large AET events following large precipitation events. All large water loss events occurred during the 24 hr.–96 hr. period following a rainfall event, and most of them occurred after the largest precipitation events.

Although these relatively large precipitation events are important in the water budget of the grassland, small events may be particularly significant in the drier regions. In the semiarid region of the Great Plains, precipitation events of 10 mm or less accounted for 41% of the growing season rainfall and 83% of the rainfall events (Sala and Lauenroth 1982). The humid tallgrass prairie may be largely unaffected by small rain showers, but lysimeter experiments in the shortgrass steppe have demonstrated that blue grama sod responds to water applied in 5 mm increments following drought conditions. Specifically, Sala and Lauenroth (1982) found that addition of only 5 mm of water significantly ($P < 0.05$) increased leaf water potential during the next day, when at noon the values were typical of nonstress conditions. Furthermore, this effect continued through the second day after watering. Though the population and ecosystem responses to these events have not been fully examined, it appears that semiarid grasslands take advantage of even small amounts of rainfall.

A close relationship exists between annual evapo-transpiration (AET) and aboveground net primary production (ANPP) in grasslands and several other vegetation types (Rosenzweig 1968; Webb et al. 1978). This relationship (Fig. 6.2) is nonlinear, with the rate of increasing net primary production decreasing at higher levels of actual evapo-transpiration. As shown below, within the shortgrass steppe and cold desert ecosystems, both the AET of the previous year and that of the current year were related to ANPP.

Year of AET Relative to Year of ANPP	Variance (R^2) of ANPP Accounted for by AET (%)
Current year	55
Previous year	44
Current and previous years combined	97

The highly significant increase in R^2 for the combined AET values of the current and previous years indicates that in these water limited shortgrass systems, moisture conditions of the previous year contribute to the ANPP of the current year (Cable 1975; Hyder et al. 1975). The effects of the moisture regime of the previous year may be the result of several mechanisms; e.g., higher water content of the soil at the beginning of the second year, greater seed crop to begin the second year, and increased root system and carbohydrate reserves of perennials at the beginning of the second year.

Water use efficiency (WUE) can be calculated as the aboveground yield of biomass or production per unit of evapo-transpiration per unit time, or g of ANPP per kg of transpired water. For shortgrass steppe–cold desert systems, the range of WUE was 0.2–0.7 (Webb et al. 1978). This regression between ANPP and AET was used to calculate the minimum annual precipitation needed to sustain zero net primary production. In the case of the shortgrass steppe, this value is approximately 60 mm per year (Lauenroth 1979). Above this minimum in the shortgrass steppe, ANPP increases with increasing precipitation in the range of 0.5 g/m2/mm. However, as precipitation increases toward an environmental situation where the ecosystem is no longer water-limited, the rate of increase in ANPP is a decreasing function of water use

P. G. Risser

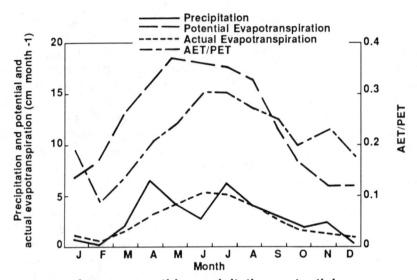

Average monthly precipitation, potential evapotranspiration, actual evapotranspiration, and the ratio of AET to PET.

Figure 6.3. Average monthly precipitation, potential evapotranspiration, actual transpiration, and the ratio of AET to PET on a shortgrass steppe in northeastern Colorado. From W. J. Parton, W. K. Lauenroth, and F. M. Smith (1981). Water loss from a shortgrass steppe. Agricultural Meteorology 24:97–109. © Elsevier Scientific Publishers.

(Fig. 6.2). Water use efficiencies that range from 1.25–0.37 indicate that when soil water is more available, vegetation is unable to capitalize on the additional water at the same rate (Webb et al. 1978).

Risser et al. (1981) related several abiotic variables to five methods of calculating net primary production on the tallgrass prairie. In all cases, more than 85% of the variance could be accounted for by one or two variables, including cumulative-potential evapo-transpiration, heat sum or temperature, and solar radiation (Table 6.3). For these predictions, the independent variables describe annual net primary production, including differences of within season growth and estimates of cumulative biomass. Thus, it is clear that several easily measured abiotic variables can be used to predict several measures of primary production.

Regression models such as those just discussed are useful for developing general relationships between easily measured abiotic factors and components of ecosystem productivity (Hake et al. 1984; Rogler and Haas 1947; Rosenzweig 1968; Smoliak 1956, 1986; Sneva and Hyder 1962). These models frequently use annual or monthly abiotic measurements to account for within-season differences or conditions at the beginning of a growing season, or include current and long-term abiotic values (Hanson et al. 1983; Sneva and Hyder 1962). In general, however, these models are site specific, or are limited to

Table 6.3. Predictors of Primary Production on the Tallgrass Prairie[a]

Calculations of Annual Net Primary Production	Abiotic Variables	r^2
(1) Cumulative standing crop of current live	Cumulative potential evapotranspiration (PEV)	0.88
(2) Cumulative current live plus recent dead	Cumulative number of days above 4.4°C (TEMP)	0.96
(3) Cumulative positive increments in total standing crop	Cumulative solar radiation (SOLR)	0.94
(4) Cumulative positive increments in current live plus positive increments of recent dead	SOLR + TEMP	0.88
(5) All cumulative positive increments at 90% confidence level	SOLR + TEMP	0.87

[a] From Risser et al. (1981), by permission of Van Nostrand Reinhold Publishing Company.

sites where the measured variables and interactions are similar to those at the site for which the model was developed. Because of this limitation, more mechanistic models relating climate and yield or production have been developed (Innis 1978; Risser and Mankin 1986; Wight and Hanks 1981; Wight et al. 1984).

One illustrative forage production model, ERHYM (Wight and Neff 1983), uses water content in the rooting zone soil at the beginning of the growing season, daily precipitation, and amounts and estimates of potential evapotranspiration as input variables, and then calculates daily water content of the soil, actual transpiration, and potential transpiration. Total herbage yield is then calculated from the following relationship: actual yield (Y)/potential site yield (Yp) = actual transpiration (T)/potential transpiration (Tp).

The results of this equation provide an index of the growing season's climatic effect on yield, since the model and index account for the amount and distribution of precipitation, for the associated evaporative demand, and for the water content of the soil (Wight et al. 1984). When water is nonlimiting, actual transpiration (T) and potential transpiration (Tp) are equal, $T/Tp = 1.0$. Previous work in the northern Great Plains suggests that site potential herbage yield is about 1.5 times the long-term site average yield; thus, the yield index can be expressed in terms of the long-term average site yield: yield index = $1.5T$.

Wight et al. (1984) used this model for an upland range site in eastern Montana. To forecast the current year's forage (yield index), the model is run with the following input variables: available water content in the soil at the beginning of the growing season and daily precipitation, mean air temperature, and solar radiation from weather records. Each model run generates a yield index, T/Tp, based on these variables, where only the beginning water content of the soil is year specific, and the remaining three variables are constant since they are from weather records. If the yield indices are normally distributed, the central tendency (mean) and variance (standard deviation) can be used to describe probability statements related to the forecasted yield. Using 55 yr. of weather data and 12 yr. of yield and soil water data, two-thirds of the field-measured yields were within one standard deviation of the forecasted yields for the growing season of April, May, and June.

6.2. Abiotic Factors and Ecosystem Nutrient Cycling

Topographic position along hill slopes is correlated with a number of abiotic and biotic components (Schimel et al. 1985a). Typically, lower slope positions have higher levels of organic carbon, total nitrogen, C/N ratios, total phosphorus, available phosphorus as a percentage of total phosphorus, water, total soil organic matter, fine clay particles, and annual nitrogen mineralization, but lower rates of nitrogen mineralization relative to the total amount of nitrogen. Thus, the availability of nitrogen to vegetation increased downslope, but the turnover rate of nitrogen decreased. Evidence from fertilizer studies at this site (Stillwell 1983) indicated that in most years the footslopes were N-limited, while the other sites on the slope were water-limited. Also, perennial aboveground and total belowground biomass increased downslope, as did the concentration of nitrogen in the aboveground vegetation (Table 6.4). The concentration of nitrogen in belowground biomass did not increase downslope, but this pattern of tissue nitrogen concentration may be confounded by the combination of live (nitrogen content of 1.1% nitrogen) and detrital and senescent (nitrogen content about 2.5% nitrogen [Clark 1977]) roots.

Table 6.4. Ash-free Aboveground and Belowground Vegetation Biomass and Organic N Content of a Shortgrass Steppe Catena in 1981[1]

Component	Ridgetop Grasses	Ridgetop Forbs	Backslope Grasses	Backslope Forbs	Footslope Grasses	Footslope Forbs
Aboveground Biomass (kg/ha)[2]	702.0	1270.0	859.0	1427.0	1160.0	425.0
N (%)	1.1	1.1	1.2	1.0	1.3	1.7
N (kg/ha)	7.7	14.0	10.3	11.7	15.1	7.2
Roots + Detritus (kg/ha)	19,160.0		18,460.0		27,390.0	
N (%)	1.9		1.9		1.7	
N (kg/ha)	364.		350.		465.	

[1] See Schimel et al., Ecology, (1985a) for standard deviations.
[2] Annual and biennial forbs were important in 1981, but not in other years between 1979 and 1983.

Although phosphorus may cause an increase in productivity of the tallgrass prairie by as much as 50% (Reardon and Huss 1965), nitrogen is usually far more effective in increasing primary production (Moser and Anderson 1965). However, increases in yield caused by nitrogen in the tallgrass prairie usually occur only when precipitation is normal or above normal (Owensby et al. 1970), e.g., supplemental nitrogen and water increased yields by 80%. In the shortgrass steppe (Dodd and Lauenroth 1979), additional water, and especially water plus nitrogen, cause an increase in plant production (Fig. 6.4). Additional nitrogen also raises the crude protein content of the forage and, frequently, the water use efficiency, although nitrogen additions may reduce soil water content because of increased transpiration by the more rapidly growing plants (Launchbaugh and Owensby 1978; Seastedt 1985b).

Both general and specific relationships between abiotic factors and primary production of grassland ecosystems, are, as we have seen, well-described. How-

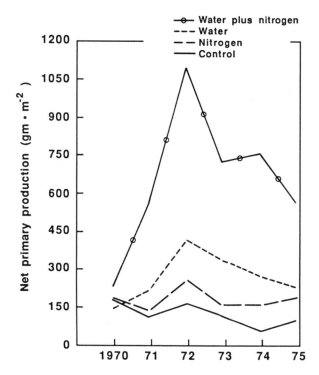

Aboveground net primary production for water and nitrogen stress experiment, 1970–1975, Dodd and Lauenroth, 1979.

Figure 6.4. Aboveground net primary production for water and nitrogen stress experiments on a northeastern Colorado shortgrass steppe, 1970–1975. From Dodd and Lauenroth (1979), in N. R. French, *Perspectives in Grassland Ecology*, Springer-Verlag, New York.

ever, the relationship between abiotic factors and the consumer function is less well understood. Little information exists concerning the responses of invertebrates to precipitation patterns, though a succession of dry years is usually followed by increased grasshopper numbers (Smith 1954). Drought is frequently associated with increases of total available carbohydrate in plant tissues (> 50%) and leaf nitrogen (> 80%) in some grasses (White 1976), but the relationship between invertebrate responses and tissue quality is not conclusive. In the tallgrass prairie, various rainfall patterns did not demonstrate consistent changes in either number or biomass of invertebrates (Risser et al. 1981), though in the shortgrass prairie, microarthropod densities and biomass increased about fourfold with the addition of water and nitrogen (Dodd and Lauenroth 1979).

No studies have been conducted to evaluate the responses of the small mammal community to fertilizer or irrigation on the tallgrass prairie, though in terms of large domestic herbivores, increases in livestock production usually accompany fertilizer applications to the rangeland (Launchbaugh and Owensby 1978). Grant et al. (1977) found that supplemental nutrients had little or no effect on small mammal populations on the shortgrass steppe; adding water caused some increases in small mammal populations, but the addition of both water and nutrients had marked positive influences on small mammals, especially because of the invasion by prairie voles (*Miocrotus ochrogaster*).

In an effort to comprehensively evaluate the response of the grassland ecosystem to abiotic variables, Risser and Parton (1982) used a large simulation model to evaluate the response of a grazed (2.4 ha/animal unit) tallgrass prairie to annual increases in nitrogen (10 g/m^2) and phosphorus (3 g/m^2). Throughout these model calculations, supplemental irrigation was added to maintain soil water tension above 0.5 MP_a. Adding water to the system increased the rate of nitrogen flows between most compartments of the model, particularly nitrogen uptake, root uptake, shoot death, and transfers to litter (Table 6.5). The model depicted increased rates of mineralization of N from plant material and soil organic material as well as increased formation of soil organic material. Thus, the available nitrogen pools increased, making increased plant uptake rates consistent with soil nitrogen availability. The addition of water stimulated aboveground primary production (+ 80%) more than did plant nitrogen uptake (+ 13%). Because of the increased plant growth with relatively less nitrogen uptake, the concentration of nitrogen in plant tissue was lowered. This reduction, in turn, decreased the consumer uptake rates of nitrogen.

Adding nitrogen to the system also increased the rate of nitrogen flows and increased the average nitrogen concentration in the shoots over the growing season from 1.2 to 1.6%. After 3 yr., 39.4% of the added nitrogen was found in the live plus the dead aboveground and belowground plant material components. The fixed nitrogen pool (exchangeable NH_4+) contained 37.7%, and the only losses from the system were 2.3% by volatilization of nitrogen and 1.7% by animal removal.

Tallgrass prairie grasses use soil water from the upper soil horizons first, and evaporative stresses are, of course, concentrated at the soil surface. As the

Table 6.5. Simulated Nitrogen Flows in the Tallgrass Prairie Ecosystem (Tabular Values Represent the Last Year of 3 Yr. Computer Simulations)[a]

Nitrogen Flows	Ungrazed	Moderate Seasonal Large Herbivore Grazing and Supplemental Fertilizer	Moderate Seasonal Large Herbivore Grazing and Supplemental Fertilizer and Water
Root uptake	5.4	6.9	8.1
Shoot uptake	3.4	4.4	5.0
Shoot death	3.5	3.3	5.2
Transfer to surface litter	3.5	3.6	6.1
Root death	2.3	3.2	3.3
Mineralization from plant material	3.1	4.6	5.3
Ammonification from soil organic N	1.7	1.8	2.5
Formation of soil organic N	1.7	2.2	2.8
Consumer uptake	0	1.2	1.0
Return by consumers	0	0.6	0.5
Volitilization by consumers	0	0.3	0.2
Removal by cattle	0	0.2	0.27.

[a] From Risser and Parton (1982), Ecology.

soil surface layers become drier, soil moisture is increasingly utilized from deeper horizons. As a result, the deep roots of the perennial tallgrasses confer an advantage during drought years (Adams and Wallace 1985; Weaver et al. 1935). The rate of nitrogen mineralization increases with soil water content, as do rates of uptake of nitrogen and decomposition. However, rates of mineralization continue at lower levels of moisture content than do rates of root uptake of nitrogen (Smith et al. 1977).

Although nitrogen in plant tissues usually increases during drought, nitrogen levels in leaf tissues may be reduced when the drought begins early in the season (Hayes 1985). Under drought conditions, herbivory may increase, perhaps because of increased nitrogen concentrations in the plant tissue. Further, the profusion of flowering the year after drought may take place because nitrogen mineralization has occurred during drought years, but not nitrogen uptake. Consequently, more nitrogen may be available in the soil the year after drought.

6.3. Discussion

The general or coarse relationships of abiotic factors with grasslands and savannas are well known. Furthermore, much of the variance in ecosystem production can be accounted for with various combinations of abiotic variables.

The following generalizations describe the correlations between abiotic and biotic variables that have been documented in the preceding text:

Biotic	Abiotic
Aboveground primary production	Precipitation and water use
Belowground primary production	Solar radiation and temperature
Cool and warm season growth habit	Temperature and water

Between 60–90% of the variance of most measures of standing crop and net primary production of grassland ecosystems can be accounted for by 1–3 abiotic variables, though these combinations vary according to the dependent variable selected for analysis.

Because these regression models are likely to be limited to the conditions under which they were developed, yield production models and whole system models have been developed that relate ecosystem processes to abiotic variables. These models have permitted an evaluation of how specific energetic processes respond to abiotic variables, particularly climate and nutrients.

Grassland primary production responds positively to nitrogen, especially when soil water is adequate, but usually demonstrates little response to other nutrients. In the shortgrass steppe, additions of nitrogen cause only slight changes in standing crop; additions of water increase the proportion of legumes and microarthropod predators; additions of water and nitrogen cause increases in exotic plant species, certain species of small mammals characteristic of lowland habitats, and macroarthropod and microarthropod predators (Dodd and Lauenroth 1979; French 1979b). In the tallgrass prairie, responses are similar, though the magnitude of response is smaller (Moser and Anderson 1965; Launchbaugh and Owensby 1978).

Although older studies on the productivity of savannas exist (Ovington et al. 1963), and recent studies outside North America can be found (McNaughton 1985), relatively few attempts to relate abiotic and biotic factors in savannas have been undertaken, except for characterizations of climatic conditions (Risser et al. 1981) and evaluations of fire and soils (Curtis 1959). This paucity of information is particularly unfortunate since understanding the interplay between trees and the herbaceous layer would be valuable, and because the climatic transitional zone should lead to powerful tests of ecosystem processes operating at the margins of both grasslands and forests.

6.4. Future Perspectives

Despite rather accurate descriptions of how abiotic factors relate to such general ecosystem characteristics as primary production and standing crop, a number of promising research directions are open to further experimentation. Of primary importance is understanding the relationships between abiotic variables and the distribution of nutrients in the soil–plant parts of the ecosystem. Simulation models indicate that rates of transfer between ecosystem compartments are altered by simple increases in soil moisture (Risser and Parton 1982), but little is known about these responses under variable abiotic conditions. These transfers and storages affect rates of nutrient cycles, so coupling the abiotic factors with nutrient cycles is particularly important.

On the one hand, plant nutrient content may significantly affect herbivory, and herbivory, in turn, influences primary production. On the other, abiotic factors influence primary production and nutrient content. As a result, the controls between and among abiotic variables and herbivory require description. For example, it is crucial to understand how decreasing soil moisture sequentially affects the availability of soil nitrogen for plant uptake, microbial transformations, and microfauna grazing. Aboveground production frequently increases with higher levels of available soil moisture, but the nitrogen content of the forage tissue decreases. The consequences of this decreased forage quality and presumed decrease in mineralization rate are not well defined with respect to ecosystem behavior.

Most studies of the relationships between abiotic variables and nutrient cycles of the grasslands have failed to account for hill slope and topographic differences. Since substantial differences exist between ridges and lower slopes, especially in the water limited environment of the grasslands (Schimel et al. 1985a), differentiation of these topographic positions will be important for understanding ecosystem responses to changing and variable abiotic factors. Furthermore, in some ecosystems, outputs from the heterogeneous landscape may influence global processes such as N_2O and methane production. Thus, the challenge is first to understand the landscape differentiation in the relationship between ecosystem process and abiotic variables, and then to translate these relatively small-scale phenomena to regional and global scales.

Finally, the North American savannas offer exciting opportunities for explaining relationships between abiotic variables and ecosystem properties. Intricate relationships have been demonstrated among climate, vegetation structure, forage quality, and herbivory in African savannas (McNaughton 1985), and similar processes are undoubtedly to be found in North America.

7. Grasslands and Savannas: Regulation of Energy Flow and Nutrient Cycling by Herbivores

James K. Detling

Although the importance of herbivores in some terrestrial ecosystems has been questioned, results of research conducted in grasslands and savannas suggest that herbivores may have pervasive effects on ecosystem structure and function. In contrast to many terrestrial ecosystems that support relatively low herbivore loads and annually lose 5–10% or less of their net primary production (NPP) to herbivores (Chew 1974; Wiegert and Evans 1967; Owen and Wiegert 1976), grasslands and savannas typically support heavy herbivore loads. As discussed below, these herbivores may consume half or more of the annual aboveground net primary production (ANPP) and one fourth or more of the annual belowground net primary production (BNPP). As a result of these relatively high rates of consumption, it has been suggested that the structure of grasslands is the result of numerous interactions, many of which are either direct effects of, or mediated by, herbivores (McNaughton 1983). A simplified conceptual model of carbon, mineral nutrients, and water, and a few of the factors affecting their flows in a grazing system, is shown in Figure 7.1. In addition to direct consumption of plant biomass and consequent alteration of standing crop, the following effects have been attributed to grazing herbivores in grasslands and savannas: alteration (both increases and decreases) in rates of NPP through consumption (Hutchinson 1971; Vickery 1972; McNaughton 1976, 1979b; Detling et al. 1979, 1980) and fertilization (Schimel et al. 1986); alterations in plant–soil water relations (Archer and Detling 1986); changes in species composition and plant life form (Glendening 1952; Ellison 1960;

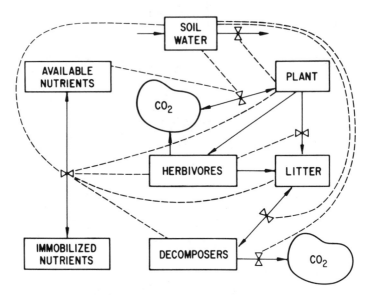

Figure 7.1. A simplified conceptual model illustrating several of the direct and indirect effects herbivores may have on the flows of carbon, nutrients, and water.

Caughley 1976; Laws 1970); alteration of gene frequencies and selection of prostrate growth forms in populations of grazed plants (Brougham and Harris 1967; McNaughton 1979a; Detling and Painter 1983; Detling et al. 1986); enhanced nutrient retention in ecosystems (Botkin et al. 1981); possible increased (Woodmansee 1978) or decreased (Schimel et al. 1986) loss of nitrogen via ammonia volatilization; nutrient translocation and concentration (Senft 1983; Ellis and Detling 1983); changes in nutrient cycling rates and total nutrient supply (Dean et al. 1975; Floate 1981); enhanced nutrient uptake by grazed plants (Ruess et al. 1983; Chapin and Slack 1979; Jaramillo 1986); and facilitation of energy and nutrient flow to other herbivores of the same or different species (McNaughton 1976, 1979b, 1984; Coppock et al. 1983a, 1983b; Coppock and Detling 1986).

7.1. Annual Rates of Consumption

Grasslands and savannas represent the potential natural vegetation of 3.3 × 10^7 km^2 or about 25% of the Earth's land surface (Lauenroth 1979). Although they occur in areas where soil water availability falls below the requirements for forests during part of the year, they receive enough precipitation to sustain the grasses as the dominant (or at least a major) component of the vegetation. Annual rates of ANPP in these areas typically lie in the range of 100–600 g/m^2/yr, although values above 1000 g/m^2/yr have been recorded (Lauenroth 1979; Sims and Singh 1978). Although estimates of belowground net primary production are more difficult to obtain, data from the U. S. IBP program summarized by Sims and Singh (1978) suggest that BNPP ranged from about 100–1000 g/m^2/yr in the eight North American grasslands examined.

Estimates of consumption and wastage of aboveground net primary production by a variety of native and domesticated mammals and insects are given in Tables 7.1–7.3. The estimates of both production and consumption were conducted in almost as many ways as there were investigators, and this undoubtedly contributes to the wide variation in these estimates. Nevertheless, some generalizations can be made from the data summarized here. For example, insects and small mammals (Table 7.3) usually consumed less than 10–15% of ANPP. By contrast, consumption by domesticated herbivores such as cattle (Table 7.2) and native large mammals (Table 7.1) was usually more than this. Although annual consumption by such herbivores was often in the range of 20–50% of ANPP, numerous examples of annual consumption rates above 50% were found. In many of the cases where such high consumption values were recorded, however, it is likely that the recorded values represented localized areas of quite high consumption rates rather than high mean consumption rates over entire landscapes (e.g., McNaughton 1985).

Among the many belowground herbivores of grasslands, nematodes, which typically consume from 5–15% of the annual BNPP (Table 7.4), may be the most important group of consumers. Calculations by Scott et al. (1979) indicate

Table 7.1. Estimates of Percent ANPP Consumed and Wasted by Large Native Mammalian Herbivores

Location	Community	% Consumed	Reference
Uganda	Savanna	30–40	Strugnell and Pigott (1978)
Uganda	Savanna	28	Lamprey (1964)
Tanganyika	Savanna	60	Petrides and Swank (1965)
Serengeti	Savanna	17–94 ($\bar{x} = 66$)	McNaughton (1985)
Serengeti	Savanna	15–39	Norton-Griffin (1979)
Serengeti	Long–shortgrass	19–34	Sinclair (1975)
South Dakota	Mixed grass	1–16	Coppock et al. (1983b)
New Mexico	Desert grassland	1	Pieper (1981)

Table 7.2. Estimates of Percent ANPP Consumed and Wasted by Cattle

Location	Community	% Consumed	Reference
Nigeria	Savanna	45	Ohiagu (1979)
South Africa	Broad leaf savanna	19	Gander (1982)
Western USA	12 rangelands	40–60	Lacey and Van Poollen (1981)
Western USA	5 IBP sites	19–30	Sims and Singh (1978)
Idaho	Mountain grassland	0–47 ($\bar{x} = 16$)	Leege et al. (1981)
Texas	Shortgrass	29–40	Heitschmidt et al. (1982)
Denmark	Pasture	60–80	Bülow-Olsen (1980)
Japan	Grassland	56–74	Akiyama et al. (1984)

Table 7.3. Estimates of Percent of ANPP Consumed and Wasted by Invertebrate Herbivores

Location	Community	Herbivore	% Consumed	Reference
Serengeti	Savanna	Insects and small mammals	4–9	Norton-Griffin (1979)
Serengeti	Short-long grass	Grasshoppers	4–8	Sinclair (1975)
Kenya	Savanna	Termites	6	Coughenour et al. (1985)
South Africa	Broad leaf savanna	All insects	7–17	Grunow et al. (1983) Gander (1982)
Saskatchewan	Grassland	Grasshoppers	7–14	Bailey and Riegert (1973)
Western USA	Short-tallgrass	All arthropods	3–7	Scott et al. (1979)
Tennessee	Grassland	All arthropods	10	Van Hook (1971)
Colorado	Shortgrass	Grasshoppers	7	Hilbert and Logan (1983)
Poland	Meadow	Grasshoppers	8–14	Andrzejewska and Wojcik (1970)

Table 7.4. Estimates of Percent of BNPP Consumed and Wasted by Below-ground Invertebrates

Location	Community	Herbivore	% Consumed	Reference
Western USA	Short-tallgrass	All plant feeders	13–41	Scott et al. (1979)
South Dakota	Mixed grass	Nematodes	8–15	Smolik (1974) Sims and Singh (1978)
South Dakota	Mixed grass	Nematodes	6–13	Ingham and Detling (1984)
Montana	Mixed grass	Nematodes	7	Coughenour et al. (1980)

that these consumers accounted for 46–67% of total belowground consumption of primary production. Most of this consumption represents ingestion of cell contents and not cell wall materials. Consequently, such consumption rates may be equivalent to consumption rates several times higher by plant chewing herbivores which also ingest cell walls.

7.2. Herbivore Effects on Primary Producers and Primary Production

7.2.1. Changes in Species Composition

As indicated above, it has long been recognized that heavy grazing by herbivores can lead to changes in species composition and plant life form in grasslands and savannas. Sims et al. (1978) reported changes in plant functional groups as influenced by light to heavy cattle grazing in a variety of North American grassland types examined during the U. S. IBP. Data from four grassland types are summarized in Figure 7.2. Although changes in plant functional groups occurred in response to cattle grazing in each of the four grassland types considered, the changes were not consistent across all grasslands. In the shortgrass prairies, there was a very slight reduction in grasses, forbs, and shrubs, but a doubling in the proportion of succulents (cacti) in response to grazing. Cattle grazing in the northern mixed grass prairie resulted principally in a shift from a C_3 (cool season) grass dominated system to a C_4 (warm season) grass dominated system. The tallgrass prairie, which was dominated by C_4 grasses, showed an increase in forbs in response to cattle grazing, while the desert grassland (also C_4 grass dominated) showed an increase in both forbs and shrubs in cattle-grazed areas.

When grazing intensity is exceptionally heavy, as in the case of overgrazing by domesticated ungulates, or in areas where native herbivores gather in dense populations for prolonged periods, vegetational changes may occur quite rapidly. For example, in a northern mixed grass prairie in Wind Cave National Park, South Dakota, replacement of grasses by forbs was observed within 3–

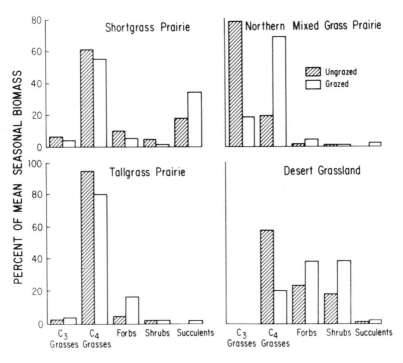

Figure 7.2. The effect of cattle grazing on percentage contribution of five plant functional groups to mean season-long aboveground biomass at U.S. IBP grassland sites studied 1970–1972. Calculated from data summarized by Sims et al. (1978).

8 yr. from the time the area was colonized by prairie dogs (Figs. 7.3 and 7.4). By 25 yr. following colonization by prairie dogs, forbs and dwarf shrubs had almost completely replaced grasses as the dominant life forms in this area (Coppock et al. 1983a). As has been observed in other studies of herbivory in grasslands, plant species diversity was maximized in moderately disturbed (i.e., young prairie dog towns colonized for 3–8 yr.) areas. By the time forbs and shrubs had replaced graminoids as the dominant life forms, plant species diversity had been reduced to a level comparable to that of native uncolonized grassland (Coppock et al. 1983a).

7.2.2. Herbivore-Induced Changes in Plant Morphology and Population Structure

Populations of plants from relatively heavily grazed areas are frequently characterized by individuals which are shorter and more prostrate than those from ungrazed or lightly grazed areas, and these morphological differences often appear to be genetically controlled (Gregor and Sansome 1926; Stapledon 1929; Kemp 1937; McNaughton 1979a; Detling and Painter 1983; Detling et al. 1986). This principle is illustrated in the morphological data from North American (Detling and Painter 1983) and African (McNaughton 1979a) grasses collected from permanent grazing exclosures and nearby heavily grazed areas (Table 7.5). In the case of the African grasses, the taller and more erect populations differentiated within twelve yr. or less following construction of a grazing ex-

Figure 7.3. Bison grazing in a prairie dog town, Wind Cave National Park, South Dakota. J. K. Detling photo.

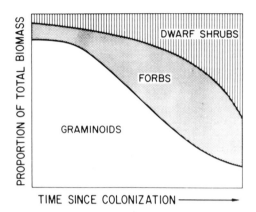

Figure 7.4. Changes in relative proportions of graminoids, forbs, and dwarf shrubs as a function of time since colonization by prairie dogs. The rate of change in dominance of plant functional groups is partially dependent upon density of prairie dogs and frequency and duration of visitation and utilization by other herbivores, particularly bison. Thus, while Coppock et al. (1983a) observed that forbs and dwarf shrubs had only begun to replace graminoids as the dominant life form after 3–8 yr. on one prairie dog colony, forbs had completely replaced graminoids after only 3 yr. of occupancy at another colony (Archer et al., in press).

closure in an area which was periodically heavily grazed by native ungulates. In contrast, the shorter, more prostrate forms of *Agropyron smithii* from North America had differentiated within twelve yr. following colonization and relatively heavy grazing by prairie dogs (*Cynomys ludovicianus*), in an area which previously had only been lightly grazed by native ungulates. Because productivity of the shorter, more prostrate plant growth forms may be reduced proportionately less by defoliation than productivity of the taller, more erect forms (Detling and Painter 1983; Detling et al. 1986), such grazing induced morphological changes (Table 7.5) have been interpreted as adaptations which allow those populations to better withstand subsequent heavy grazing pressure (Detling and Painter 1983; Detling et al. 1986). In other instances, however, it appears that the shorter populations may evolve as a grazing avoidance, rather than a grazing tolerant, evolutionary strategy. For example, Jaramillo (1986) observed that although heavily grazed populations of the North American shortgrass species *Bouteloua gracilis* were shorter than lightly grazed conspecifics, similar intensities of defoliation reduced productivity by similar amounts in the two populations. He hypothesized that grazing selected for shorter individuals not because they were more grazing tolerant, but because they were less frequently or less intensively grazed than their taller, more readily apparent and available conspecifics. Finally, in still other cases (Quinn and Miller 1967), the dwarf prostrate growth forms of plants from heavily grazed populations appear to represent plastic responses of existing genotypes to repeated heavy defoliation.

Table 7.5. Mean Internode Length, Leaf Length, and Culm or Leaf Angle in Five Grass Species at Least Eight Months After Being Collected From Areas Which Were Heavily Grazed by Native Herbivores or Within a Grazing Exclosure, and Transplanted to a Uniform Environment[a]

| | Population Origin | | |
| | Within | Outside | |
Species	Exclosure	Exclosure	p[b]
	Leaf blade length (cm)		
Agropyron smithii	23.1	13.3	0.001
Sporobolus marginatus	7.7	9.6	n.s.
Cynodon dactylon	4.3	2.1	0.05
Harpachne schimperi	9.2	4.3	0.05
Themeda triandra	9.5	9.4	n.s.
	Mean internode length (cm)		
Agropyron smithii	18.0	4.4	0.001
Sporobolus marginatus	6.1	6.3	n.s.
Cynodon dactylon	8.1	3.5	0.001
Harpachne schimperi	3.6	2.7	n.s.
Themeda triandra	17.1	7.9	0.001
	Leaf angle (° from horizontal)		
Agropyron smithii	62.2	29.9	0.01
	Culm angle (° from horizontal)		
Harpachne schimperi	54.4	27.5	0.05

[a] Data from *Agropyron smithii* are from the South Dakota grassland studied by Detling and Painter (1983); all other data are from the Serengeti, Africa study of McNaughton (1979a).
[b] n.s. = not significant at 5% level.

The consequences of such genetic or plastic responses to grazing at the population level have not been fully evaluated with regard to their impact on ecosystem processes in grasslands and savannas. For example, results of laboratory studies suggest that the shorter growth forms characteristic of heavily grazed populations are less productive, even if relatively more grazing tolerant, than are the taller growth forms (Detling and Painter 1983; Detling et al. 1986). There is, however, little evidence to suggest that primary production is lower in heavily grazed natural or managed grassland and savanna ecosystems dominated by these populations. Indeed, as discussed below, some of these ecosystems may be more productive than similar ungrazed or lightly grazed systems dominated by taller, more erect growth forms. Furthermore, investigations into the significance of such grazing induced population differentiation on processes related to nutrient uptake and cycling in grazing systems are in their infancy (Jaramillo 1986).

7.2.3. Grazing Optimization Hypothesis

One of the most interesting and perhaps controversial (Belsky 1986, 1987; McNaughton 1986) topics concerning the effects of grassland herbivores or ecosystem function in recent years concerns the ways in which these consumers affect primary production and related processes. Figure 7.5 shows three hypothetical primary production responses that producers may exhibit as grazing intensity increases (adapted from Tieszen and Archer 1979; Hilbert et al. 1981; McNaughton 1979b). Traditionally, primary production has been considered to generally decrease with an increase in grazing intensity (curve *A* of Fig. 7.5), or to be relatively unaffected under light grazing, but decrease under heavier grazing (curve *B*, Fig. 7.5). More recently, the notion that net primary production is maximized at some optimal grazing intensity before decreasing at higher levels of grazing intensities (curve *C*, Fig. 7.5) has become quite popular. This so-called "grazing optimization hypothesis" (Hilbert et al. 1981) was advanced on the basis of indirect evidence by Dyer and Bokhari (1976) and on the basis of grassland production data from the Serengeti by McNaughton (1976; 1979a; 1979b). For example, McNaughton (1979b) found that aboveground net primary production in the Serengeti shortgrass plains was maximized at an optimal grazing intensity of about 25% removal of green biomass by migratory wildebeest (Fig. 7.6).

Although McNaughton and colleagues (McNaughton 1983; Wallace et al. 1985; Coughenour 1985) have shown in a variety of laboratory experiments that some species of Serengeti graminoids exhibited increased production following defoliation, the universality of the "grazing optimization hypothesis" is not supported by a large empirical data base from field studies. Moreover, it is likely that the specific weather conditions govern whether or not net primary production is enhanced by grazing. For example, Heitschmidt et al. (1982) found that short duration cattle grazing resulted in an increase in ANPP in a relatively dry year and a decrease in ANPP during a relatively moist year (Fig. 7.7). In a recent review, Lacey and Van Poollen (1981) collected data from 12 studies which reported a moderate level of use (40–60% of current

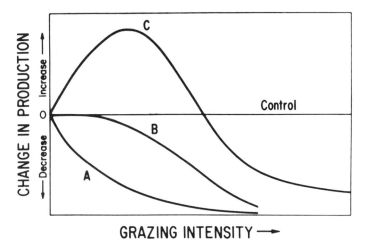

Figure 7.5. Hypothetical ways in which primary production may be affected by grazing. See text for explanation.

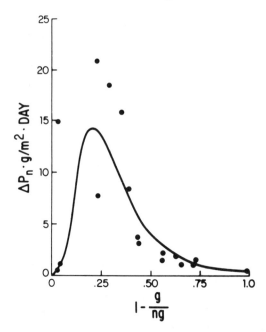

Figure 7.6. Relationship between intensity of grazing $(1 - g/ng)$ by wildebeest (*Connochaetes taurinus Albojubatus* Thomas), where g was plant biomass in recently grazed areas and ng was biomass within a nearby permanent grazing exclosure, and stimulation of aboveground net primary productivity (ΔP_n) above that of ungrazed control plots in the Serengeti Plains of Africa. Data of McNaughton (1979b).

year's ANPP) in western North American rangelands. Results of their review (Fig. 7.8) indicated that ANPP was either unaffected or reduced in these grasslands at these levels of utilization. Such an analysis, however, is not necessarily an adequate evaluation of the "grazing optimization hypothesis," since "moderate utilization" from the perspective of a rancher may be above optimal utilization from the perspective of the primary producers. Specifically, amount of plant material consumed, duration of individual grazing events in a given location, and seasonal grazing patterns of domesticated herbivores may all differ markedly from conditions under which plants in these grasslands evolved with native herbivores. Nevertheless, results of laboratory defoliation experiments on North American grassland species (e.g., Detling et al. 1979, 1980; Detling and Dyer 1981; Detling and Painter 1983) have failed to show enhanced plant growth following a variety of defoliation treatments. This suggests the possibility that there may be some fundamental differences in plant growth patterns in response to defoliation between the North American and Serengeti graminoids which have been studied in detail.

Results of two independent ecosystem level modeling efforts provide additional evidence that grazing adapted graminoids from temperate and tropical grasslands respond somewhat differently to grazing. In the first study, Capinera

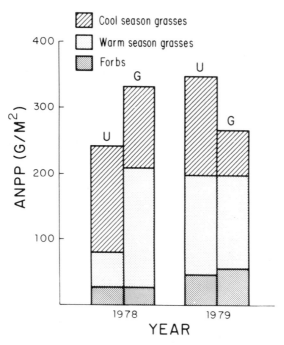

Figure 7.7. Effects of short duration cattle grazing on aboveground net primary production in a Texas, U. S. A., grassland during relatively dry (1978) and moist (1979) years. U = ungrazed plots; G = grazed plots. Data of Heitschmidt et al. (1982).

et al. (1983) evaluated the response of pure swards of the dominant North American shortgrass species, *Bouteloua gracilis*, to defoliation by different sized populations of three size classes of grasshoppers. In the second study, Coughenour (1984) simulated the response of Serengeti graminoids to defoliation at different heights and frequencies by large mammalian herbivores. I have extracted previously unpublished ANPP and BNPP results from the simulations of Capinera et al. (1983) and compared them with Coughenour's (1984) results for the Serengeti shortgrass plains (Fig. 7.9). The studies differed from one another in several important ways, including model structure; weather conditions; type of herbivores simulated; frequency, intensity, and seasonality of defoliation; and method of expressing grazing intensity (see legend of Fig. 7.9 for explanation). Nevertheless, the results may be instructive in several ways. First, in spite of having similar levels of ANPP under ungrazed conditions, simulated ANPP was increased by over 100% at some grazing intensities in the Serengeti graminoids, and little (< 15%) enhancement of ANPP was simulated in the North American grass. Second, because of the greater grazing induced stimulation of ANPP, far more forage was available to, and subsequently removed by, herbivores in Coughenour's (1984) Serengeti simulations. Third, in both studies, precipitous declines in BNPP occurred at lower grazing intensities than those with inhibited ANPP. Thus, consistent with results of

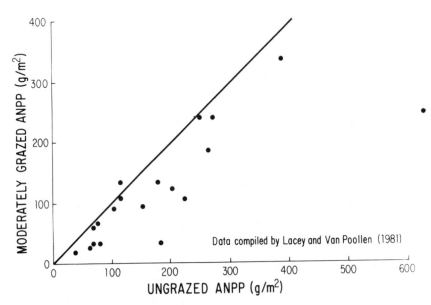

Figure 7.8. Relationship between aboveground net primary production of ungrazed and moderately grazed (40–60% removal of current year's ANPP) western North American rangelands. Data compiled by Lacey and Van Poollen (1981).

many laboratory and field studies cited below, one major effect of grazing was to increase the proportion of the total energy flow through the primary producers which went through the aboveground part of the system. Unfortunately, accurate assessment of ANPP and BNPP, especially under a wide range of grazing intensities in the field, is difficult, time consuming, and expensive. Consequently, few reliable comparative data are available to validate the results from these models. This clearly represents an area in which more research effort is needed.

Results of a number of experiments indicate that within a few days following defoliation there is an enhancement in the photosynthetic rate of remaining undamaged or newly produced tissue (Detling et al. 1979, 1980; Detling and Painter 1983; Wallace et al. 1984; Nowak and Caldwell 1984). In addition, a high proportion of the newly produced photosynthate is allocated to the production of new photosynthetic machinery and less is allocated to root production (Coughenour 1984; Detling et al. 1979). As a result, root production and standing crop are frequently less in grazed grasslands than in ungrazed grasslands (Weaver 1950; Schuster 1964; Ingham and Detling 1984).

As was discussed previously, herbivory by root feeding nematodes and other belowground herbivores may constitute a significant portion of total plant consumption in grasslands and savannas (Table 7.4). In spite of our growing recognition of these relatively large amounts of consumption by belowground herbivores, relatively little is known of their effects on rates of primary production. In controlled laboratory experiments, Stanton (1983) and Ingham and

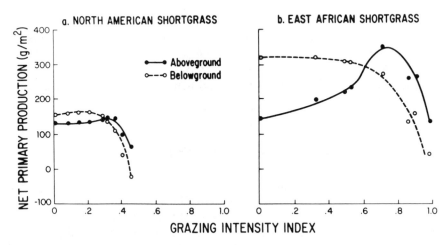

Figure 7.9. Results of simulation modeling studies of the effects on aboveground and belowground NPP of (a) grasshopper grazing on North American shortgrass prairie (from Capinera et al. 1983); and (b) large mammals grazing on the shortgrass Serengeti plains of East Africa (Coughenour 1984). The indices of grazing intensity (GI) were calculated differently in the two studies. In the North American study, GI was calculated as amount consumed divided by ANPP of ungrazed grassland. In the African study, GI was $1 - g/\mu g$, where g and μg were mean seasonal shoot biomass (i.e., current season's standing crop) of grazed and ungrazed grasslands, respectively.

Detling (1986) found that root feeding nematodes at densities equivalent to those observed in the field caused statistically significant but relatively minor reductions in both shoot and root production in two species of *Bouteloua* from North American grasslands.

Few field studies involving manipulations of populations of root feeding nematodes have been made, and results from these studies have been variable and inconclusive. In the growing season following application of a nematicide to a North American mixed grass prairie, Smolik (1974) reported 28–59% increases in ANPP. By contrast, Stanton et al. (1981) found no significant increase in aboveground standing crop in the two growing seasons following nematicide treatment of a North American shortgrass prairie. They did, however, report that root biomass was about 25% greater in the nematicide treated plots. Results of a recent nematicide study in a northern mixed grass prairie in South Dakota were variable and did little to clarify our understanding of nematode influences on temperate grassland productivity (Ingham and Detling, in prep.). At one site, aboveground net primary production in an area in which nematodes had been killed was increased by 30–50%, but in a nearby replicate site, the nematicide had no significant effect on ANPP.

Considering the amount they consume, and that they may have a disproportionately large impact on the fine root system (Ingham and Detling 1986), phytophagous nematodes might well be expected to significantly alter primary productivity and energy flow in grasslands. The variable nature of the results of the few studies done on this topic, however, makes it difficult to generalize concerning the extent of nematode impact. Moreover, interpretation of results may be further complicated by potential effects of nematicides on nontarget organisms (Stanton et al. 1981) such as bacterial or fungal feeding nematodes which are important in nutrient cycling, and hence in regulating primary production (Ingham et al. 1985).

7.3. Alteration of Nutrient Cycling by Herbivores

The relatively large consumption rates by herbivores in many grasslands (Tables 7.1–7.4) suggest that sizeable quantities of nutrients are removed by grazers, and that grazing animals may substantially influence nutrient cycling. Indeed, nutrient recycling via grazing animals has been implicated as an important factor in the maintenance of soil fertility (Floate 1981). As suggested by the diagram for the grassland nitrogen cycle (Fig. 7.10), grazers may have a variety of direct and indirect effects on various pathways of nutrient cycles. Floate (1981) considered the direct effects to include herbage consumption, trampling of soil, transfer of standing live and dead vegetation to litter, excretion of nutrients via urine and feces, and removal of nutrients by the har-

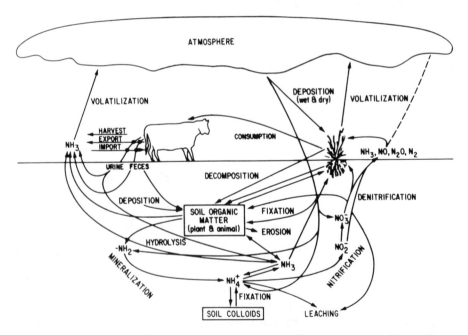

Figure 7.10. Conceptual diagram of the nitrogen cycle, illustrating some of the pathways which can be directly or indirectly affected by grazers in grasslands. Courtesy of D. G. Milchunas.

vesting of animal products. He considered the indirect effects to be a secondary result of these primary effects. Thus, indirect effects may include modification of botanical composition and sward structure, and alteration of the proportion and distribution of nitrogen returned to the soil via the plant litter and animal pathways. These will further affect decomposition and mineralization rates, and the extent of losses.

Botkin et al. (1981) hypothesized that cycling of nutrients through herbivores may help to maintain a pool of nutrients near the soil surface, and that these nutrients would be readily available to the vegetation. Furthermore, because ANPP may be stimulated by grazing (Fig. 7.6), they further suggested that herbivores may stimulate uptake of nutrients through the consumption of vegetation, and that this would result in more rapid turnover for the various elements. Consistent with such an hypothesis, it is commonly observed that the nutrient concentrations in aboveground plant biomass of grazed areas are higher than nutrient concentrations in plants from similar ungrazed areas (McNaughton 1984). Thus, it has been observed that shoot nitrogen concentrations of graminoids, forbs, and shrubs from heavily grazed prairie dog colonies were consistently higher than nitrogen concentrations in plants from adjacent uncolonized areas (Coppock et al. 1983a; Krueger 1986). Results of controlled environment experiments on the North American shortgrass species *Bouteloua gracilis* (Jaramillo 1986) and the Serengeti sedge *Kyllinga nervosa* (Ruess et al. 1983; Ruess 1984) indicate that uptake of nitrogen per unit of root biomass is frequently enhanced by defoliation.

In considering additions and losses of nitrogen in North American grassland ecosystems, Woodmansee (1978) concluded that animals had a significant effect on loss of nitrogen from the systems. He estimated that 50% of the nitrogen excreted by ungulates was lost via ammonia volatilization, and he calculated that such losses were equivalent to 30–100% of the annual nitrogen inputs via wet and dry deposition in the seven grassland sites he examined. However, in a more detailed field study in a Colorado shortgrass prairie, Woodmansee and his associates (Schimel et al. 1986) later calculated that such ammonia volatilization losses were not nearly as great as originally calculated. Thus, while Woodmansee (1978) calculated annual losses via volatilization of ungulate waste products to be 0.2 g $N/m^2/yr$, Schimel et al. (1986) calculated that such losses ranged from 0.004 in relatively moist lowland sites to 0.016 g $N/m^2/yr$ in more xeric upland sites. By contrast, Woodmansee (1978) calculated that annual ammonia volatilization losses from plant and animal residues would be on the order of only 0.05 g $N/m^2/yr$, but calculations based on the data and assumptions of Schimel et al. (1986) indicate that ammonia volatilization losses from vegetation should be 0.20–0.24 g $N/m^2/yr$. Because such losses are a function of aboveground plant biomass, grazers might be expected to reduce such losses in direct proportion to their effects on aboveground biomass. Thus, in a grazed grassland whose live aboveground biomass is half that of a similar ungrazed grassland, nitrogen losses by ammonia volatilization may be only about half those of similar ungrazed grasslands.

7.4. Grazing Facilitation Among Herbivores

The final point to be considered here regarding herbivore regulation of energy flow and nutrient cycling in grasslands and savannas concerns the concept of grazing facilitation. Specifically, a number of studies have shown that grazing by one species of herbivore may enhance feeding by the same or another species of herbivore in the same area. For example, McNaughton (1976) reported that one month following heavy grazing by dense concentrations of migratory wildebeest in the Serengeti plains of Africa, Thomson's gazelles were significantly associated with areas previously grazed by the wildebeest. Similarly, Coppock et al. (1983b) and Coppock and Detling (1986) have observed that North American bison feed preferentially on areas subjected to heavy grazing by prairie dogs. The phenomenon has been confirmed by Krueger (1986), who also observed that pronghorn feed preferentially on colonized areas as well (Fig. 7.11). Such examples of grazing facilitation have been attributed at least partially to higher concentrations of nutrients, particularly nitrogen, in the foliage of previously grazed areas (Coppock et al. 1983a, 1983b; McNaughton 1984). Interestingly, grazing facilitation may not be limited to aboveground herbivores. For example, Smolik and Dodd (1983) observed that a number of species of root feeding nematodes had higher population densities in the soil of shortgrass prairie which had been heavily grazed by cattle than in the soil of nearby ungrazed grassland. Similarly, Ingham and Detling (1984) reported that some species of both plant parasitic and nonparasitic nematodes are found

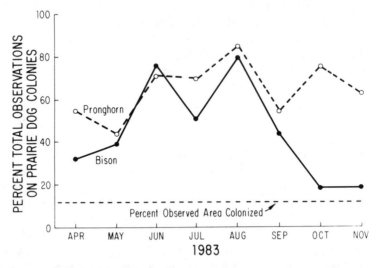

Figure 7.11. Utilization of heavily grazed prairie dog (*Cynomys ludovicianus*) colonies for feeding by North American bison (*Bison bison*) and pronghorn (*Antilocapra americana*) in mixed grass prairie. Both pronghorn and bison foraged on the colonies a greater proportion of the time than would be expected on the basis of proportion of the study site which was colonized (horizontal dashed line). Data from Krueger (1986).

in greater densities in the soil of heavily grazed prairie dog colonies than in adjacent lightly grazed, uncolonized areas. Likewise, Roberts and Morton (1985) observed that biomass of soil dwelling larvae of Scarabaeid beetles was greater in pastures receiving moderate grazing pressure by sheep than in very lightly grazed or heavily grazed pastures (Fig. 7.12). Seastedt (1985a) has postulated that such increases in root feeding invertebrates around the root systems of plants experiencing foliar herbivory occur as a result of higher nitrogen concentrations in the roots of these plants.

7.5. Conclusions

Grassland herbivores typically consume 15–60% of the annual aboveground net primary production and 5–15% of the belowground production. Although light to moderate levels of grazing have been reported to enhance NPP, particularly in the case of large ungulate grazing in some tropical African grasslands, few well-documented cases of this exist. By contrast, available data from temperate North American grasslands indicate that there is usually a decrease or no effect of grazing by either aboveground or belowground herbivores on NPP. Regardless of its effect on NPP, defoliation invariably results in an increase in the proportion of carbon and nutrients to aboveground structures. Aboveground grazers tend to feed preferentially in areas grazed previously by the same or other species, and root feeding invertebrates are often more abundant in the rooting zones of plants which have experienced moderate grazing pressure aboveground. In both cases, increased nutrient concentrations (particularly N) in shoots and roots have been implicated as possible partial causative factors. Grazing may further affect nutrient cycling by keeping nutrients

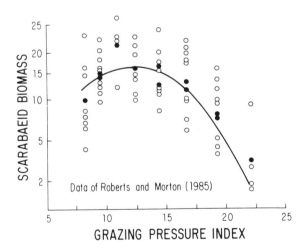

Figure 7.12. Total biomass of Scarabaeid larvae in a *Phalaris aquatica/Trifolium repens* pasture grazed by sheep in Australia. Data of Roberts and Morton (1985).

in a relatively readily available, rapidly cycling pool near the soil surface, and it may reduce ammonia volatilization losses from leaf surfaces.

Acknowledgements

Development of this paper was enhanced by discussions with many of my students and colleagues at Colorado State University. I am particularly indebted to April Whicker for assistance in compiling data summarized in Tables 7.1–7.4; David Schimel for sharing unpublished information and ideas concerning nutrient cycling; and Michael Coughenour for the insight and information he provided concerning African grazing systems. Daniel Milchunas graciously permitted me to use his previously unpublished figure of the nitrogen cycle of a grassland. Manuscript preparation was supported by NSF Grant No. BSR-8406660.

8. Agroecosystems Processes

David C. Coleman and Paul F. Hendrix

8.1. Agroecology—Agriculture from an Ecological Viewpoint

Modern agriculture has evolved from a long tradition of basic and applied research in a variety of disciplines, including agronomy, botany, zoology, soil science, climatology, chemistry, economics, engineering, and many others. Indeed, agricultural problems have been crucibles for theoretical developments in many of these disciplines. The science of ecology is no exception, having benefited from and contributed to agricultural theory and practice. Liebig's observation in 1840 that crop growth is limited by the nutrient element in shortest supply relative to demand became a cornerstone in modern ecology, known as the Law of the Minimum (Odum 1983). On the other hand, the widely used practice of integrated pest management, which combines chemical and biological controls, was developed from a knowledge of predator–prey interactions, which are central issues in population ecology (Flint and van den Bosch 1981). Thus, the roots of agriculture and ecology are deeply intertwined historically and, in fact, are older than either discipline as formal bodies of knowledge. When humans first cultivated plants and domesticated animals, they were building on at least a rudimentary understanding of interactions between living organisms and their environment. The extent of this understanding and the success of early agriculture were undoubtedly limited, but as understanding increased so did agricultural productivity. Techniques such as crop rotation (planting crops in different areas in alternate years to avoid buildup of pathogens and negative allelopathic effects), intercropping (growing selected crops together, to the mutual benefit of each), and multicropping (growing crops successively throughout the year to prevent erosion and provide green manure) could not have come about without some knowledge of interactions among species and of their basic environmental requirements.

Although linkages between ecology and agriculture continue, different priorities and lines of research have occupied these disciplines in recent years. Traditionally, ecological research has been concerned with populations, communities, and ecosystems under natural conditions, and relatively less with applications of ecological knowledge. Over the past two decades, there has been increasing interest in "stress ecology" (Odum et al. 1979), which focuses on effects of human and natural disturbances on ecological systems. Interestingly, this has been one avenue by which ecologists have become interested in agricultural systems. In this context, land clearing, cultivation, and intensive management are seen as forms of perturbation driving ecosystems away from some natural state. Agricultural systems once again serve as systems for increased understanding of nature.

The primary focus of agricultural research throughout much of this century has been increasing plant and animal productivity and developing technologies to do so. Spectacular increases in food and fiber production have been achieved as a result of this research. However, the large energy requirements and potential for environmental degradation inherent in modern agricultural practices have raised questions about long-term sustainability of food production, in both developed and developing countries. Consequently, research efforts are beginning to focus on low input, high efficiency agriculture rather than on maximum production (Tangley 1986a), and this is creating renewed interest in ecological aspects of food and fiber production. Those aspects that can reduce farming costs and preserve or increase soil fertility will undoubtedly receive greatest attention.

Utilizing ecological principles in agriculture does not mean abandoning the technological innovation that has improved human living standards worldwide. On the contrary, streamlining old technologies and developing new ones within the context of functioning agroecosystems presents an exciting challenge for agricultural research and development. An ecosystem perspective has much to offer toward these goals, as we will show in the following pages.

8.1.1. What is an Agroecosystem?

Several definitions of "agroecosystems" are in use, including: ecosystems containing a significant component of anthropogenic inputs and outputs necessary for regulation, with some fraction of the system harvested (House et al. 1983). The definition which we prefer to use is: an agroecosystem is an ecosystem manipulated by frequent, marked anthropogenic modifications of its biotic and abiotic environments.

Odum (1984a) recognizes four major distinctions between natural ecosystems and agroecosystems, all resulting from human management: (1) in addition to solar power, auxiliary energy from human and animal labor, fertilizers, pesticides, irrigation water, and fuel-powered machinery are added as energy subsidies to agroecosystems; (2) biotic diversity in agroecosystems is reduced to maximize economic yields of desired products; (3) artificial selection rather than natural selection produces the dominant plants and animals; and (4) agroecosystems are under external, goal oriented control rather than internal control mediated by "subsystem feedback," as in natural ecosystems.

As a result of these factors, human management tends to decouple interacting components and processes in agroecosystems, both in the fields where they occur and in our perception of food production systems. For example, plowing represents a large energy input which, among other presumed benefits, aerates the soil, buries and fragments plant residue, and promotes residue decomposition. Plowing also reduces the abundance of earthworms, which perform the same functions in natural ecosystems. In addition, however, earthworms excrete nutrient-rich feces which stimulate soil microbial activity, nutrient turnover, and plant production (Stout 1983). Thus, minimum tillage techniques not only reduce energy input costs, but also may yield added benefits associated with increased earthworm activity. Insecticides, another managerial input, also can decouple agroecosystem components by destroying natural predators which prey on pest insect species. Integrated pest management procedures represent an attempt to overcome this problem by using a combination of biological and chemical controls on pest populations. Likewise, chemical fertilizers are a widely used input to agroecosystems. However, added nitrogen reduces the fixation by legumes of atmospheric dinitrogen (Lynch 1983), an important process in many natural ecosystems. Introducing legumes into cropping systems couples an important biotic function into agroecosystems, thereby reducing the need for chemically fixed nitrogen inputs.

All ecosystems are subject to some influence, direct or indirect, from human activities (Bormann 1986). Perhaps it is best to drop the now timeworn distinction between natural and human managed, or applied, ecosystems. The extent and timing of the disturbances is what sets managed systems apart from nonmanaged ones. As noted by Odum (1971), there are some important differences along the continuum between managed and nonmanaged ecosystems. Principal among these is the imposition of immediate, short-term goals. Human managed ecosystems are managed for net production, (usually primary production), whereas most natural systems are much closer to a balance of production and respiration, or a P/R ratio approaching 1 (Table 8.1, Coleman et al. 1976). Intermediate levels are reached, of course, in early successional stands.

It is useful to consider terrestrial ecosystem function within the framework of the major factors which are involved in soil formation: climate, organisms, parent material, and relief (Jenny 1941). These factors, operating over time, lead to various system properties which are the result of key processes, such as mineralization and immobilization, leaching, decomposition, etc. (Table

Table 8.1. Annual Production and Respiration as kcal/m^2/year in Growth-type and Steady State Ecosystems (Portions after Odum 1971 b). Systems Arranged in Decreasing Magnitudes of Net Ecosystem Production (NEP).[a]

Production and Respiration	Alfalfa Field (U.S.A.)	Young Pine Plantation (Great Britain)	Lightly Grazed Short-grass Prairie (Colo.)	Mesic Deciduous Forest (Tenn.)
Gross primary production (GPP)	24,000	12,200	5,230	27,976
Autotrophic respiration (R_A)	9,200	4,700	1,778	18,200
Net primary production (NPP)	14,800	7,500	3,452	9,776
Heterotrophic respiration (R_H)	800 (3,000– 4,400)	4,600	2,379	9,172
Net ecosystem production (NEP) (NPP − R_H)	14,000 (12,200– 13,600)	2,900	1,073	604
Net ecosystem respiration (R_E) ($R_A + R_H$)	10,000 (12,200– 13,600)	9,300	4,157	27,372
Ecosystem metabolism (P/R) (GPP/R_E)	2.44 (2.00–1.80)	1.13	1.26	1.02

[a] Numbers in parentheses include calculated microbial respiration. From Coleman et al. (1976), Agroecosystems. © Elsevier Science Publishers.

8.2). Using this framework, we are then able to compare and contrast system behaviors. Agroecosystems, being among the more heavily perturbed systems, are near one end of a continuum of ecosystems, ranging, on the other end, to relatively undisturbed natural areas such as national parks and wilderness areas.

8.1.2. Agroecosystems as Components of the Biosphere

Human habitations currently occupy less than 2% of the Earth's land surface, yet our activities influence nearly 40% of global terrestrial primary productivity. Agriculture accounts for 20% of terrestrial net primary productivity and approximately 30% of land surface area (10% cultivated and 20% grazing land) (Odum 1983; Vitousek et al. 1986). Until the mid-1900s, increases in global cropland area followed growth in the human population, new lands being claimed from natural ecosystems through clearing of forests, drainage of wetlands, irrigation of arid lands, and plowing of grasslands. Although new lands are still being cultivated, the rate of growth in cropland area is now decreasing and is expected to do so for the remainder of this century. By the year 2000, total cropland area may increase by about 4%, whereas human numbers will increase by 40% (Brown 1985). Obviously, the intensity of food and fiber production on existing agricultural lands will have to increase substantially. As a result, the global impact of agriculture is certain to increase well into the next century (National Research Council 1986).

Table 8.2. Factors Influencing Soil Development (Controlling Factors Affecting Processes over Time, Influencing Ecosystem Properties)[a]

Controlling Factors			
Parent material	Climate	Vegetation	Relief
	Man		
Rangeland	Cropland	Forests	
Grazing	Cultivation	Seeding and planting	
Species	Fallow	Site preparation	
Nutrient input	Crop selection	Watershed management	
Fire	Residue management	Fire	
	Nutrient inputs	Harvest	
	Water management		
	Fire		
	Harvest (removal)		

Processes	Development of Ecosystem Properties
Energy inputs and transformations	Soil
Radiation	Base status
Primary production	Texture
Decomposition	Organic matter
Nutrient cycling	Phosphorus
Immobilization	Sulphur
Mineralization	Nitrogen
Weathering	Salinity
Translocation	Consumers
Transport	Vegetation
Erosion	
Gaseous	
Leaching	

[a] From Coleman et al. (1983).

Agroecosystems are the focus of large fluxes of matter and energy, especially in technologically advanced countries, and therefore their influence on global biogeochemical cycles may be significant. For example, chemical fertilizer use has increased nearly tenfold since 1950, amounting to 121 million metric tons (MT) in 1984 or 25 kg per capita per year (Fig. 8.1, Brown 1985). The amount of nitrogen fertilizer used (61 million MT) is equivalent to nearly 30% of the atmospheric nitrogen fixed by all terrestrial biota (Rosswall 1983). Similarly, the amount of phosphorus applied to agroecosystems as fertilizer (14 million MT) is equivalent to approximately 10% of that taken up from the soil by terrestrial biota (Richey 1983).

Clearing of forests, particularly in the tropics, and plowing of grasslands may release substantial amounts of carbon into the atmosphere. Palm et al. (1986) estimate that deforestation in the tropics accounts for nearly 90% of global net release of CO_2 from terrestrial biota into the atmosphere (total released from soil and biota in 1980 was $1.8–4.7 \times 10^{15}$ gC). Southeast Asia accounts for 17% of the global release, which results primarily from conversion of forest to shifting or permanent agriculture (Table 8.3). Stewart et al. (1983) estimate

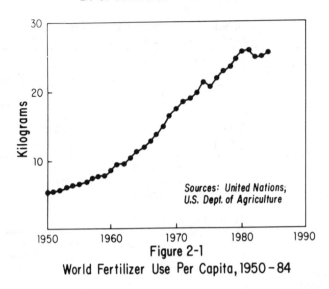

Figure 2-1
World Fertilizer Use Per Capita, 1950 – 84

Figure 8.1. World fertilizer use per capita 1950–1984, from L. R. Brown, State of the World, 1985. © W. W. Norton & Co., New York

Table 8.3. Components of the Flux of CO_2 From Southeast Asia in 1980 (Values Given $\times$ 10^{15} gC)[a]

		Vegetation		Soil	Total
	Burned	Decay on Site	Regrowth	Soil	Total
Permanent agriculture	0.0567	0.0549	0	0.0259	0.1375
Shifting agriculture and land impoverishment	0.2036	0.1925	−0.2258	0.0330	0.2033
Forest harvest and regrowth	0.0412	0.0553	−0.0874	0.0044	0.0135
Total	0.3015	0.3027	−0.3132	0.0633	0.3543
				Products Decay off site	0.0760
					0.4303

[a] From Palm et al. (1986) in R. Lal, P. A. Sanchez and R. W. Cummings, Jr., Eds., Land clearing and development in the tropics. Proceedings of a conference organized by the International Institute of Tropical Agriculture, Ibadan, Nigeria, 1986. A. A. Balkema, Rotterdam.

that continuous cultivation over the past 50–100 yr. of the North American Great Plains has released 50% of the carbon and nitrogen from soil organic matter. Interestingly, the atmospheric "pulse" of CO_2 resulting from plowing of these virgin prairies between 1850 and 1890 was detected by Wilson (1978) in the tissues of bristlecone pine trees in California (Coleman et al. 1984).

Because of the magnitude of these fluxes of matter and their potential global effects, our view of the biogeochemistry and energetics of the biosphere is incomplete without an understanding of the processes that occur within agroe-cosystems. In addition to primary production, such processes include herbi-vory (consumption of plant materials by humans, livestock, and pests—see Detling, Chapter 7) and decomposition of organic matter (soil organic matter and plant residues remaining after harvest). Decomposition and mineralization of organic matter (OM) are of particular interest, both in ecology and agron-omy, because they influence the fertility and aeration of soils and thus the availability of nutrients and oxygen to primary producers. These processes are the focus of our discussion of agroecosystem research.

Although our chapter is not concerned with forestry in detail, the impacts of agricultural activity in forested regions, particularly in developing countries, must be considered. Clearing of forested lands in Central and South America, and in parts of Asia and Africa is proceeding rapidly (Table 8.4), representing millions of hectares per year added to regions of human utilization. Much of this land is being used for row crop cultivation and grazing lands (Lal 1986; Vitousek et al. 1986). Effects on the global carbon cycle have already been indicated. For further viewpoints on forest ecology, refer to Chapter 5.

8.2. Agroecosystem Studies

8.2.1. Ecosystem Processes and Conservation Tillage

Process studies, for example, those concerning crop growth, yield, and fertilizer use efficiency, have a long history in agricultural research. Studies with an ecosystem emphasis have been funded for only the last 8–10 yr. in North America, although aspects of agroecological research (especially soil science) have been carried out for more than a century in other countries; e.g., at Rothamsted, in Great Britain (Russell 1973). Key processes involved in nu-trient cycling (decomposition, immobilization, mineralization, and nutrient losses due to leaching and volatilization) have been a focus of much of this work. The ecosystem view emphasizes interactions among these processes and the system components which they influence (Fig. 8.2, Frissel 1977).

As with other ecological studies, agroecosystem research has been carried out at various levels of resolution, ranging from laboratory microcosm, growth

Table 8.4. Global Deforestation Rates of Tropical Forests Between 1980 and 1985 for 76 Countries[a]

| Forest Category | Deforestation Rates (% of the Remaining Forest) | | | |
	Tropical America	Tropical Africa	Tropical Asia	Total
Closed forest	0.64	0.62	0.60	0.62
Open forest	0.59	0.48	0.61	0.52
All forest	0.63	0.52	0.60	0.58

[a] Reprinted with permission from Ambio, 13, The uncertain energy path—energy and third world development, O'Keefe and Kristoferson, Pergamon Journals, Ltd., Copyright 1984.

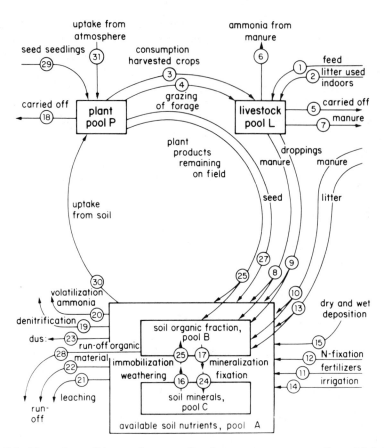

Figure 8.2. Flow chart of the nutrient transfers for an agroecosystem. From M. J. Frissel 1977, Cycling of mineral nutrients in agricultural ecosystems. Agroecosystems 4: 1–354. © Elsevier Science Publishers.

chamber, and greenhouse studies to field studies. These studies have been mostly "site specific," but, as will be discussed later, efforts are being made to understand interactions among sites at different positions in the landscape.

Two processes of major importance to agronomists and ecologists are soil erosion and organic matter (OM) degradation, which can reduce both soil fertility and depth of rooting by crops. Both of these processes are exacerbated by conventional agricultural practices such as moldboard plowing and other tillage methods, crop residue removal, and bare fallowing (Anderson and Coleman 1985). Plowing in particular can cause erosion, and, over time, progressive mixing of surface and subsurface horizons, thereby altering water relations, soil organic matter status and nutrient availability (Fig. 8.3). A number of agroecosystem studies have focused on alternative practices which can mitigate these processes. For example, in long-term studies of a minimum tillage technique (stubble–mulch) vs. conventional clear–fallow regimes, Bauer and Black

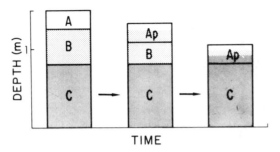

Figure 8.3. Changes in soil profiles that result from simultaneous cultivation and erosion, with ultimate incorporation of subsurface material into the plow layer. From Schimel et al., Geoderma. © 1985 Elsevier Science Publishers B. V.

(1981) found that stubble–mulching (cultivating the top 1–2 cm, leaving the residues on the soil surface) could reduce OM losses by as much as 50% in a 25 yr. period.

A further motivation for the conservation trend is the savings realized by no tillage, as compared to conventional tillage regimes (Lowrance et al. 1984a; Phillips et al. 1980; Anderson and Coleman 1985; Phillips and Phillips 1984). No tillage regimes employ herbicide application and careful timing of seeding to reduce fuel-energy inputs, enhance water retention and infiltration, improve crop growth, and enhance nutrient retention in the effective rooting zone (Lamb et al. 1985). Recent research in the U.S.A. and Canada has shown that trends toward reduced organic matter content and increased erosion (Larson 1981) can be stopped, and even reversed, by modifications of high fossil fuel energy inputs involved with frequent tillage (Gebhardt et al. 1985). In semiarid systems, shifting from the more usual fallow–rotation in alternate years to yearly cropping of dryland wheat gradually arrests losses of soil OM, and even increases OM over a several year period (Coleman et al. 1984). Similarly, in humid regions of the U.S., no tillage management not only reduces erosion but also increases OM in upper soil layers (Blevins et al. 1984; Langdale et al. 1984).

Barber's (1979) work is illustrative of this point. In a series of long-term experiments (10–12 yr.), he measured the impact of various residue management systems on organic matter status in Indiana corn fields. He varied treatments between removal of corn stalks, removal of stalks and roots, and return of twice the normal amount of residues. This comparative residue input or removal study showed that stalk residues introduced 11% of their C to soil organic C in the 11 yr. study. In another experiment, conducted for 12 yr., root inputs were calculated to contribute 18% of their C to formation of new OM. It is very likely that additional root inputs from annual wheat cropping, as well as more leaf residues, contributed significantly to the enhanced OM status. Thus, fine roots and mycorrhizae contribute significant amounts of carbon not usually recovered by most harvest sampling techniques used in field studies (Newman 1985; Read et al. 1985; Fogel 1985). These inputs may constitute anywhere from 20–40% more than the fibrous root input alone.

Several studies of long-term effects of "minimum" tillage vs. conventional tillage have found an initial decrease in crop yields, averaging about 10%, followed by a gradual increase, until they slightly surpass the outputs from conventionally tilled plots. Greater amounts of herbicides (a fossil fuel subsidy) are applied, but energy savings from reduced tillage requirements more than compensate for increased herbicide costs. Actual diesel fuel use may be reduced from 71 to 39 liters per ha., contributing to an overall energy output/input ratio of 5.4 for no tillage vs. 4.9 for conventional moldboard plow systems (Phillips and Phillips 1984).

8.2.2. Soil Biota and Conservation Tillage

Recent work of Holland and Coleman (1987) extends the research discussed above into biotic mechanisms of the changes induced by no tillage regimes. Drawing on the earlier work of Doran (1980a, 1980b), they examined the amounts of fungal and bacterial biomass present in the litter and soil of experimental plots of winter wheat subject to no tillage vs. bare-fallow regimes. They found significantly enhanced fungal populations (50–60% greater than in conventional tillage) in the top 0–5 cm of no tillage soils. Because fungal hyphae can translocate nutrients from soil into nutrient-poor surface residues, fungi may play a role in the initial N-immobilization in the upper levels of the profile. When compared with higher turnover and carbon losses of bacterial populations thought to dominate under conventional tillage, fungal domination in no till may favor greater activity of fungal based food web organisms, and lead to greater retention of microbial C and N in the profile. Holland and Coleman (1987) also followed the decomposition of ^{14}C-labeled wheat straw, and found significant enhancement of carbon retention in litter and microbial biomass in the surface litter, compared to buried litter which mimicked conventional tillage treatments (Table 8.5).

Detailed mechanisms of the biological and chemical changes which occur during the gradual buildup of surface litter materials in agroecosystems are currently being studied in several locations in the U.S. Studies of effects of residue placement on detrital food web dynamics are under way at Colorado State University and at the University of Georgia. Comparative studies of conventional (CT) and no tillage (NT) agroecosystems at these and other sites suggest (Hendrix et al. 1986):

1. NT soils are generally stratified physically and chemically with more nutrients localized near the soil surface (Table 8.6).
2. CT soils (moldboard plowed) show increased OM decomposition and accompanying nutrient mineralization.
3. nutrient uptake by crops and weeds is often greater in CT systems, although perhaps not over the long term (Rice et al. 1986).
4. NT residues tend toward greater nutrient immobilization, nutrients thus being less available until they are released from the microbial biomass.

Table 8.5. Straw [14]Carbon Loss and Incorporation into Microbial Biomass in Field Studies Conducted on a Nunn Clay Loam (Data are Means + SE)[a]

| | Incorporated Straw | | Surface Straw | |
	May 1984	Sep 1984	May 1984	Sep 1984
Soil organic[b] [14]C to 10 cm (mCi/ha)	315 ± 56	420 ± 25	144 ± 9	172 ± 30
Microbial biomass ^{14}C to 10 cm (mCi/ha)	57 ± 6	63 ± 2	44 ± 7	38 ± 6
Straw ^{14}C lost (mCi/ha)	735 ± 87	1062 ± 85	235 ± 69	496 ± 63
% [14]C retained as soil organic C	43 ± 6	41 ± 4	66 ± 9	36 ± 6
% [14]C retained in microbial biomass	8 ± 1	5 ± 0.4	43 ± 11	11 ± 6
% original ^{14}C remaining (final/ initial)	54 ± 5	34 ± 5	85 ± 4	69 ± 4

[a] Reprinted by permission of Holland and Coleman (1987), Ecology.
[b] Soil organic ^{14}C was measured by wet oxidation of soil separated from straw fragments.

At the University of Georgia's Horseshoe Bend agroecosystem research site, the soil fauna show strong differences in distribution and abundance, probably as a function of both residue placement and tillage practices. Earthworm biomass is about six times greater under NT than under CT (House and Parmelee 1985) (Table 8.7).

Other significant differences occur with ground beetles and mites, the latter of which are heavily dominated by fungal feeding Cryptostigmatids and Prostigmatids (Visser 1985).

Among other active groups, the nematodes are often recognized as key participants in the nutrient cycling process (Ingham et al. 1985). At Horseshoe Bend, the situation is somewhat mixed, with bacterial feeders being twice as abundant in CT as in NT during summer. The fungal feeding nematodes were twice as numerous in the no till soils, but were generally less abundant in both treatments than were bacterivores (Parmelee and Alston 1986).

A biological footnote to this comes from Yeates (1981) and Yeates and Coleman (1982), as reviewed by Lee (1985). Earthworms have a well-documented effect on plant debris through trituration and comminution activity. They have a significant effect on some of the soil mesofauna, as well. Earthworms in New Zealand markedly depressed soil nematode populations in both pot experiments and in long-term field studies. Nematode standing crops were only 40–50% of those in treatments containing identical soil types, but no earthworms (Yeates 1981). Given the results in New Zealand pasture soils, with and without earthworms, do our shifts in nematode populations reflect not only changes in food sources, but also a hitherto overlooked predator–prey relationship? If earthworms are more numerous in no tillage treatments, and indiscriminately ingest soil as well as organic debris, are they reducing nematode populations (of all types; i. e., microbivorous as well as plant parasites) in our no till sites just as happened in New Zealand?

These sorts of questions are amenable to experimental analysis and manipulations under field conditions. Currently we have work underway at the Uni-

Table 8.6. Organic Carbon and Nitrogen[a] and Double Acid Extractable Nutrients[b] (kg/ha) in Conventional Tillage (CT) and No Tillage (NT) Soils at Horseshoe Bend[c]

Depth (cm)	Organic C		Organic N		Phosphorus		Potassium		Calcium		Magnesium	
	CT	NT	CT	NT	CT	NT	CT	NT	CT	NT	CT	NT
0–5	9067 [d]	14155	849 [d]	1201	21 [d]	37	103 [d]	176	314 [d]	709	50 [d]	92
5–13	11481	10672	1306	1234	28	15	167	158	580	505	94 [d]	75
13–21	7141	4942	962	815	12 [d]	6	121	106	411 [d]	283	66 [d]	46
0–21	27689	29769	3117	3250	61	58	391	440	1305	1497	210	213

[a] Annual means from 1983.
[b] Data from June 1983.
[c] Hendrix et al. 1986 © 1986 by The American Institute of Biological Sciences.
[d] Tillage treatments differ significantly at $P = 0.05$.

Table 8.7. Numbers and Estimated Biomass of Soil Fauna in Conventional Tillage (CT) and No Tillage (NT) Agroecosystems at Horseshoe Bend

	Numbers·m^{-2}			mg dry wt·m^{-2}	
	CT		NT	CT	NT
Nematodes[a]					
Bacterivores	1836	e	909	237	117
Fungivores	227	e	500	14	31
Herbivores	945		1064	93	104
	3008		2473	344	252
Microarthropods[b]					
Mites	41081	e	78256	118	303
Collembola	6244	e	14684	17	40
Insects	2105		2548	—	—
	49430		95489	135	343
Macroarthropods[c]					
Ground beetles	7	e	33	6	30
Spiders	1	e	17	1	14
Others	6	e	28	—	—
Annelids[d]					
Earthworms	149	e	967	3129	20307
Enchytraeids	1837		520	59	17
	1986		1487	3188	20324
Total	54438		99526	3674	20962

[a] Means of samples from June–October 1983; numbers are × 10-3.
[b] Means of samples from May–December 1983.
[c] Means of samples from April–June 1983.
[d] Means of samples from April 1983.
[e] For numbers of organisms, tillage treatments differ significantly at $P = 0.05$.
From Hendrix et al. 1986. Bioscience, AIBS, Washington, D.C.

versity of Georgia using mesocosms (small field enclosures) (Odum 1984b) to ascertain the effects of reduced populations of bacteria, fungi, microarthropods, and annelids on subsequent residue decomposition and nutrient cycling under NT and CT systems.

8.2.3. Conservation Tillage in Practice

Although conservation tillage techniques yield benefits such as reduced soil erosion and increased soil organic matter, in practice they require more careful planning and skillful management than do conventional techniques. For example, without soil cultivation, weed control relies more heavily on herbicides, many of which are expensive. Cost effective use of these chemicals requires timing of application with susceptible stages of weed growth, which in turn requires some knowledge of weed biology and ecology. Also, certain properties of no till soils can alter the effectiveness of herbicides. For example, the tendency toward low soil pH decreases effectiveness and carryover of triazine herbicides, and thus more frequent liming of no till soil may be required (Phillips and Phillips 1984).

As with most pesticides, there can be undesirable "nontarget" effects of herbicides if they are used improperly. Such effects include reduced activity of beneficial soil organisms and movement of persistent chemicals (e.g., triazines) into surface and groundwater (Brown 1978; Hendrix and Parmelee 1985). Some of the newer herbicides and weed control techniques emerging from research in biotechnology may reduce these risks, but at the same time may create new ones (discussed later).

Conservation tillage also requires management skills in the use of fertilizers. For example, temporary immobilization of nitrogen fertilizers into crop residues can reduce nitrogen availability to crops at early stages of growth (Blevins et al. 1984). Nitrogen applications may need to be spaced incrementally in time or increased in total amount, depending on residue quantity in the field and the state of its decomposition (i.e., carbon/nitrogen ratio). Alternatively, subsurface applications of nitrogen and phosphorus can minimize contact with surface residues and increase their availability in the rooting zone, but this technique requires more expensive, specialized equipment.

There is considerable interest in techniques that are termed "microbial management," which would attempt to manipulate the actual decomposition process to synchronize nutrient mineralization with crop growth. This entails manipulating the timing and physical placement of residues, adjusting resource quality of the residues, or applying amendments to stimulate or inhibit activities of soil microbes and fauna (Swift 1986). Further research is needed to bring such techniques into practice. Interestingly, impetus for research on placement of residues in the field may come from recent legislation in several European countries (e.g., Denmark (Christensen 1986)) banning the time-honored practice of burning crop residues in the field. This legislation stems largely from the "nuisance" created by smoke in suburban developments as they spread into agricultural areas. Because of slower decomposition rates in north temperate regions, crop residues may tend to remain on the field for extended periods if not physically removed. As a result, soils may remain cool and wet in spring, thereby delaying the planting and germination of seeds for summer crops. Certain plant pathogens also are favored by these conditions. Finding ways to accelerate decomposition or otherwise manage crop residues without increasing erosion or sacrificing soil fertility will present an interesting challenge for researchers in cool climates.

8.3. Landscape-Level Agroecosystem Studies

Although many studies of agroecosystem processes are site specific, it is useful to consider effects of processes over larger geographic areas, or landscapes. Landscape-oriented studies provide information on both short-term (months to years) and long-term (decades to centuries) processes in a framework which can be extended over a wider, regional scale. In addition, such studies reveal the importance of linkages among managed and unmanaged ecosystems. Several recent studies have used a toposequence (upslope to downslope) framework (Gregoric and Anderson 1985; Smeck 1985; Schimel et al. 1985b).

The main objectives of Schimel et al. (1985b) were to determine the nature and extent of changes in soil OM status in matched pairs of field toposequences in Great Plains soils derived from three different parent materials. The soils in grassland and cropland (usually spring wheat) in central North Dakota were derived from sandstone, siltstone, or shale in glaciated, wind eroded landscapes in Grant County, in southwestern North Dakota. The fields, selected for detailed pedological studies by R. D. Heil, R. Aguilar, and associates, had known histories, with all tillage on the cropped half of the lands occurring since the lands were broken out of native virgin grassland in 1938. Schimel et al. (1985b) determined the changes in soil organic matter and soil nutrient status which could be attributed to mineralization in situ, and how much could be traceable to accumulation of eroded materials downslope, as a result of wind or water assisted movement downhill. The soils were sampled at several locations in the toposequence: summit, backslope, and toeslope, along the ca. 300 m transect.

The conditions of the profile were characterized in the control (rangeland) areas, and laterally across the landscape, upwind from the prevailing winds on the prairie, on similar locations to the cropland. Particularly at summit and backslope locations, the plow layer, or Ap layer, was extensively eroded, as well as having significantly lower organic C and N (Table 8.8). To address questions of the extent of changes in labile C and N, Schimel et al. (1985) carried out laboratory studies, including NH_4^+-N and NO_3^--N extractions, and also 20 day $CHCl_3$ incubations (Jenkinson and Powlson 1976) to determine microbial C and N in the soils. These soils were sampled at three depths, 0–5, 5–10, and 10–20 cm. The mineralization data showed several interesting trends (Table 8.9). Microbial biomass and mineralizable C and N concentrations were correlated with organic C and N concentrations, not with total

Table 8.8. Proportional Loss of Organic C to 100 cm After 44 Years of Cultivation From Three Soil Toposequences[a]

Position	Loss of Organic C (%)		
	Sandstone	Siltstone	Shale
Summit	−54	−45	−45
Shoulder	−40	−49[b]	−53
Upper backslope	−56	−39	−24[b]
Lower backslope	−18[b]	−49	−24[b]
Footslope	−34	−37	−24
Average weighted by area	−34	−46	−35

[a] Reprinted with permission from Schimel et al. (1985b), Geoderma.
[b] Identified as depositional by profile reconstruction (Kelly 1984).

Table 8.9. Microbial Biomass C, N, N Mineralized (0–10 d, Nonfumigated): Biomass N ratio and k_n for Surface Soils of Rangeland and Cultivated Toposequences[a]

	Microbial Biomass C (μg g^{-1})		Microbial Biomass N (μg g^{-1})		N Mineralized 0–10 d (Nonfumigated) (μg g^{-1} d^{-1})		k_n		N Mineralized 0–10 d (Nonfumigated)/ Biomass N %	
	range.	cultiv.	range.	cultiv.	range.	cultiv.	range.	cultiv.	range.	cultiv.
Sandstone										
Summit	1118	416	190	73	0.8	0.5	0.29	0.35	0.65	2.91
Backslope	1032	422	179	75	0.7	0.5	0.32	0.39	0.65	3.86
Footslope	1181	699	201	118	0.9	0.8	0.29	0.30	0.77	1.83
Mean	1110	512	190	88	0.8	0.6	0.30	0.35	0.69	2.87
Siltstone										
Summit	1455	609	247	105	0.5	0.8	0.28	0.32	0.21	1.72
Backslope	1556	342	264	60	0.7	0.6	0.29	0.38	0.37	2.67
Footslope	1215	559	205	98	0.5	0.6	0.27	0.38	0.36	1.37
Mean	1409	503	239	88	0.6	0.7	0.28	0.37	0.31	1.92
Shale										
Summit	1203	633	206	112	0.7	0.7	0.30	0.39	0.56	2.84
Backslope	1122	844	195	149	0.3	1.9	0.34	0.38	0.27	5.18
Footslope	1335	725	231	129	0.5	2.5	0.32	0.41	0.45	3.72
Mean	1220	734	211	130	0.5	1.7	0.32	0.39	0.43	3.91

[a] Reprinted with permission from Schimel et al. (1985b) Geoderma.

accumulation in the profile. Total accumulation and proportion of N mineralized was correlated with clay content. Parent materials had significant effects on microbial biomass and mineralization rates. The siltstone site had the highest microbial biomass and mineralizable C and N, and had the highest losses in laboratory incubations. The sandstone site had high soil, but low OM losses, while the shale experienced both low soil and OM constituent losses. Relating the soil types and constituents to ecosystem theory, we can say that the sandstone and shale sites (coarse and fine textured) may be considered "resistant" in regard to perturbation, whereas the siltstone site was less so. Development and exploration of this technique could be very useful in other landscape oriented agroecosystem studies.

Integrated effects of agroecosystem processes on landscapes also have been studied in watersheds with defined boundaries and known inputs and outputs of water and nutrients. Water runoff, soil transport, and total outputs of nitrogen and phosphorus all were reduced by conservation tillage practices in an experimental watershed on the Georgia Piedmont, U.S.A. (Langdale and Leonard 1983). Interestingly, N and P concentrations were higher in runoff from conservation tillage systems, but much lower total volume was exported. Higher nutrient concentrations under conservation tillage may have resulted from release of nutrients from crop residues on the soil surface, limited incorporation of fertilizer into the soil, or increased fertilizer applications during double cropping (growing crops during cool as well as warm seasons to minimize erosion).

Woods et al. (1983) found that a large watershed with mixed land uses retained nutrients added as fertilizer to cultivated systems within. Retention may have been due to storage in the soil profile or to uptake by forest ecosystems downslope from the cultivated fields. Riparian ecosystems can accumulate a sizeable fraction of nutrients lost from agroecosystems, thereby serving as nutrient "filters" for groundwater and surface streams in agricultural areas (Lowrance et al. 1984b; Peterjohn and Correll 1984). The significance of the "nutrient filter" effect is an indirect one for the farmer who has lost nutrients downslope from his fields. However, if he has a combined crop and woodlot operation, the increased nutrient "capital" can be harvested in his trees, enhancing the total capital return.

A further dimension to landscape level studies is that of time. The very process of soil formation, or pedogenesis, leads to important changes in key elements, such as phosphorus, in the soil profile. Over a 22,000 yr. chronosequence, rates of soil weathering have a major influence on soil P status and availability (Fig. 8.4). Longer-term chronosequential studies will be useful in comparing community dynamics across climatic regimes, and across soil types (Smeck 1985; Walker and Syers 1976).

8.4. Merging -*Logos* and -*Nomos* (Science and Management)

In the next 10–20 yr., it is apparent that many seemingly disparate threads of scientific activity will be brought together, as ecologists, agronomists, and other scientists continue to work on common research problems in applied ecology.

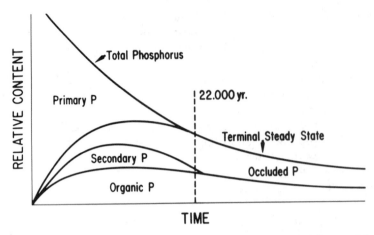

Figure 8.4. Changes in forms and amounts of soil P with time (Walker and Syers 1976; names of the P forms have been changed). From Smeck 1985. Phosphorus dynamics in soils and landscapes. Geoderma 36:185–199. © Elsevier Science Publishers

This synthesis is epitomized in the classical and heretofore irreconcilable difference between ecology and economics. For ecology, the *oikos* refers to the environment, whereas the economist tends to consider the *oikos* as man and his (chiefly financial) activities. Viewed from the perspective of the interdisciplinarian, the differences between science (*logos*) and management (*nomos*) become reduced, and the boundary lines blurred. This follows from a number of other developments in the applied environmental sciences, such as Integrated Pest Management (IPM). Rather than trying to eradicate all members of pest plant or animal populations, IPM practitioners endeavor to keep the target populations below the so-called "economic threshold," below which "pests" have no apparent economic effect on crop yield (Flint and van den Bosch 1981). The immediate result of an IPM approach to agroecosystems is that less pesticide is used, reducing application costs and enabling a more diverse biota to act, often including natural predators. Because there are more refugia and therefore potentially more predatory arthropods in NT than in CT fields (Blumberg and Crossley 1983), there may be a synergism between two complementary, progressive agricultural practices.

The notion of an "economic threshold" could be extended, as an analogy, to a perceived "ecological threshold" of processes which occur in agroecosystems. Beyond such a threshold (e.g., at higher levels of erosion or lower rates of soil OM regeneration), long-term sustainability of a system might be threatened. In this context, the ecological or holistic perspective leads the scientist or manager to consider aspects of physics and chemistry, as well as the biology of the system under study. The ties between *logos* and *nomos* then enable the researcher to ask what the effects are from conventional vs. conservation tillage management on such longer-term phenomena as soil erosion and organic matter status. The Universal Soil Loss Equation, applied on a nationwide basis, assumes an average annual loss of 1 mm topsoil per acre, or 5 tons/acre. Inasmuch as nearly 10% of U.S. croplands have erosion rates 6–8 times higher than this national average (Clark et al. 1985), changes in tax or land use policy could be implemented to encourage conservation management, and thus to mitigate these exceedingly high rates of soil loss.

More immediate objectives should focus attention on the basic mechanisms of agroecosystems. Greenland (1981), in a thoughtful review article, called for a greater understanding of how minimum tillage (or direct drilled) croplands function. Without frequent tillage, croplands are thought to become compact, yet the buildup in soil fauna, noted by Edwards (1977), Barnes and Ellis (1979), House et al. (1983), and Hendrix et al. (1986) probably mitigates this process. Soil aeration has long been considered an important fauna assisted need in temperate grassland (Darwin 1881). For the smaller, 20–50 μm diameter water storage pores, Greenland (1981) notes that the smaller oligochaetes, enchytraeids, may play a significant role in forming these pore spaces.

8.4.1. Ecosystem Optimization

An alternative approach, which should appeal to the mergers of -*logos* and -*nomos*, is the following: perhaps optimal levels of ecological processes, such as decomposition, mineralization, extent of soil invertebrate activity, etc., can be "designed in" during agroecosystem management. The scientific studies of agroecosystems noted above are attempting just such combinations, or melding, of research and management objectives. More generally, is there a range of problems on which ecologists, agronomists, and economists can collaborate to better promote the long-term sustainability of agroecosystems, and the sustainability, even survival, of the farmer–manager? The Conservation Reserve Program, enacted as part of the legislation in the 1985 U.S. Farm Bill, shows signs of heavy participation by farmers, who will get financial incentives to join the program. By fall of 1987, the U.S. Department of Agriculture (USDA) plans to place 15 million acres in the reserve program, and 40 million by the end of the decade (Guither 1986). This program is important because it marks a strong movement in North America away from "feeding the world." Current subsidized programs encourage production, but at a net cost to the taxpayers and to ecosystem stability, both costs being destructive in the long run. Similar problems exist in the European Economic Community, and in other parts of the industrially developed world.

Developments in this area have been studied by a U. S. National Research Council Board of Agriculture committee on alternative agriculture. The committee intends to evaluate the current status of modern production agriculture in all its phases, and then to make recommendations for national policy over the next several years. A number of points have arisen in the committee's work, which should be mentioned briefly in this review paper. First, and foremost, there is strong recognition and appreciation of the productivity of the majority of farms and ranches in the U.S. The crisis in which millions of farmers find themselves is one of debt capital, rather than one of ordinary financial input and outputs. To use economic parlance, the debt/equity ratio was deliberately skewed in 1979–81, when it was decided that U.S. farmers would "feed the world." Subsequent wide oscillation (increase) of the value of the dollar made the change not only short-term but also nearly fatal for the financially overextended farmers. Similar financial and overproduction problems bedevil farmers in Europe and Australasia, as well.

The principal lesson for farmers worldwide is to include financial as well as ecosystem management locally. The farmer who improves his financial status through, for example, diversification into an array of multiple crops, including animals in the total system, will be better off financially and ecologically than the monocrop farmer (Francis et al. 1986).

8.5. Environmental Biotechnology and Agroecosystems

An important new dimension to ecosystem studies is coming into being, with a broad segment of the biological sciences community becoming involved. Thus, in addition to manipulating agroecosystems, it is now possible to manipulate the genomes of the plants, animals, and microbes in these systems. This new field, environmental biotechnology, considers both the agricultural benefits and ecological consequences of releasing the altered organisms into the environment. Entire symposia and conferences have been devoted to this topic, so we can only touch on a few points, briefly (e.g., Halvorson et al. 1985).

Work on engineered livestock (e.g., microbial production of bovine growth hormone, somatotropin, for increased milk production) and engineered crop plants (e.g., herbicide resistant cultivars) continues apace, and could increase agricultural production substantially in future years. Of course, plant and animal breeding have been carried out successfully for millennia. The major contribution of biotechnology is to accelerate the process, by insertion of foreign genetic elements (e.g., fragments of bacterial chromosomes, or plasmids) into cells, achieving new and unusual traits.

Developments in engineered microbes are also proving manifold. For example, an important goal is the enhancement of nitrogen fixation in leguminous crops by *Rhizobium spp.* A better understanding of the infection process by the symbionts, and perhaps incorporation of genomes of free-living, nitrogen-fixing bacteria into crop plant roots may give us a wider range of possible nitrogen fixation avenues than we now have (Bohlool et al. 1984). Symbiotic actinomycetes, so-called actinorhizal (*Frankia*) associations, are important in some temperate and tropical crop plants and fuelwood trees (Wheeler et al. 1986).

In the area of pesticide development, plant produced insect toxins are now possible. Genes encoding the insect toxins produced during sporulation of the naturally occurring microbial pesticide, *Bacillus thuringiensis*, have been cloned and are now studied widely. Transgenic plants will be engineered to produce the toxin protein (Hauptli and Goodman 1986). Some caveats should be borne in mind, however. At a USDA Agricultural Research Service stored grain products lab at Manhattan, Kansas, several populations of grain beetles have been found to have partial or near total resistance to the *B. thuringiensis* toxin. Coping with the inherent evolutionary mechanisms involved in resistance under field conditions will prove to be an intriguing and extremely challenging problem.

An equally interesting problem is the potential "escape" of altered genetic material from improved plant cultivars into their weedy congeners. For example, a gene sequence coding for resistance to the widely used herbicide, glyphosate, has been successfully introduced into domesticated plants (Shah et al. 1986). In field crops, this trait might allow for more liberal herbicide application regimes (higher dose rates, more frequently, with less crop protection). However, cross-fertilization between an altered cultivar (e.g., grain sorghum, *Sorghum bicolor*) and a congeneric weed (Johnsongrass, *Sorghum halepense*) could transfer herbicide resistance to the latter, defeating the purpose of the genetic manipulation and possibly creating a herbicide resistant weed strain. Other engineered traits could be similarly transferred. These are problems in which ecologists and population biologists should play a leading role.

In the process of monitoring the spread of engineered organisms in the environment, it will also be of considerable importance to follow the cost of "doing business" with engineered vs. nonbioengineered organisms. Prior to releasing organisms into the environment, screening trials in laboratory microcosms and field mesocosms may be important for biotechnology "risk–benefit analyses" (Odum 1984b). The concerns of many biologists regarding the various impacts of biotechnology are both strong and real. With new engineered microbes having many new roles, physiologically and functionally, in ecosystems, what will happen if further, unintended genetic transfers (plasmid transfers) occur? The trials mentioned above are crucial to a fuller preparedness in dealing with releases of bioengineered organisms.

In addition to fears of biotechnological "monsters" unleashed in the environment, concerns have also been voiced about the effects biotechnology is likely to have on the social, political, and economic structure of agriculture and rural communities. A recent report by the U.S. Office of Technology Assessment (OTA) predicts large increases in U.S. agricultural production due to advances in crop and animal biotechnology and in information technology (reviewed by Tangley 1986b). Accompanying this increase will be the disappearance by the year 2000 of one million small and moderate size farms, leaving some 50,000 large farms accounting for 75% of agricultural production in the U.S. Because large, corporate farms tend not to rely on local communities for goods and services, substantial disruptions could occur in institutions such as stores, schools, churches, and community organizations in rural areas.

Whether or not risk–benefit analyses have been performed on the effects of biotechnology at this level, it seems generally agreed that the changes it portends are undesirable. Thus, "to assure a diverse, decentralized farm structure, where all sizes of farms have an opportunity to compete and survive," OTA recommended a set of policies individually targeting small, moderate, and large farms. These include recommendations that large farms not have access to government subsidy programs (income and price supports) and that new technologies be made available to moderate size farms.

It is desirable for agronomists, economists, and ecologists facing the 21st century to address all aspects of biotechnology in a holistic context. There are many opportunities present in this burgeoning new field, ones which will undoubtedly come about in spite of transitory budgetary constraints.

8.6. Conclusions

As human populations and demands on natural resources continue to grow over the next several decades, it is important that we recognize the global implications of our activities. Agriculture, an activity common to all people, places heavy demands on resources in both developed and developing countries. Although critical agricultural problems (desertification, deforestation, salination, etc.) may be localized in certain regions, they are likely to have impacts on a larger scale through economic, political, and social systems, and possibly through effects on hydrologic and biogeochemical cycles. Dealing with these problems, while at the same time devising highly productive, sustainable food production systems, presents a great challenge to our scientific and technological abilities.

We are certainly in need of not only holistic, but also transdisciplinary perspectives in ecosystem studies, particularly in agricultural and basic ("pure") disciplines. As we continue to study system processes, we must increase our priorities in bringing adequate "people power" to bear on what the organisms are, and how they function. When one adds the need to consider conjoint effects of physics, chemistry, hydrology, and economics of applied ecosystem studies, the problems become greater, but not insurmountable. There is a groundswell of public recognition for new, interdisciplinary approaches to study big system problems.

Acknowledgements

Support for development of ideas in this chapter was provided by NSF Grant BSR 8506374 to the University of Georgia. M. Beare and D. Wright provided helpful critiques of the manuscript.

9. Ocean Basin Ecosystems

Richard T. Barber

Natural scientists have for 100 years wrestled with the difficulty of formalizing the concept of a functional entity of biological and physical organization. In a review of the concept, Evans (1956) discusses Möbius's (1877) use of *biocoenose*, Forbes's (1887) use of *microcosm*, Markus's (1926) use of *naturkomplex*, Friederich's (1927) use of holocoen, and Thienemann's (1939) use of *biosystem*; but Evans prefers Tansley's (1935) use of *ecosystem* to describe the basic unit of ecology. Evans's thesis is that there is a functional entity of biological and physical organization (i.e., an ecosystem), and that level of organization should be the fundamental unit of ecology. Two particularly important discussions of the ecosystem concept are those of Odum (1969; 1977). Odum (1969) provides our current working definition of *ecosystem* as the unit of biological organization interacting with the physical environment such that the flow of energy and mass leads to a characteristic trophic structure and material cycles. This definition emphasizes that the existence of a *characteristic trophic structure* and *characteristic material cycles* provides the means to determine the boundary between one ecosystem and another. The 1969 article goes on to suggest that ecological succession is a central tenet of ecosystem theory. This line of reasoning causes concern for me as an ocean ecologist, or at least as a pelagic ocean ecologist. The decade to century time scale of ecological succession that is a central feature of terrestrial and benthic marine ecology is difficult, or even impossible, to detect in the fluid medium of pelagic ocean ecosystems. In the pelagic realm, wind and current systems appear to reset the successional time clock to zero each year or perhaps even with each storm or variation in winds. There is by no means agreement on this issue. Steele and Henderson (1984) modeled the low frequency changes in species abundance which take place in ocean ecosystems at the 10 to 100 yr. scale. However, the changes they describe are not functionally analogous to succession as defined by Odum (1969) as a process that is: (1) orderly and directional; (2) resulting from biological modification of the physical environment; and (3) culminating in a stabilized ecosystem. Succession is observed on the day to month scale in pelagic systems, particularly in phytoplankton (Smayda 1963, 1980; Levasseur et al. 1984; MacIsaac et al. 1985), after spring stratification or after an episode of upwelling, but evidence for directional succession involving the decade to century time scale is lacking. Smayda (1980) specifically compared phytoplankton species succession to terrestrial plant succession as defined by Odum (1969) and characterized oceanic succession as being circular with a successional cycle that usually reiterates itself annually or after each storm. It is clear from Smayda's (1980) review and Hutchinson's (1967) characterization of planktonic succession that, while there are cyclic changes in planktonic species, these changes are not orderly and directional on the decade to century time scale, and they do not lead to a climax community or stabilized ecosystem.

If long-term directional succession is central to the ecosystem concept, but is not detectable in the ocean, then is the ecosystem concept useful for describing the pelagic ocean? An answer to that question is given in Odum (1977), where hierarchical theory is related to the ecosystem concept. Odum suggests that one consequence of hierarchical organization is that as components, or subsets, are integrated or combined to produce a larger functional whole, new properties emerge that were not evident at the next lower level. This consequence, the emergence of new properties, makes the ecosystem concept useful, and even necessary, for describing the physical and biological behavior that characterizes the ocean. If all of the characteristics of a component (or population) can be accounted for by analysis of the component (or population), then clearly no higher level of functional organization is present. Furthermore, if analysis of the "larger functional whole" explains nothing new about the population, then no properties have emerged, and it is necessary to conclude that there was no valid functional organization at the higher level. The central question becomes what, if any, new understanding does the ecosystem concept provide that cannot be obtained from a study, no matter how thorough, of the separate regions of the ocean. The answer to that question is quite dramatic and of practical societal interest.

9.1. An Example

From my dissertation research two decades ago (Barber 1967) to the present (Barber and Chavez 1986), many colleagues and I have worked on coastal upwelling, an ocean phenomenon that occurs in only a few restricted locations around the world (Barber and Smith 1981). Originally, this work (which was carried out in the spirit of interdisciplinary but reductionist science) concentrated on analysis of the specific physical and biological processes taking place in the limited domain where the upwelling phenomenon occurred (Barber 1977). As the work progressed, it became increasingly apparent that certain aspects of the physical and biological variability of coastal upwelling could not be explained when the analysis was limited to atmospheric and physical processes actually occurring in the upwelling region. We interpreted these observations to mean that coastal upwelling was part of a higher level of organization (Barber and Chavez 1983), which turned out to be the basinwide thermal dynamics of the ocean and the global wind system (Fig. 9.1; Wyrtki 1975; Philander 1986a). In essence, one can say that coastal upwelling is a component of the global heat budget. When the global heat flux is perturbed, the physical and biological components of coastal upwelling necessarily change, as shown in Fig.s 9.2–9.4.

This coupling of the global heat flux to living resources extends through economics to the global social system, as shown in Fig. 9.3. When ocean productivity decreased because of a perturbation of the global heat budget, an important world commodity, Peruvian fish meal, became scarce. This drove up the value of a substitute commodity, soybean meal. The increased demand and value of soy meal changed both land development and agricultural practices (Freivalds 1976; Barber et al. 1980) in many parts of the world. Sensitivity of biological (and economic) components of coastal upwelling to large-scale ocean and atmospheric processes is interesting, because the major physical process (wind driven coastal upwelling) and the major biological process (phytoplankton blooms) are inherently mesoscale processes that involve relatively short time scales of 1–10 days and small space scales of 5–200 km (Barber and Smith 1981). The character of both physical (R. L. Smith 1983) and biotic variability (Barber and Chavez 1983) forced us to recognize that coastal upwelling is a component of a larger functional whole, as well as an inherently mesoscale phenomenon. The character of this larger functional whole was revealed clearly in 1982 and 1983, when nature provided a signal in the form of the 1982–83 El Niño (Barber and Chavez 1986). This extraordinary natural event demonstrated the basinwide coupling between components, such as the low latitude gyre and coastal upwelling, that were previously thought to be relatively independent of each other.

9.2. The Other Point of View

An interesting case can be made for the point of view that there exists in the ocean a set of more or less independent ocean "ecosystems." For several years, I have taught a course entitled "Analysis of Marine Ecosystems", which examines in detail the following "ecosystems":

1. Coastal upwelling
2. Low latitude gyre

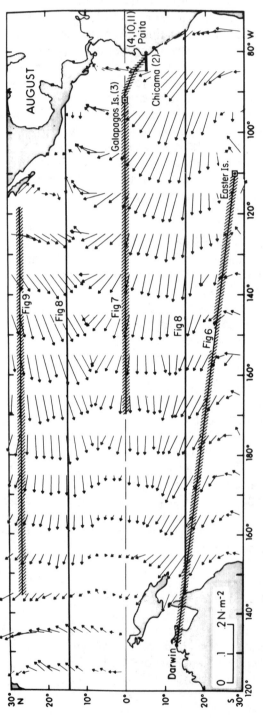

Figure 9.1. Mean surface wind stress over the tropical Pacific for August. The orientation of the arrow indicates the direction of the wind stress and the length of the arrow indicates magnitude of the stress in newtons m⁻². This figure is modified from Gill (1982) and Wyrtki and Meyers (1975, 1976 (© Journal of Applied Meteorology, American Meteorological Society)). The locations of sections given in Figures 9.6, 9.7, and 9.9 are shown by crosshatched lines; the numbers in parentheses indicate the figures that present data from that location. The solid lines at 15°N and 15°S indicate the limit of the tropical area where the upper layer volume anomaly (Fig. 9.8) was calculated.

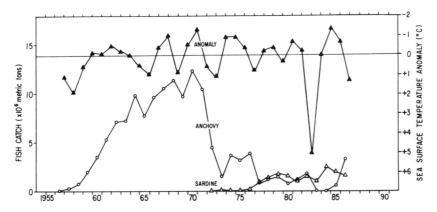

Figure 9.2. Physical and biological coupling in the basinwide ocean ecosystem shown by the relationship between the annual anomaly of surface temperature and the annual Peruvian catch of anchovy and sardine. The anomaly temperature scale is inverted: upwards indicates cooler temperatures; downwards indicates warmer temperatures (El Niño). The annual temperature anomaly is relative to the 26 yr. mean temperature of 16.9°C at Chicama, Peru (see Fig. 9.1). The anomaly is calculated for the southern hemisphere thermal year from July to June spanning half of two calendar years; the value is plotted in the middle of the thermal year in January. The fish catch for each calendar year is plotted in the middle of the calendar year in July. Each downward deflection of the temperature anomaly line indicates a year when less nutrient-rich water upwelled along the Peru coast because of the El Niño phenomenon. The El Niño episodes in 1972–73, 1976–77, and 1982–83 were accompanied by progressive stepwise decreases in anchovy. From 1975–1985, when anchovy were at very low levels, sardines became increasingly abundant. While this relationship looks like direct competition between the two fish stocks, there is evidence that subtle changes in basinwide thermal conditions caused enhanced sardine recruitment during the 1975–85 decade. Analysis of this relationship is still being investigated. From Barber and Chavez, Vol. 222, No. 4629, pp. 1203–1210,) 1983 © by A. A. A. S.

3. Equatorial upwelling
4. Subarctic gyre
5. Southern ocean
6. Eastern boundary current

The approach used is to define the characteristic trophic structure and material cycles by applying the same analysis to each system. Each of the six systems has a different combination of stratification, nutrient supply, primary productivity, and productivity regulation (Table 9.1). This analysis indicates that there are distinct ocean regions that appear to meet Odum's criteria of having (1) characteristic trophic structure, (2) characteristic material cycles, and (3) recognizable boundaries. When one examines in detail the taxonomic character or other community ecology properties, as several authors have (McGowan 1971, 1974; Reid et al. 1978), the differences between regions are reinforced.

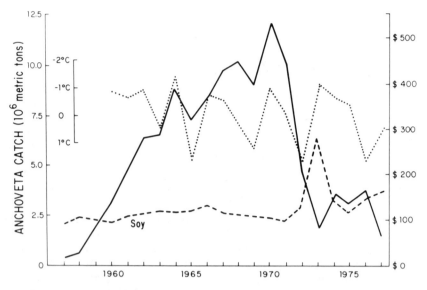

Figure 9.3. Physical, biological, and economic connections during the 1972 El Niño. The heavy solid line is the annual anchovy catch by Peru; the dotted line is the annual sea surface temperature anomaly of the eastern Pacific as measured in the Galapagos Islands (see Fig. 9.1); the broken solid line is the cost of soy meal on the world market in 1972 U. S. dollars per metric ton. The temperature anomaly scale is inverted, with upwards indicating enhanced upwelling; i.e., colder temperatures. The Peruvian anchovy industry was the world's largest supplier of protein meal, and in the 1960s, the availability of relatively cheap fish meal held down the cost of protein meal throughout the world. When the supply of Peruvian fish meal decreased in 1973, the value of other protein meals, especially soybean, increased. As a result of the increase in value, soybean production was rapidly increased throughout the world. In Brazil, several million acres of Amazon forest were cleared (Freivalds 1976), and in the United States, large-scale clearing and drainage of wetlands took place, starting in 1974, in response to the 1973 increase in value of soybeans From Barber et al. (1980), with permission from American Water Resources Association.

For example, Table 9.2 presents McGowan's (1974) comparison of five biotic provinces of the North Pacific.

What, then, is the difficulty with saying that there are numerous relatively well-defined ocean ecosystems whose boundaries are determined by the large-scale ocean circulation? The problem is that during the 1982–83 El Niño, relatively rapid transitions from one set of conditions in Table 9.1 to another took place in a single location, as the ocean, as well as the large-scale pressure and wind fields, progressed through an El Niño Southern Oscillation (ENSO) cycle (Philander 1983). The transitions were coherent and, to some degree, complementary, in that while the eastern Pacific became much less productive, the western Pacific became slightly more productive over a large area.

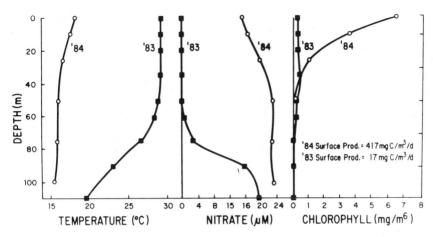

Figure 9.4. Vertical profiles of temperature, nitrate, and chlorophyll during the peak of the 1982–83 El Niño in May 1983, and during normal conditions in May 1984. The station sampled in 1983 and 1984 was at 5°S latitude and 81°30'W longitude, about 25 km off the coast of Peru, near the fishing port of Paita, where the information in Figures 9.10 and 9.11 was collected. The location of Paita is shown in Figure 9.1. Conditions in 1984 were "normal" in the Pacific in that the seawater was cool (16 to 18°C), nutrient-rich (about 15 μM NO_3 at the surface), and high in phytoplankton abundance (about 6.5 mg chlorophyll m^{-3}). During the mature phase of the 1982–83 El Niño in May 1983, the water at this location was warm (29°C), and had no detectable nitrate, and had greatly reduced phytoplankton abundance. The conditions in May 1983 resemble normal conditions in a low latitude gyre (Table 9.1).

During the peak of the anomaly in April, May, and June 1983, upwelling centers along the coast of Peru had surface temperatures of over 29°C, nutrient-depleted water in the surface mixed layer, and very low primary productivity with a tropical oceanic phytoplankton species assemblage (Ochoa et al. 1985). By any ecosystem or community ecology criterion, an observer would identify the ecosystem as being a low latitude gyre (see Table 9.1, No. 2) or a central water mass biotic province (see Table 9.2, No. 2), as defined by McGowan (1974). Fig. 9.4 compares vertical profiles of oceanographic properties in May 1983, at the peak of the mature El Niño, and May 1984, when normal conditions prevailed. The station is at 5°S latitude about 25 km offshore, and during normal conditions is representative of the very productive Peruvian upwelling centers (MacIsaac et al. 1985). Large scale changes in surface temperature for the Pacific and Atlantic in June 1983 and June 1984 are compared in Fig. 9.5. The June 1984 temperature field shows "normal" or climatological mean conditions, with the cool eastern boundary current sweeping up the west coast of South America along Peru and turning westward into the equatorial current system. In marked contrast, during June 1983, the cool water signature of the boundary current and equatorial upwelling are not present in waters off South America. The May 1984 profiles (Fig. 9.4) and June 1984 large-scale temperature field (Fig. 9.5) show normal or climatological mean conditions of

Table 9.1. Summary of Stratification, Nutrient, and Productivity Characteristics of Six Regions of the Ocean Basin Ecosystem

Region	Stratification[a] Strength	Stratification[a] Duration	Nutrients[b] Level[c]	Nutrients[b] Source	Primary Productivity per unit area[d]	Process Regulating[e] Productivity	Reference
1. Coastal upwelling	Mosaic of weak and strong	Continuous	High	Advection	Medium to high	Space[f]	Barber and Smith (1981) MacIsaac et al. (1985) Ryther (1969)
2. Low latitude gyre	Strong	Permanent	Low	Mixing	Low to medium	Nutrient supply	Blackburn (1981) McGowan and Hayward (1978) McGowan (1971)
3. Equatorial upwelling	Strong	Permanent	High	Advection	Medium	Physical process/grazing	Vinogradov (1981) Walsh (1976) Cowles et al. (1977)
4. Subarctic gyre	Strong	Seasonal	High	Mixing	Medium to low in winter	Grazing in summer, mixing and low light in winter	McGowan and Williams (1973) McGowan (1974) Miller et al. (1984)
5. Southern Ocean	Weak	Seasonal	Very high	Mixing	High to low in winter	Mixing in summer, low light in winter	Nemoto and Harrison (1981) Smith and Nelson (1985) Marra and Boardman (1984)
6. Eastern boundary current	Medium	Permanent	Medium	Advection	Medium	Grazing/nutrient supply	Abbott and Zion (1985) Huyer et al. (1984) Bernstein et al. (1977)

[a] Stratification is the resulting balance between the buoyant heat flux and the wind driven turbulence.

[b] Nutrients here refers to "new" nutrients in the sense used by Dugdale and Goering (1967).

[c] Level is relative to the uptake kinetics of phytoplankton as described by Dugdale (1967). "High" is a concentration level where the uptake vs. concentration hyperbola is always saturated; "low" is a concentration which is on the slope of the hyperbola and concentration is always regulating the uptake rate; "medium" refers to nutrient concentrations that vary from saturating to regulating.

[d] low < 0.1 gC/m²/day; medium ≈ 0.5 gC/m²/day; high > 1.0 gC/m²/day

[e] Regulation here can mean regulating the intensity of primary production (as in No. 2) or regulating the abundance of primary producers (as in No. 3).

[f] This comment emphasizes that the domain where coastal upwelling provides optimal nutrients and optimal stratification is very limited in area.

cool, nutrient-rich surface waters. As shown in Fig. 9.4, during the anomaly in 1983, surface primary productivity and standing crop of phytoplankton were reduced twentyfold; therefore, it is not surprising that the large-scale physical changes had profound biological consequences throughout the upwelling food web (Barber and Chavez 1983; Arntz et al. 1985; Kelly 1985).

At the same time that coastal upwelling, eastern boundary, and equatorial current systems became much less productive in 1983, the low latitude gyres and the warm pool region of the Pacific became somewhat richer over a very large area (Dandonneau and Donguy 1983). Changes also took place in the high latitude North Pacific subarctic gyre (Wooster and Fluharty 1985) and in the Southern Ocean (Heywood et al. 1985), with both regions becoming less productive. The nature of the dynamic connection from the tropics to the midlatitude and high latitude ocean gyres is not at all clear, but the direction of change is well-established, with the tropical western Pacific becoming more productive, and the high latitude gyres less productive.

Particularly important in an analysis of the basinwide connection is the observation that local atmospheric forcing off central Peru did not change significantly during El Niño. In April, May, and June 1983, the coastal region became oligotrophic and gyrelike in character, despite upwelling favorable winds that continued to blow along the central Peru coast at strengths near to or greater than the climatological mean conditions (Enfield and Newberger 1985). Our interpretation of this sequence is that wind driven coastal "upwelling" was occurring through much of the nine month warm anomaly (October 1982 through June 1983) along the central Peru coast, but, because of the increased depth of the thermocline and nutricline, warm, nutrient-depleted water was entrained into the upwelling circulation (Barber and Chavez 1983). If this interpretation is correct, the basinwide processes regulating the vertical thermal and density structure of the Pacific completely overrode the mesoscale processes that normally maintain high nutrient levels in the upwelling centers, and determine the biological character of the region. The low latitude gyre and coastal upwelling are connected parts of a larger system. In particular, they are linked by the equatorial and eastern boundary current systems, and probably by the western boundary currents as well. When we observe the whole system, a new property, the relationship of local productivity, species abundance, and species distribution to the global heat flux, becomes evident. It seems appropriate to call this single, interconnected system "the basinwide ocean ecosystem" and to try to describe how this ecosystem functions.

9.3. The ENSO Cycle

The phenomenon that structures the basinwide ecosystem is the El Niño/Southern Oscillation (ENSO) cycle (Philander 1983). The cool phase, or normal portion, of the ENSO cycle is responsible for the mean regional characteristics of the basinwide ecosystem that are summarized in Table 9.1; the El Niño, or warm phase, drives the variability of the ecosystem. The El Niño component

Table 9.2. A Comparison of Five Biotic Provinces of the North Pacific Modified by McGowan from McGowan (1974). (The provinces are numbered to correspond with the regions in Table 9.1.)

Biotic Province	Percent of Endemic Species	Number of Species	Evenness[a]	Variability Zooplankton[b]	Primary Productivity (gC·m⁻²·yr⁻¹)[c]	Productivity Peak	Degree Stratified[d]	Range to top of Thermocline (m)[e]	Leakage[f]	Type of Phytoplankton Limitation[g]
1. Eastern tropical Pacific	Low	High	Moderate	Moderate 2×	75	Two peaks, spring and fall	Complex	10–60, but complex	Moderate	N
2. Central (low latitude gyre)	Low	High	High	Low	About 40	None	High, but halocline shallow in summer, deep in winter	30–110	Very low	N and P
3. Equatorial	Moderate	High	High	Low	100–200	None	Both high and very low	35–65 and 60–140	Moderate	?
4. Subarctic	High	Low	Low	High > 2×	75–80	Spring	Low, but deep halocline	20–130	Very high	Light

Biotic Province	Percent of Endemic Species	Number of Species	Evenness[a]	Variability Zooplankton[b]	Primary Productivity (gC·m⁻²·yr⁻¹)[c]	Productivity Peak	Degree Stratified[d]	Range to top of Thermocline (m)[e]	Leakage[f]	Type of Phytoplankton Limitation[g]
6. Transition (Eastern boundary current)	Moderate	Moderate	Moderate	Low < 2×	40–75	Late spring to early summer	Low, no halocline	20–120	Very high	Probably N

[a] Refers to the "evenness" of the rank order of zooplankton species abundance and may or may not be true of the phytoplankton and nekton.

[b] Variability of zooplankton standing crop refers to temporal variability on a seasonal or longer basis.

[c] Primary productivity per year has been measured over an entire year only in the Subarctic and Eastern Tropical Pacific. The other values are estimates based on short-term (a few days or weeks) measurements and extrapolated over the entire year. These extrapolated values were weighted by factors derived from what we know of the seasonality of chlorophyll standing crops.

[d] Degree stratified refers to the "strength" of the thermocline in terms of the steepness of the temperature gradient. It is related to the degree of vertical mixing of deeper, nutrient-rich water.

[e] Range to top of thermocline is a measure of the thickness of the mixed layer, but south of 40°N the thermocline "topography" is much more complicated than indicated here.

[f] Leakage refers to the proportion of the standing crop of plankton and nutrients estimated to be carried out of the system of its origin per unit time.

[g] Type of phytoplankton limitation refers to the fundamental factors regulating turnover rates and specifically excludes grazing. N means nitrate limitation; P means phosphate limitation.

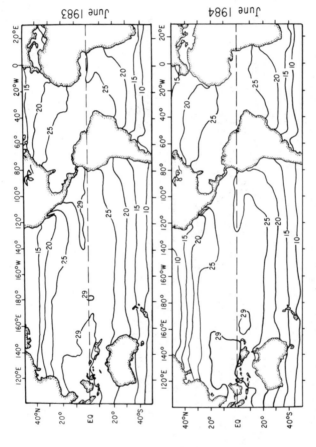

Figure 9.5. Sea surface temperature fields in June 1983 and June 1984 in the Pacific and Atlantic. This analysis blends satellite and shipboard temperature measurements and was provided by Vernon Kousky of the Climate Analysis Center of NOAA. The June 1984 temperature field in the Pacific is close to the climatological mean condition, with a well-developed "cold tongue" of water cooler than 25°C reaching out from the coast of South America. In June 1983, the "cold tongue" in the Pacific is missing and there are large regions of anomalously warm water off Central and South America. In June 1984, the "cold tongue" of Benguela Current is missing and the eastern equatorial Atlantic shows a dramatic warm anomaly (Philander 1986b; Weisberg and Colin 1986; Hisard et al. 1986). This comparison shows that large-scale interannual variability can modify the oceanographic character of traditionally defined ocean regimes, and the changes appear to be coherent. The sequential changes that occurred in the Pacific and Atlantic in 1983 and 1984 suggest that the world's ocean basin ecosystems are interconnected. Courtesy of the Climate Analysis Center, National Oceanic and Atmospheric Administration, Washington, DC.

of the name refers to the redistribution of water and heat in the tropical Pacific Ocean, while the Southern Oscillation component of the name refers to a coupled oscillation of the South Pacific atmospheric high pressure system and Indonesian atmospheric low pressure system (Fig. 9.6). The coupled oscillation of the two pressure systems causes changes in the basinwide trade winds. In each ocean basin, the combination of low latitude trade winds (Fig. 9.1) and midlatitude westerlies sets up a basinwide slope in sea level, thermal structure, density structure, and, most importantly for the ecosystem, nutrient structure.

Interactions of the ENSO cycle in relation to the west vs. east temperature gradient and the basinwide thermocline and nutricline tilt are shown in Fig. 9.7. As Panel A indicates, the Pacific is warmest in the west. Most of the solar heat falling on the warm water (28°C) of the western Pacific is transferred to the atmosphere by convection, back radiation, and evaporation. Satellite measurements of outgoing long wave radiation (OLR) (Weickmann 1983; Lau and Chan 1985) show that the major convective center of the Pacific is located over the Indonesian "maritime continent" (Ramage 1969). The tremendous center of convection and evaporation over the warm pool of the western Pacific causes air to rise, creating the Indonesian low pressure system, and feeds into the upward branch of the east–west atmospheric circulation cell shown in Fig. 9.7 (Newell and Gould-Steward 1981). On the eastern portion of the basin, equatorward trade winds force upwelling of colder subsurface waters, as shown in the right side of the panels in Fig. 9.7. At the surface, this upwelled water extracts heat from the atmosphere, causing subsidence and creating the South Pacific high pressure system (Figs. 9.6 and 9.7). Trade winds moving around the region of high pressure push surface water westward in the Equatorial current system (Fig. 9.1). As the water moves westward, the upper layer gains heat from the sun, so the wind driven circulation transports heat to the western Pacific, which keeps it warm and the eastern Pacific cool. Thus, the west vs. east temperature gradient in the tropical Pacific sets up a pressure gradient (Fig. 9.6) that forces the large-scale trade winds (Fig. 9.1), and at the same time, the trade winds are responsible for forcing the circulation that maintains the west vs. east temperature gradient (Rasmusson and Wallace 1983; Cane 1983).

When normal trade winds persist for several years, they accumulate surface water in the west, and the volume of upper layer water is increased. The western Pacific warm pool expands towards the east and the atmospheric heating spreads eastward as well. Eastward migration of atmospheric heating and its associated low causes trade wind reversals, and reduces the fetch of the easterly trade winds, as shown in Panel B of Fig. 9.7. As a result of the reduced easterly wind stress and trade wind reversals, equatorially trapped internal waves that propagate eastward are excited (Busalacchi and O'Brien 1981), and warm surface water surges eastward (Wyrtki 1975). Both advective transport of warm surface water eastward and equatorially trapped waves result in thermocline depression, so less cool water is upwelled by local processes in the central and eastern portion of the equatorial Pacific. Hence the eastern equatorial Pacific becomes still warmer, and the zone of atmospheric heating migrates further east. When

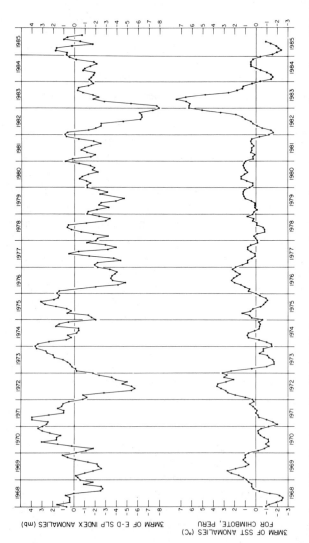

Figure 9.6. Time series of the Southern Oscillation Index anomaly and sea surface temperature anomaly at Chimbote, Peru, showing the coupled ocean–atmosphere relationship. The Southern Oscillation Index expresses the magnitude of the atmospheric pressure gradient between the South Pacific High, measured at Easter Island, and the Indonesian Low, measured at Darwin, Australia. The index is the pressure in millibars at Easter Island minus the Darwin pressure. See Figure 9.1 for the location of Easter Island and Darwin; Chimbote is south of Chicama on the central Peru coast. The anomaly is the pressure difference for a given month minus the long-term climatological pressure difference for that month; the sea surface temperature anomaly is computed the same way. The values plotted are three month running means of the index and temperature anomalies. When the index anomaly is positive, the pressure gradient is stronger than average; when the anomaly is negative, the gradient is weaker than average and consequently the trade winds are weaker. Variations in the trade winds set in motion the ENSO process described in Figure 9.7. This figure was provided by Dr. William H. Quinn of Oregon State University; see Quinn and Neal (1983) for more details on the Southern Oscillation Index.

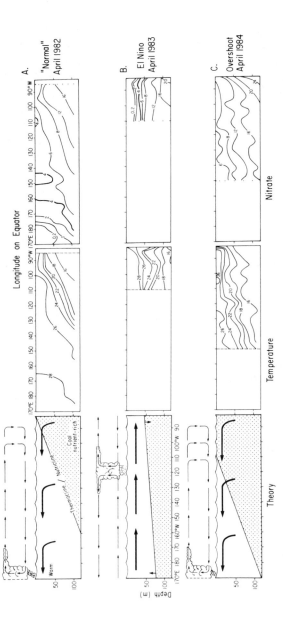

Figure 9.7. Interactions of the ENSO cycle in relation to the west vs. east temperature gradient and the basinwide thermocline and nutricline tilt. The nitrate and temperature profiles are from equatorial cruises (Barber and Chavez 1986) before, during, and after the 1982–83 El Niño. The April 1983 cruise (B) did not extend far to the west, but global sea surface temperature observations, such as those shown in Figure 9.5, showed that warm (> 28°C) water covered the equatorial region from the coast of South America to Southeast Asia in April 1983. As shown in (B), water > 28°C is nutrient depleted; that is, it contains < 0.2 µM NO_3. The center of the convective heating during normal conditions is usually further to the east than shown in A; Figure 5 shows that the > 29°C warm pool is between 120°E and 160°E; the convective center is usually also in this region (Gage and Reid 1985). During the mature phase of the 1982–83 El Niño in April 1983, the center of convective activity was around 120°W (Lau and Chan 1985), as shown in Panel B. After El Niño, resonance in the basinwide thermocline tilt results in more cold water in contact with the atmosphere along the eastern boundary, as shown in Panel C. This forces more subsidence, which strengthens the pressure gradient, causing stronger than average trade winds. The overshoot is evident in Figure 6 as a cool anomaly in 1984 and 1985. From Barber and Chavez (1986), reprinted by permission from Nature, Vol. 319, No. 6051, pp. 279–285 © 1986 Macmillan Magazines Limited.

this process gains enough momentum to lock the atmosphere and ocean into the anomalous phase (Panel B of Fig. 9.7), it is the phenomenon known as El Niño (Cane et al. 1986). The May 1983 portion of Fig. 9.4, the June 1983 scene of Fig. 9.5, and the April 1983 portion of Fig. 9.7 show different aspects of the mature El Niño phase.

The normal trade wind phase (Fig. 9.7A) or the El Niño phase (Fig. 9.7B) of the ENSO cycle appear to be self-sustaining conditions, but instead they oscillate from one phase to another. Observations (Wyrtki 1985) and modeling (Pares-Sierra et al. 1985) indicate that there is anomalous poleward transport of heat in the form of warm water to the midlatitudes from the tropics during the El Niño phase of the ENSO cycle. The aperiodic oscillation of the two phases was modeled by Cane and Zebiak (1985), who suggest that at the peak of El Niño (Fig. 9.7B) water drains poleward, reducing the eastern portion of the equatorial reservoir of warm surface water. When there is no longer enough warm water in the tropical Pacific to sustain above normal surface temperatures in the east, the El Niño begins to decay. Cane and Zebiak (1985) postulate that this decay results in an overshoot that leaves the eastern Pacific colder and more nutrient-rich, as shown in Panel C of Fig. 9.7. Comparison of nitrate and temperature profiles from equatorial cruises before (April 1982) and after (April 1984) the 1982–83 El Niño provides evidence of the overshoot. Work along the Peru coast in early 1985 found evidence of increased denitrification (Codispoti et al. 1986). It is assumed that increased nutrient availability caused enhanced productivity, which in turn resulted in increased denitrification.

The volume of upper layer water in the tropical Pacific is monitored through a network of island sea level stations by Klaus Wyrtki of the University of Hawaii. Fig. 9.8 shows how the volume of upper layer water in the tropical Pacific between 15° N and 15° S has varied in the last decade. The increase in volume in 1982 and rapid decrease in 1983 show clearly the signal of the 1982–83 El Niño. The draining of this warm tropical water poleward of the tropics during 1983 is apparently the physical process responsible for the high latitude effects in both the subarctic (Wooster and Fluharty 1985) and the Southern Ocean (Heywood et al. 1985).

9.4. Ecosystem Consequences of the ENSO Cycle

The normal basinwide tilt in the thermal and nutrient structure (Fig. 9.7) makes the entire eastern side of the ocean basin inherently richer than the central or western portions, because the subsurface nutrient pool penetrates to the sunlit layer where enough radiant energy is available for the photosynthesis of new organic material (Barber and Chavez 1983). The shallower nutricline in the eastern boundary current and in the eastern portion of the equatorial regions also means that any physical process that causes additional turbulent or advective vertical transport (island wake mixing, tidal mixing, wind driven upwelling, shelf–break mixing) will supply more nutrients to the surface layer. Therefore, the eastern portion of the ocean basin is, during climatological mean

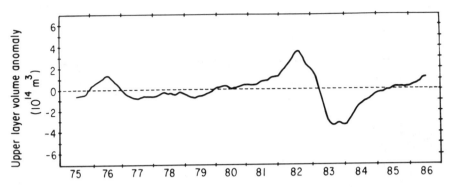

Figure 9.8. Upper layer volume in the tropical Pacific between 15°N and 15°S in 10^{14} m^3, relative to the climatological mean upper layer volume of 70×10^{14} m^3. The upper layer volume is a measure of the amount of warm water above the thermocline; as such, it is an index of ocean heat content. This figure shows that a dramatic change in the tropical Pacific heat content is associated with the ENSO cycle. This figure is provided by Dr. Klaus Wyrtki of the University of Hawaii, and is updated through January 1987. See Wyrtki (1985) for a description of the sea level net and data processing procedure.

conditions, always richer than the central or western portions. Fig. 9.9, from Joseph Reid's (1965) atlas, shows that an east–west basinwide tilt in nutrients is present across the Pacific at midlatitudes 30–40° off the equator. The east vs. west generalizations apply to the temperate ocean as well as to the tropical ocean.

Potential energy accumulated and stored in the high sea level and deep thermocline of the warm pool region of the western Pacific is available to force a rapid redistribution of water back across the basin when the winds weaken or reverse. The redistribution of warm, nutrient-depleted water forces the nutricline down in the entire eastern boundary current and eastern portion of the equatorial regions, causing a transition towards more oligotrophic character (Fig. 9.7B). At the same time in the western Pacific, the thermocline and nutricline must rise as surface layer water "drains" to the east (White et al. 1985). Direct evidence of the rise in the nutricline in the western Pacific was obtained in December 1986 during the weak ENSO episode that began in late 1986 and early 1987.

In the eastern boundary current region off South America and in the equatorial current system, the El Niño phase of the ENSO cycle results in large-scale redistributions, some direct thermal effects, and widespread food stress because of the productivity decrease (Chavez and Barber 1985). The numerous biological effects are well described for Peru in Arntz et al. (1985), and for Chile in Kelly (1985). The three classes of biological effects (migration, thermal shock, and starvation) reduce the abundance of species indigenous to the cool eastern boundary current and equatorial upwelling, but the most pervasive biological effect is the disruption of reproduction. The portion of an organism's

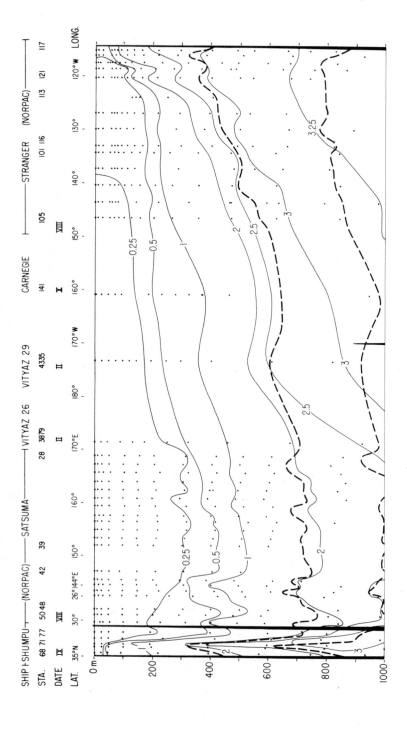

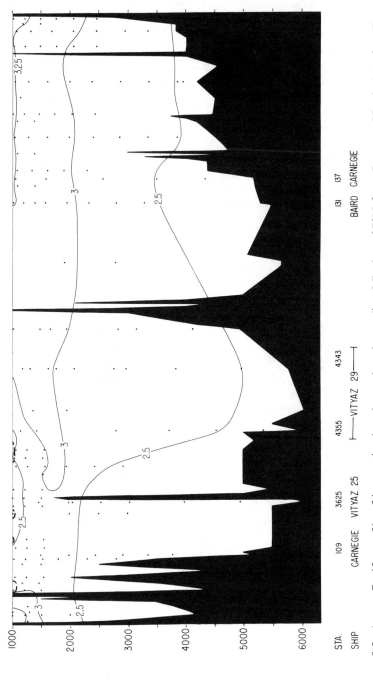

Figure 9.9. A transPacific profile of inorganic phosphate–phosphorus (in μM) along 27°N from Japan to North America. From Reid 1965 © Johns Hopkins University Press. The west to east upward tilt of the nutrient gradient at 27°N emphasizes that basinwide tilt is important in determining the midlatitude, as well as lowlatitude, character of the ocean basin ecosystem.

life cycle most sensitive to environmental variability is the early life history, involving eggs, larvae, and juveniles. The entire 1983 cohort was lost for anchovy (Santander and Zuzunaga 1984), as well as for certain species of birds and pinnipeds (Arntz et al. 1985).

While the typical upwelling species decreased in abundance, tropical oceanic assemblages of species flourished in the region during the anomaly. Tropical fish and invertebrates were the most obvious new organisms during the anomaly, but during El Niño, the coastal waters supported phytoplankton assemblages characteristic of low latitude gyres (Ochoa et al. 1985; Rojas de Mendiola et al. 1985). As shown in Fig. 9.4, there was a twentyfold reduction in surface primary productivity and phytoplankton standing crop during May 1983, when the tropical assemblage was present, but assimilation numbers in 1983 and 1984 were similar at 6.5 mgC mgChl^{-1}hr^{-1}, indicating that the sparse tropical phytoplankton were photosynthesizing at a specific rate equal to the upwelling assemblage.

The collapse of the upwelling food web and the rapid establishment of tropical species in the coastal waters were impressive biological phenomena in the 1982–83 El Niño, but equally impressive was the immediate reestablishment of a highly productive phytoplankton community of typical upwelling species when the anomaly suddenly ended in mid-July 1983. Fig. 9.10 shows that by July 20, 1983, classical coastal upwelling temperature, nutrient, and chlorophyll profiles with extremely high values next to the coast were reestablished at the 5°S upwelling center. The surface and 60 m temperature records in Fig. 9.11 indicate that while the warm anomaly slowly increased in strength from September 1982 to May 1983, the cooling was fast, with the normal upwelling conditions rapidly appearing in July and August 1983, along the coast of Peru.

Rapid transitions of the pelagic community in this region are not unique to the strong 1982–83 El Niño. At the onset of the 1976 El Niño, a sudden bloom of the dinoflagellate phytoplankter, *Gymnodinium splendens*, took place in the coastal waters along Peru when a sudden warming of the nutrient-rich surface layer increased the static stability (Barber and Chavez 1983). Anecdotal accounts from 1972 and 1976 suggested that other biological consequences of El Niño showed up almost simultaneously along the entire Peru coast, persisted for months, and then rapidly disappeared. The relationship of productivity and species distribution to a global atmospheric oscillation is an emergent property in the sense used by Odum (1977). The existence of this new property requires that we acknowledge a basinwide unit of biological and physical organization, that is, a basinwide ocean ecosystem which determines the trophic structure and material cycles of the ocean basin.

9.5. What is the Utility of Understanding the Ocean Basin Ecosystem?

Determining emergent properties of a system as large as the basinwide pelagic ecosystem is a difficult undertaking. The spatial domain of even a small ocean basin, such as the North Atlantic, is difficult to resolve with slow ships and

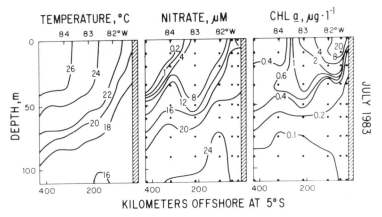

Figure 9.10. Profiles of temperature, nitrate, and chlorophyll along a 5°S latitude section that extends offshore 400 km from Paita, Peru, on July 20, 1983, one month after the June 1983 temperature survey shown in Figure 9.5. The June 1983 survey showed warm El Niño conditions in the eastern Pacific; the July 1983 survey found cool, nutrient-rich, high phytoplankton biomass in the coastal upwelling region adjacent to the Peruvian coast. The rapid reestablishment of strong coastal upwelling, with its characteristic high nutrient and chlorophyll concentrations, emphasizes the rapid transitions that occur in pelagic communities when the basinwide physical regime changes. From Barber and Chavez, Vol. 222, No. 4629, pp. 1203–1210, © 1983 by A. A. A. S.

scarce satellites. Are there substantive benefits for society that will justify efforts to achieve a new level of understanding about how the ocean ecosystem functions? One well-defined benefit, elaborated by Mann (Chapter 15), deals with traditional fisheries management. For the great majority of ocean species, fish stocks are regional in nature. The traditional approach to management of these regional resources has involved analysis of the component populations and the local environment. It has become clear that ecosystem properties, especially physical conditions, operating on scales much larger than the range of the population, may be responsible for changes in reproductive success and adult abundance (Cushing 1982 and Fig.s 9.2 and 9.3). In this context, a benefit of understanding the basinwide ecosystem will be an improved ability to predict local variations in living resources such as those shown in Fig. 9.11. This benefit, while important to resource managers, is probably not the major societal benefit that will accrue from a more accurate understanding of the ocean ecosystem.

Odum (1977) said, "It is the properties of the large-scale integrated systems that hold solutions to most of the long-range problems of society." While few of us in the scientific community recognized the wisdom of this comment a decade ago, the validity of Odum's prediction has been demonstrated. Acid rain and carbon dioxide modification of the planet's climate are examples of problems that can be solved only if the emergent properties of large-scale biogeochemical integration are understood. For both of these issues, the lack

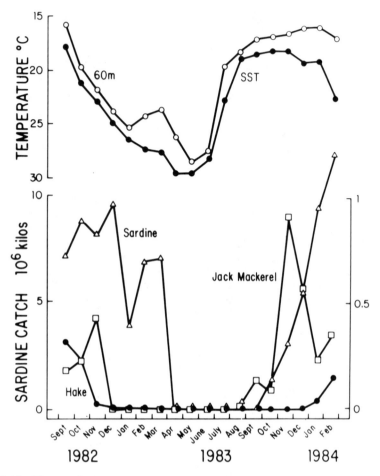

Figure 9.11. Changes in ocean temperatures and the monthly catch of three important commercial species at Paita, Peru (5°S) during the 1982–83 El Niño. The temperature scale is inverted, and shows both the surface and 60 m temperature. Each fish species responded differently to the habitat changes that took place as the 1982–83 El Niño developed. Note that in July 1983, cool, upwelling conditions were rapidly reestablished (see Fig. 9.10), but hake, sardine, and jack mackerel did not return in large numbers for several months. The response of higher trophic level organisms to El Niño is distinct from that of the phytoplankton and zooplankton. See Arntz et al. (1985) and Barber and Chavez (1986) for further discussion of the biological responses to El Niño. From Barber and Chavez, Vol. 222, No. 4629, pp. 1203–1210, © 1983 by A. A. A. S.

of consensus in the scientific community about the nature of higher order integration has been used by the U.S. government as a reason for taking no action. Somewhat ironically, the U.S. government is making a very strong case for the need to understand the ecosystem.

Odum has emphasized repeatedly that ecology and economics are parallel and closely related disciplines, and expressed regret that they are not perceived as such, particularly by the economic community (Odum 1975). Resource development provides a good example of how ecosystem understanding relates to economics. A fisheries manager or a fisherman needs information on where the fish are now, and how many there will be next year; but if we move up one hierarchical level in the social structure the needs for information change. Investors and political leaders making policy decisions on resource development need an understanding of the probability, frequency, and intensity of resource variability for realistic economic decision making. In this context, economics and the ecosystem concept are closely related; understanding the variability of the ocean ecosystem will show economic decision makers that the variation of living marine resources is a normal and inevitable characteristic that must be accommodated in economic plans.

Recognition of the inevitability of natural fluctuations, such as the ENSO cycle, would in itself be a large step toward improving economic development related to natural resources. Odum, in his seminal 1969 paper, "The Strategy of Ecosystem Development," provided a conceptual plan for "resolving man's conflict with nature by recognizing the central role played by succession." That plan is indeed appropriate for terrestrial affairs, where succession is of paramount importance, but for harmonious ocean affairs, we need to accommodate variability. I suggest that Odum now write a corollary paper on "The Strategy of Ecosystem Variability" to contribute to the resolution of nature's conflicts with man that arise from the natural variability of the ocean.

10. The Cycling of Essential Elements in Coral Reefs

Christopher F. D'Elia

When the Odum brothers arrived at Enewetak in 1954 to do the research that resulted in their widely cited paper on the trophic structure and productivity of a windward coral reef (Odum and Odum 1955), both ecosystem science and coral reef ecology were in their infancy. Thanks to the Odums' extraordinary efforts during their short sojourn in the Marshall Islands, and to their prolific careers, we have seen a marked growth in interest in ecosystem science and in the study of reef ecology. In that context, this chapter considers the role of ecosystem science in assessing the cycling of essential elements in coral reefs.

Ecosystem science is the systematic study of the structure and function of an ecosystem. Such study normally involves the measurement of metabolic features such as the energetics and nutrient fluxes of the ecosystem and is typically practiced in an observational, not an experimental fashion (cf. Chapter 1). Essential to the paradigm of the ecosystem according to E. P. Odum (1971b) is the concept that there are hierarchies of organization in ecological systems, and that each level displays emergent properties not characteristic of lower levels of organization. This concept is often termed "holism." It follows that only holistic approaches are capable of identifying emergent properties at a given level of organization, for such properties could not be deduced from interactions of components at lower organizational levels (Odum 1979; Mann 1982a). These properties would not be evident using the traditional reductionistic approach of much of today's science. In essence, the notion that emergent properties exist implies that the whole, at a given level of organization, is greater than the sum of the parts. However, we rarely see specific examples of emergent properties evaluated in the scientific literature. The ecosystem approach of examining system functional and structural integration forms a basis for derivative "reductionistic"[1] studies at lower levels in the organizational hierarchy. In my opinion, such reductionistic studies are necessary, but not sufficient, for understanding specific processes, mechanisms, and structural attributes that contribute to system integration at higher levels. However, a dynamic tension between practitioners of holistic and reductionistic approaches often interferes with communication between the two groups (cf. Patten and Odum 1981).

The present understanding of the dynamics of essential nutrient elements in coral reefs has traditionally drawn from a suite of techniques employing the holistic approaches of ecosystem science and biological oceanography, as well as the more traditional reductionistic (to the ecologist at least) approaches of organismal biology, physiology, and biochemistry. I will devote particular attention to studies aimed at the synecological level that should help identify such emergent properties at the ecosystem level of organization, and then proceed to the autecological level. Although I am reviewing what is known about the dynamics of essential elements on coral reefs, I do it from the perspective of this volume, which hopes to evaluate the validity of the ecosystem approach. Accordingly, much of the focus here is on the holistic level.

In the course of reviewing the literature germane to this paper, three things became apparent: (1) the principal pathways of cycling of essential elements on coral reefs are neither well defined nor well understood; (2) nutrient related processes on reefs vary in time and space much more than is widely appreciated; and (3) few, if any, attempts have been made to identify emergent properties of coral reef nutrient cycling as such. Although advanced mathematical system theories are being developed for ecosystems, which could help greatly in defining important features of nutrient cycling and energetics, application of these theories has primarily focused on systems simpler than coral reefs (Ulanowicz 1986).

10.1. General Features of Coral Reefs

Odum and Odum (1955) soon discovered in their Enewetak study that coral reefs are among the most productive of all ecosystems. Coral reef primary productivity is considered in detail in Chapter 11. Suffice it to say here that while the gross primary productivity of coral reefs is high, net community primary productivity is low. The widely drawn inference from this high gross, but low net primary productivity, is that the essential elements supporting productivity must cycle tightly; net diurnal production is thus offset by net nocturnal consumption with nutrients being repeatedly assimilated and re-

---

[1] Reductionism may be defined alternatively as the antonym to holism, as used here, or in the sense of reducing complex information or processes to simple terms.

generated. Most studies addressing coral reef nutrient productivity relationships have been conducted on reefs in the Pacific distant from the influence of land masses. Whether such sites are the norm or the exception is uncertain, and much effort could be expended arguing about what constitutes a "typical" coral reef.

The Odums' Japtan reef study site at Enewetak was the site of a subsequent expedition, Symbios (Johannes et al. 1972), and shares many features with other Pacific sites studied (Fig. 10.1). Water flow is wave driven over the reef in a unidirectional fashion. Distinct physiographic zones from windward to leeward include the windward buttress zone with extensive coral cover, the algal ridge, an encrusting zone, a zone of smaller coral heads, a zone of larger heads, a zone of sand and shingle, and the leeward lagoon. Kinsey (1983b) has emphasized how differently the various zones of coral reefs function and has warned that one of the most metabolically active areas of reefs, the "reef flat" (an area of intense macroalgal productivity), is often viewed as synonymous with "coral reef" in the discussion of material fluxes. In fact, the windward buttress zone that is characterized by the most luxuriant coral cover is actually outside the boundaries of most reef studies (Smith and Harrison 1977). Accordingly, coral reefs are in many aspects algal reefs, and the generalizations that apply to "coral" reefs are derived from ecosystems in which macrophytes are important, if not dominant, primary producers.

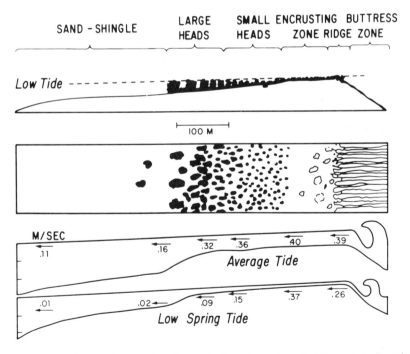

Figure 10.1. Physiographic zonation for a windward coral reef at Enewetak. Reprinted by permission from Odum and Odum (1955).

10.2. Coral Reefs vs. Other Shallow Water Marine Systems

Table 10.1 is a crude attempt to compare and contrast key features of coral reefs with those of other shallow water marine and estuarine ecosystems (see also Pomeroy 1970 and Johannes and Betzer 1975 for other comparisons). Of the shallow water ecosystems listed in Table 10.1, it is instructive to make a brief comparison between coral reefs and the other two major tropical shallow water ecosystems: seagrass and mangrove communities, of which the former is often in proximity to coral reefs. Temperate salt marshes are also included. A key to why these communities are as they are may rest in the prevailing nutrient regimes they encounter as well as other obvious considerations such as water depth, sediment type, and salinity. To my knowledge, there have been no extensive attempts to classify these shallow water communities according to nutrient regimes, although such an attempt may prove worthwhile. Vernberg (1981) has discussed the need for more such comparative studies to improve our understanding of ecosystem dynamics and how they respond to stress.

Unlike their temperate, shallow water, marine counterparts, coral reefs are characterized by high levels of benthic, not pelagic, productivity. Nutrient concentrations in coral reef environments are often vanishingly low—typically an order of magnitude less than for shallow water temperate ecosystems. Nonetheless, concentrations vary more widely than is often perceived. Table 10.2 provides selected examples of nutrient concentrations at a variety of coral reef

Table 10.1. Characteristics of Selected Shallow Water Marine and Estuarine Ecosystems That Relate to the Cycling of Nutrient Elements (the Comparison is Necessarily Subjective and Generalized)

Feature	Relative Amount or Rate			
	Coral Reef	Sea Grass	Mangrove	Salt Marsh
Benthic productivity				
Gross	High	Moderate–high	High	High
Net	Low	Moderate	High	High
Water column productivity				
Gross	Low	Low	Variable	Variable
Net	Low	Low	Variable	Variable
Biomass	Moderate	Moderate–high	High	Moderate–high
Optimal nutrient				
Concentration	Low	Low	Variable	High
Inputs	Low–moderate	Low–moderate	Moderate–high	Moderate–high
Recycling	High	Moderate	Low	Moderate
Resistance to overenrichment	Low	Moderate	High	High
Water clarity	High	High	Variable	Variable

sites around the world. The extremely low nutrient concentrations limit pelagic net productivity and phytoplankton standing stocks; accordingly, phytoplankton biomass is low and water clarity is high.

Despite the low concentrations of nutrients in waters impinging on coral reef ecosystems, large amounts of nutrients may be conserved or supplied to them by a variety of internal mechanisms. For example, coral reefs are typically characterized as sites of high rates of nitrogen fixation (Capone 1983) and nutrient cycling (Johannes et al. 1972). Differences in nutrient supply and retention rates are largely unquantified, but they represent a potentially important determinant of productivity and community structure. Few studies of any shallow water marine tropical system have attempted to construct the nutrient and energy budgets necessary to evaluate nutrient-productivity relationships. Table 10.3 lists major factors affecting the rate of allochthonous nutrient inputs to shallow water benthic systems that should be considered in developing such budgets. While we might surmise that differences in allochthonous nutrient inputs substantially affect the productivity of coral reefs and seagrasses, few studies have documented differences in productivity between tropical benthic systems with different inputs (Zieman 1975; Odum and Johannes 1975). Clearly, such comparative studies would be worthwhile.

10.3. Major Sources and Sinks of Nutrients

Nutrients supporting productivity on coral reefs must ultimately come from one or more of the sources listed in Table 10.3. These sources are considered in greater detail below.

10.3.1. Advection

Although the concentrations of dissolved nutrients in seawater impinging on reefs are typically low, the advection of water to the reef has the potential of delivering large quantities of dissolved nutrients simply as a function of volume transport multiplied by the concentration (Atkinson 1981; Hatcher 1985). Accordingly, autotrophic organisms with high transport affinities for nutrients at low concentrations would be expected to flourish in such an environment. Conversely, autotrophs without such capabilities would be expected to be poor competitors. Similarly, reefs may also receive nutrients in the form of particulate materials advected to them (e.g., Glynn 1973), but few quantitative studies exist.

10.3.2. Upwelling and Endo-upwelling

Elevated concentrations of nutrients in some places may be the result of equatorial upwelling (e.g., Smith and Jokiel 1976; Kimmerer and Walsh 1981). Reefs at other latitudes may also experience upwelling. Andrews and Gentien (1982) have shown that nutrient-rich water from the East Australian Current is upwelled to the continental shelf with a period of 90 days and is advected inshore

Table 10.2. Selected Examples of Mean Surface Oceanic Nutrient Concentrations Impinging on Different Coral Reef Sites Throughout the World[a]

Site	Season	NH_4^+	NO_3^-	NO_2^-	PO_4^{3-}	DON	DOP	H_4SiO_4	Comments	Source
Canton Atoll, Phoenix Islands	Summer	1.5	2.5	NA	0.6	NA	NA	2.3		Smith and Jokiel (1976)
Discovery Bay, Jamaica	Mean, all seasons	0.2	0.26	0.13	0.2	21.4	NA	3.2		D'Elia et al. (1981) D'Elia (unpublished)
Enewetok, Marshall Islands	Spring	0.240	0.109	NA	0.174	1.79	0.152	NA		Johannes et al. (1972)
Great Barrier Reef	Summer	<0.2	0.47	< 0.02	0.31	NA	NA	1.5		Risk and Müller (1983)
Great Barrier Reef	Winter	0.32	0.54	0.16	0.26	6.0	NA	1.30	DON uncertain	Crossland and Barnes (1983)
Great Sound, Bermuda	Mean, all seasons	1.0	<0.5	NA	0.025	NA	NA	1.1		Morris et al. (1977)
Guam	Mean, all seasons	NA	0.86	NA	0.18	NA	NA	NA		Marsh (1977)

Location	Season								Notes	Reference
Houtman Abrolhos Islands	Winter	0.296	1.17	0.037	0.375	10.7	0.35	2.32		Johannes et al. (1983b)
	Summer	0.024	0.9	0.058	0.21	8.7	0.36	2.21		"
Kavaratti Atoll, Lakshadweep Archipelago		1.74	0.46	NA	0.34	NA	NA	NA		Wafar et al. (1985)
Peros, Banhos	Summer	NA	0.40	NA	0.43	NA	NA	NA		Johannes et al. (1983b)
Salomon Atolls	Summer	NA	0.98	NA	0.43	NA	NA	NA		Rayner and Drew (1984)
Sesoko Island, Okinawa	Spring	NA	<0.3	<0.1	<0.7	60–70	<0.3	1.2	DON uncertain	Crossland (1982)
St. Croix, Virgin Islands	Mean, all seasons	NA	0.283	NA	0.08	NA	NA	NA		Adey and Steneck (1985)
Tarawa Lagoon, Rep. of Kirabati	Fall	0.34	2.68	NA	0.33	4.58	NA	NA		Kimmerer and Walsh (1981)

[a] All concentrations given in μmol/L and with number of decimal places reported by authors.
NA = not available; ND = not detectable.

Table 10.3. Factors That Determine the External Rate of Nutrient
Inputs to Shallow Water Benthic Systems

Factors Promoting Low Rates of Allochthonous Nutrient Inputs
Proximity to low islands
Proximity to dry islands
Low advection of sea water

Factors Promoting High Rates of Allochthonous Nutrient Inputs
Proximity to high islands
Proximity to wet islands
High groundwater inputs
High anthropogenic inputs
High advection of sea water
Proximity to upwelling of nutrient-rich water

in a bottom Ekman layer. These authors speculate that nutrient elements so derived may constitute a substantial factor supporting the productivity of the Great Barrier Reef Shelf. In a recent paper, Rougerie and Wauthy (1986) proposed that "geothermal endo-upwelling" occurs in some atolls. These authors characterized the concept as a "vertical ascent of deep ocean water, driven by geothermal heat flow, through the atoll's internal structure; nutrient-rich upwelled water supplies the reef-building biocenosis at the surface." Endo-upwelling could supply allochthonous nutrients that counterbalance exports from reefs. This topic is largely unstudied; accordingly, the relative importance of endo-upwelling awaits further evaluation.

10.3.3. Migrations of Large Organisms

To paraphrase Pomeroy (1970), large organisms are usually a result of productive conditions rather than a cause of them. However, within reef ecosystems, large grazers like fishes and sea urchins may be responsible for the displacement of nutrients from one area to another (see review by Ogden and Lobel 1978), and grazing may affect turf algal community structure and detrital particle sizes in a way that can potentially affect important food webs and nutrient transformations such as nitrogen fixation (Wilkinson and Sammarco 1983). Indeed, coral reefs may be systems in which large herbivorous organisms play a "much more important role . . . than in almost any other system" (Mann 1982a, p. 178). In an interesting series of papers, Meyer et al. (1983) and Meyer and Schultz (1985a, 1985b) have examined the effects of fish schools on local nutrient enrichment and the growth rate of corals. These authors found that grunt schools doubled the amount of ammonium available to corals and affected the tissue nitrogen content and growth rate of the coral *Porites furcata*. Rates measured varied with both season and time of day.

10.3.4. Seabirds

Guano deposits have also been proposed as potential sources of nutrient enrichment for coral reefs, although I am unaware of any detailed quantitative study of the phenomenon. Allaway and Ashford (1984) have shown that seabird droppings can account for substantial nutrient inputs to islands in the vicinity of coral reefs and speculated that much of the phosphorus is retained in a slowly exchanging soil compartment; rates of delivery of nutrients to coral reef areas were not assessed. Carsin et al. (1985) hypothesized that Clipperton Lagoon is fertilized by guano—enough to cause symptoms of overfertilization. Ancient phosphorus deposits (i.e., phosphorites) exist near reefs (Veeh 1985) and may serve as sedimentary phosphorus reserves, but are probably rarely significant nutrient sources to reefs.

10.3.5. Groundwater, Precipitation, and Runoff

Marsh (1977), D'Elia et al. (1981), Crossland (1982), and Lewis (1985) have observed significant groundwater inputs in the vicinity of coral reefs. Groundwaters in most carbonate sediments tend to be nitrogen-rich but virtually devoid of P, although Lewis (1985) did observe significant P in groundwater at one site in Barbados. Lewis (1987) has presented limited quantitative information on the enrichment effect of groundwater discharges near coral reefs.

Like groundwater, rainwater and runoff from the land potentially deliver important nutrient inputs to reefs, particularly near high islands in regions with high rainfall (Marsh 1977). Such inputs would typically be expected to be a rich source of nitrogen, but not of phosphorus.

10.3.6. Burial, Resuspension, and Diffusion

Ultimately, all nutrients accumulated by coral reefs must either be carried away by advective flux or be buried in the sediments. Substantially higher nutrient concentrations are contained in reef sediments than in the water (e.g., Entsch et al. 1983a; Williams et al. 1985b).

High concentrations of nutrients have been measured in coral reef sediments (Entsch et al. 1983a) and coral heads (Risk and Müller 1983). Unfortunately, estimates of *rates* and the relative importance of regeneration of nutrients from such sources are lacking. In any case, these nutrients did not arise de novo—that nutrients are buried implies net input over the long term such as occurs in phosphorite deposits (Veeh 1985). One cannot invoke diffusion or resuspension irrespective of net fluxes as a mechanism to drive net productivity.

10.4. Other Factors Affecting Nutrient Availability

In addition to the fluxes that comprise the sources and sinks of nutrients, physical features of some environments either trap, concentrate, or dilute nutrients in waters near coral reefs, and thereby affect their concentration and availability.

10.4.1. Structure and Residence Time

The morphology and flushing characteristics of atolls, microatolls, and patch reefs may greatly affect the accumulation of nutrients, especially in lagoons

(Hatcher and Frith 1985). For example, Ricard and Delesalle (1981) found that Scilly Lagoon exhibited flushing characteristics related to morphological features that trap materials and account for elevated concentrations of particulate matter in comparison to oceanic waters. Welsh et al. (1979) concluded that Whalebone Bay and other coastal Bermuda waters were nutrient sinks. Smith (1984) has hypothesized that water residence time over coral reefs is also an important determinant of gross system nutrient dynamics to the extent that it affects nutrient limitation at the ecosystem level.

10.4.2. Net Evaporation and Precipitation

Particularly in atolls like Canton Atoll, where equatorial upwelling provides a source of nutrients, and where confined lagoon waters exchange poorly with the open ocean, nutrients may be concentrated slightly when evaporative losses exceed rainwater inputs (Smith and Jokiel 1976; Kimmerer and Walsh 1981). Conversely, in upwelling regions with high precipitation, dilution by precipitation can also affect nutrient concentrations (Smith 1984). Although changes in nutrient concentration from evaporation and precipitation do occur, such effects are small, being limited to the proportional change in salinity observed between the lagoon and the ocean. Accordingly, biological and chemical exchanges with the benthos usually alter nutrient concentrations more significantly. Smith and coworkers have used such salinity changes to estimate water residence time and nutrient fluxes. Fig. 10.2 gives an example of the evaporative concentration combined with utilization of phosphorus in Canton Atoll.

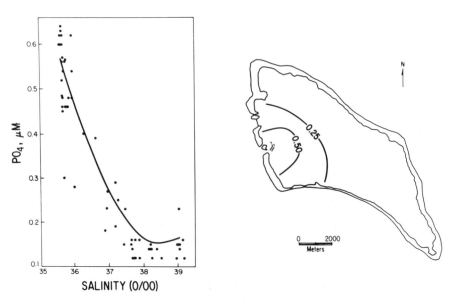

Figure 10.2. Salinity isopleths and phosphate vs. salinity plots for the lagoon of Canton Atoll. From Smith and Jokiel (1976).

10.5. Approaches to Understanding Nutrient Cycling

Research approaches to understanding ecosystems, like the subjects themselves, are often hierarchically structured. As stated above, holists and reductionists often strongly disagree about approaches to comprehending different levels of organization, and clearly such dispute exists in the coral reef research community. Some of the approaches used to understand the cycling of essential elements in coral reefs, and the major results of studies following such approaches, will be discussed in this section.

10.5.1. Synecology and Holistic Truths

Eulerian Measurements

A characteristic of many metabolically active "coral reefs" throughout the world is a strong wave driven or current driven flow of water across the reef. Sargent and Austin (1949) were the first to take advantage of this characteristic and apply what might be considered the "Eulerian" approach to the measurement of reef community metabolism. Eulerian measurements establish the flux of materials by taking discrete samples at given boundaries, and thereby allow the calculation of net changes that have occurred between those boundaries. This approach, often referred to as "flow respirometry," uses changes in concentration of nutrients, oxygen, or other parameters in water masses as they move across a reef to estimate the fluxes between the sediments and water column (see review by Marsh and Smith 1978). Planktonic processes are assumed to have an inconsequential effect on these parameters during the transit of a water mass. Results from such studies have been used to develop a holistic understanding of coral reef community nutrient processing, because the net fluxes of nutrient elements between the reef and water column are determined.

The majority of flow respirometric studies have been made in the Pacific. Such studies have provided much information about nutrient fluxes and transformations on windward atoll coral reefs in particular. The Odums' (1955) study at Enewetak is perhaps the best known and most widely cited of these studies. They measured the metabolism of the intact reef by monitoring oxygen changes in the water over the reef and constructed a detailed energy budget for the system; although they measured nutrient concentrations over the reef, they did not attempt to measure the nutrient flux.

Participants on the Symbios Expedition to Enewetak, the first major expedition to study both coral reef productivity and nutrient fluxes, focused on nutrient fluxes on the same transect the Odums had used in 1954 (Johannes et al. 1972). The Symbios team unexpectedly observed a net export of dissolved nitrogen from the reef to the water column in the forms of nitrate and dissolved organic nitrogen (Webb et al. 1975). That observation could only be plausibly explained by non-steady state phenomena, endo-upwelling (Rougerie and Wauthy 1986), or high rates of nitrogen fixation (Wiebe 1976), of which the latter seemed most likely. It remained for experiments at the organismal level to verify that nitrogen fixation did indeed occur and to identify organisms responsible for it (Wiebe et al. 1975). The net export of nitrogen prompted Webb et al. (1975) to characterize the nitrogen cycle of the reef at Enewetak as an "open" one, not characteristic of mature ecosystems (cf. Odum 1969), which tend to retain nutrients. Similar results were also found in a subsequent expedition to a site on the Great Barrier Reef (LIMER 1976).

In a complementary study on phosphorus fluxes at Enewetak, Pilson and Betzer (1973) found very little exchange of P between the benthos and the water column (Fig. 10.3A). No discernible relationship existed between short-term oxygen production or consumption rates and the uptake of dissolved inorganic ("reactive") phosphorus or release of organic phosphorus for the two transects studied (Fig. 10.3B). Atkinson (1981), who collected data over an entire year at the Kaneohe Bay (Hawaii) barrier reef system, was able to show a significant net uptake of phosphorus by the community, which he used to estimate net primary productivity, assuming a C/P ratio of 640 that he measured for reef autotrophs (Fig. 10.4). This suggested that the stoichiometric requirement of reef organisms for nutrients is substantially less than expected from the Redfield (1934) ratios (cf. Chapter 16). Henderson (1981) found similar results in microcosm studies of reef flat communities in Hawaii. Atkinson (1983) has emphasized that there is no a priori reason why the uptake of C and P should be closely coupled over the short term and that only long-term measurements could be expected to yield estimates of net productivity.

Until recently, seasonal variability in function of coral reefs has been overlooked because of a widespread perception that tropical communities are aseasonal, and because few centers of ecological investigation are located near coral reefs so as to perform continuous studies in which such variability would become apparent. We now know that higher latitude reefs may undergo striking seasonal variations in primary productivity (Kinsey 1977) and nutrient dynamics (Hatcher and Hatcher 1981), variations that seem unexpected on the basis of temperature fluctuations alone and probably relate to insolation and variations in oceanic water masses. Similarly, Johannes et al. (1983a) observed different patterns of nutrient flux that may be linked to seasonal differences in nutrient regime. This seasonality greatly complicates the interpretation of flow respirometry results and should be accounted for in future studies.

Johannes et al. (1983a) observed that dissolved inorganic nitrogen (DIN) concentrations impinging on reefs vary with season. Using the flow–respirometric approach, they reported three patterns of nutrient flux: (1) concentration-dependent; (2) diel cycle, not concentration dependent; and (3) neither. They observed a net DIN uptake in the summer, but not in the winter. Fig. 10.5 shows the concentration dependence they observed during the summer period. This is the only report I know of which shows concentration dependent nutrient uptake by an entire benthic community.

Even low latitude coral reefs may display more seasonality than we might expect, owing to possible seasonal changes in nutrient inputs. Atolls near regions of equatorial upwelling are known to have considerably higher nutrient regimes than those at higher latitudes (Smith and Jokiel 1976; Kimmerer and Walsh 1981). It is well known that the strength of upwelling varies with season (and

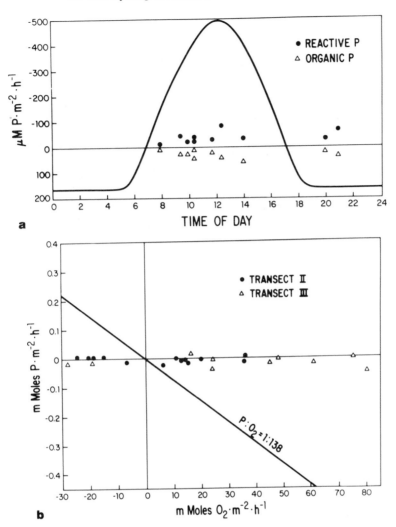

Figure 10.3. a. Productivity, respiration, and flux of phosphorus into and out of the water flowing across a windward reef transect at Enewetak. Reprinted by permission from Pilson and Betzer (1973). b. Relationship between oxygen fluxes and phosphorus fluxes at Enewetak. Phosphorus fluxes are not as would be expected from Redfield ratio. Reprinted by permission from Pilson and Betzer (1983).

indeed with year). However, not enough is known about the coupling of nutrients with reef productivity to predict what effects variations in the strength of upwelling have on productivity. Johannes et al. (1983b) have speculated that nutrient-rich upwelled waters may, in fact, be deleterious to coral reefs by causing enhanced growth rates of benthic macroalgae that compete with corals. Moreover, Kinsey and Davies (1979b) have postulated that elevated

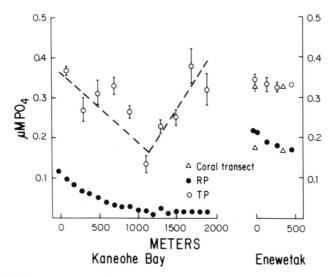

Figure 10.4. Phosphorus concentrations vs. distance down a reef transect in Kaneohe Bay, Hawaii and Enewetak, Marshall Islands. Reprinted by permission from Atkinson (1982).

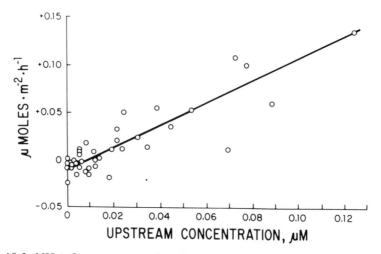

Figure 10.5. NH_4^+ flux over a coral reef transect at Houtman Abrolhos Islands vs. upstream NH_4^+ concentration. From Johannes et al. (1983a).

nutrient levels during the first half of the Holocene transgression suppressed calcification by phosphate crystal poisoning (cf. Simkiss 1964).

Glynn (1983) has hypothesized that El Niño events may influence some coral reefs. Although there are no detailed studies of the effect of El Niño on reef-living resources and nutrient supplies as for other ecosystems (Barber, Chapter 9), such events may affect the delivery of nutrients to coral reefs.

Lagrangian Measurements

Recently, D. J. Barnes (1983) has developed what might be termed a Lagrangian approach to the measurement of productivity over reef transects. Barnes's approach follows a given water mass and allows inferences about fluxes that occur in the transit of an instrumented drogue containing appropriate sensors for O_2 and pH. Fig. 10.6 presents an example of data obtained from a deployment of this drogue. Barnes found three metabolic zones within a transect on the Great Barrier Reef: (1) a windward zone of moderate productivity and moderate to low calcification; (2) a central zone, with low productivity and low calcification; and (3) a leeward zone with high productivity and high calcification. While similar instrumentation for nutrient determination has not been developed, zone to zone differences in nutrient flux will undoubtedly be found when such instrumentation becomes available.

In theory, Barnes's instrumented drogue enables one to make more refined determinations of the function of different components of reefs than can be made by setting arbitrary boundaries. His approach maximizes the possibility of discerning different rates of function from different zones of reefs.

As I indicated previously, Smith and coworkers have taken advantage of net evaporation or precipitation in confined water bodies to estimate residence time and flux related concentration changes in nutrients (Smith and Jokiel 1976; Smith 1984; Smith et al. 1984). This approach, which is in essence

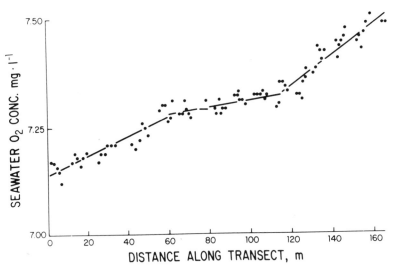

Figure 10.6. Change in oxygen concentration with distance along a transect on a reef flat recorded by an oxygen probe on an instrument drogue. From Barnes (1983).

Lagrangian, allows one to calculate elemental mass balances on the time scale of the water residence time. Accordingly, it integrates over longer periods of time than are possible with the flow respirometric or the drogue approaches, but it does not enable equivalent fine scale spatial resolution. The approach is an important one that will inevitably see greater future application.

Others (e.g., Remote Sensing)

The technique of remote sensing has not seen wide application to coral reefs. In any case, its applicability to nutrient cycling and dynamics can only be indirect, such as by determining high areas of productivity or in locating up-welling induced sea–surface temperature anomalies.

10.5.2. Autecology and Reductionistic Realities

The reductionistic approach also has its place in ecosystem science. Although Odum and Odum (1955) were able to learn that gross productivity was high, that net productivity was low, and that coral reef nutrient conservation and acquisition mechanisms were specialized, physiological experiments at the organismal level of organization have been necessary to understand what roles individual organisms play and to identify what the specialized features of a given reef are. For example, the Odums were apparently incorrect in believing that the chlorophyte, *Ostreobium*, which is found in bands in coral skeletons, was important, and while they recognized that zooxanthellae have specialized characteristics for nutrient metabolism, they could not infer much about their trophic role in the coral reef ecosystem.

The reductionistic arguments that total system function reflects, in some sense, the component functions, is not without merit. Accordingly, the reductionist often finds himself doing the important "dirty work" of determining what the holists could not determine in studies at higher levels of organization. A recent review by Gladfelter (1985) stresses some of the recent developments in research at the organism level. She states that by knowing, for example, the mechanisms by which key factors such as light or nutrients limit productivity, we may gain a predictive advantage that is now unavailable.

10.5.3. Nutrient Fluxes of Corals

Reef corals themselves have been the primary focus of most autecological studies of nutrient exchange on coral reefs. Thus, such studies serve to illustrate the application of the autecological approach and the complexity involved in trying to understand nutrient exchange from the reef benthos.

The earliest studies of coral nutrient fluxes were conducted during the Great Barrier Reef Expedition of the 1920s. Yonge and Nicholls (1931) found that coral animals that contain endozoic symbiotic dinoflagellates, known as "zooxanthellae," were capable of removing phosphate from seawater, even at very low concentrations. Kawaguti (1953) found that corals with zooxanthellae could remove ammonia from seawater at low concentrations. Concern about experimental methods in such early studies caused subsequent skepticism about their results and great uncertainty about whether the coral–zooxanthellae symbiosis displayed autotrophic or heterotrophic characteristics (Pomeroy and Kuenzler 1969).

Research in the last three decades has conclusively confirmed both the photoautotrophic nature of corals (Muscatine 1980a) and the Yonge and Nicholls (1931) findings that coral–zooxanthellae associations behave as autotrophs in removing net quantities of dissolved nutrients from solution, even at the extremely low concentrations characteristic of coral reefs (Franzisket 1973; D'Elia 1977; Muscatine and D'Elia 1978; Szmant-Froelich et al. 1981; Webb and Wiebe 1978; Pomeroy et al. 1974). Moreover, we now have extensive information on the kinetics of dissolved nutrient uptake by reef corals (reviewed by Muscatine 1980b). Fig. 10.7, drawn from Muscatine (1980b), shows the major patterns of uptake kinetics found for reef corals.

D'Elia et al. (1983) compared nutrient uptake curves for an even lower hierarchical level: that of isolated zooxanthellae. They found a close comparison between the fluxes of isolated algae and those of intact associations (Fig. 10.8). This implies that the zooxanthellae enable corals to appear as functional autotrophs in at least some cases.

Despite the specialized photoautotrophic and nutrient uptake capabilities of coral–zooxanthellae associations, we still are uncertain about the relative importance of photoautotrophic and heterotrophic modes of nutrient acquisition for corals in nature. The most attractive explanation has been proposed by Porter (1976), who has suggested that corals with large polyps and a low surface to volume ratio are adapted to zooplankton capture, whereas corals with small polyps and a high surface to volume ratio are adapted to photoautotrophy. This hypothesis awaits further testing.

10.6. Nitrogen Cycle

Autecological studies of processes and organisms involved in the nitrogen cycle have been useful in understanding some aspects of the net benthic elemental fluxes measured in flow respirometry studies. Nonetheless, much remains to be done to quantify different pathways and to establish detailed nitrogen budgets for reefs. Fig. 10.9, taken from Fenchel and Blackburn (1979), shows the major components of the nitrogen cycle. These components are individually discussed below.

10.6.1. Nitrogen Fixation

Probably the most significant discovery about coral reefs in the last two decades has been that nitrogen fixation occurs at rates uncharacteristically high for a marine ecosystem (Capone 1983). Nitrogen fixation is now known to be a key feature of the nitrogen cycle of most coral reefs, regardless of location (Wiebe et al. 1975; Mague and Holm-Hansen 1975; Burris 1976; Hanson and Gun-

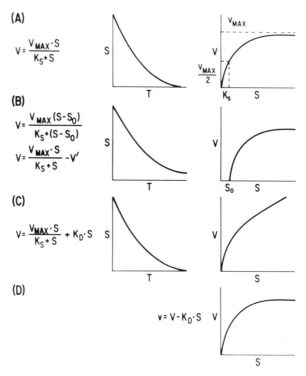

Figure 10.7. Major kinetic equations and curves applicable to nutrient uptake by intact coral/zooxanthellae symbiotic associations (from Muscatine 1980b). Equations, depletion and uptake curves for (A) Theoretical Michaelis–Menten equation; (B) Michaelis–Menten equation modified for threshold (NO_3^-) or bidirectional fluxes (PO_4^{3-}); (C) Michaelis–Menten equation modified for diffusional component (NH^{4+}). Reprinted from "Uptake, retention, and release of dissolved inorganic nutrients by marine alga-invertebrate associations," by L. Muscatine, in *Cellular Interactions in Symbiosis and Parasitism*, edited by Clayton B. Cook, Peter W. Pappas, and Emanuel D. Rudolph. © 1980 by the Ohio State University Press. All rights reserved.

dersen 1977; LIMER 1976; Potts and Whitton 1977; Wilkinson et al. 1984; Goldner 1980; see also reviews by Paerl et al. 1981; Capone 1983). Rates on reefs may approach levels achieved in managed legume agriculture (Wiebe et al. 1975), although activity is spatially variable (Goldner 1980). Capone and Carpenter (1982) estimated that coral reef nitrogen fixation rates reach 25 g · m^{-2} · yr^{-1}, higher than any other marine ecosystem listed (Table 10.4).

   Why such high rates occur on reefs, but not in most other marine ecosystems, is unknown. Corals themselves do not have the capacity to fix nitrogen, although endolithic organisms in coral skeletons probably can (Crossland and Barnes 1976). The autotrophic blue-green alga (cyanobacterium), *Calothrix crustacea*, seems to be a particularly important nitrogen fixer on coral reefs (Wiebe et al. 1975). Recent work has shown that fish grazing may be important

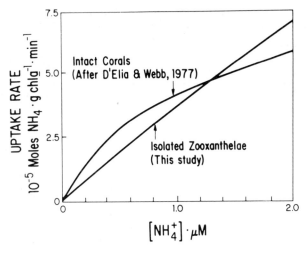

Figure 10.8. Kinetic curves for ammonuim uptake by isolated zooxanthellae composed with curves for intact symbiotic associations. From D'Elia et al. (1983).

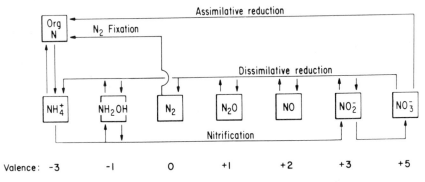

Figure 10.9. Major components of the nitrogen cycle. Reprinted by permission from Fenchel and Blackburn (1979), *Bacteria and Mineral Cycling.* © Academic Press, Harcourt, Brace, Jovanovich.

in maintaining high rates of nitrogen fixation by cropping other benthic algae that compete for space with cyanobacteria (Wilkinson and Sammarco 1983).

The nitrate exported from reefs presumably derives from the nitrogen fixation (Webb et al. 1975), but the major pathways have not been completely elucidated. Webb and Wiebe (1975) have shown that active nitrification occurs, and that the rate is affected by light, but detailed quantitative investigations have not examined the link between nitrogen fixation, the release of fixed N by blue-green algae, and the production of the ammonium substrate for ammonium oxidation. One assumes that N release by nitrogen fixers provides substrate, since light is apparently required for nitrification; this would be

Table 10.4. Estimate of the Total Annual Contribution of Combined
Nitrogen to the Global Nitrogen Cycle by Nitrogen Fixation in the
Benthic Environments of the Ocean[a]

Environment	Area (km^2 × 10^6)	N$_2$ Fixation g N·m^{-2}·yr^{-1}	N$_2$ Fixation Tg N·yr^{-1}
>3000 m	272	0	0
2000–3000 m	31	0.0007	0.022
1000–2000 m	16	0.001	0.016
200–1000 m	16	0.01	0.16
0–200 m	27	0.1	2.7
Bare estuary	1.08	0.4	0.43
Seagrass	0.28	5.5	1.5
Coral reefs	0.11	25	2.8
Salt marsh	0.26	24	6.3
Mangroves	0.13	11	1.5
Total	363		15.4

[a] Reprinted from Capone and Carpenter, © 1982 by the A. A. A. S.

consistent with close coupling between nitrogen-fixing organisms and nitrifying
ones. Another plausible alternative is that the substrates for nitrification result
from the metabolism of organisms feeding on the nitrogen fixers.

Wiebe (1976) has proposed the following schematic representation of the
nitrogen cycle on coral reefs:

The routes by which fixed nitrogen enters the rest of the community have been
summarized by Mann (1982a), and certainly include: (1) the grazing food chain
in which fishes and invertebrates release N in excreta; (2) the detritus food
chain; and (3) the direct release of dissolved inorganic and organic nitrogen
by the algae. The relative importance of these pathways remains unquantified.

10.6.2. Denitrification

Nitrate can serve as the terminal electron acceptor of respiration in a variety
of facultatively anaerobic bacteria and in a few strictly anaerobic bacteria
(Hattori 1983). The nitrate is not used as an N source, hence the products of
what is sometimes termed "dissimilatory" nitrate reduction (i.e., N$_2$ and N$_2$O)
are excreted by the bacterial cells. Because these gases are generally lost to the
atmosphere, this process is usually referred to as *denitrification*.

Denitrification is an important process to ecosystems because it offsets nitrogen fixation: when denitrification exceeds nitrogen fixation, bound dissolved
inorganic nitrogen is lost to the atmosphere as gas. Thus, "net ecosystem

nitrogen fixation" (i.e., nitrogen fixation less denitrification) might be conceived of as analogous to net ecosystem productivity (i.e., gross primary productivity less respiration). Such a rate would seem difficult to derive using autecological techniques.

Initial studies of nitrogen metabolism of coral reefs suggested that denitrification was insignificant because large amounts of nitrate were exported, and the reef surface was viewed as highly aerobic (Wiebe 1976). While flow respirometric studies at Enewetak indicated that nitrogen fixation does appear to exceed denitrification (Webb et al. 1975), the importance of denitrification in coral reef sediments may be greater than was originally thought. Smith (1984) used mass balance calculations to show that significant rates of denitrification probably occur in some reef communities. His synecological approach seems to be an effective way of estimating net ecosystem nitrogen fixation, but comprehensive nutrient budgets for reefs will require more information on the rates of denitrification in different habitats on coral reefs.

Two recent autecological studies have employed the acetylene blockage technique to inhibit N_2 formation and enhance N_2O accumulation as a presumptive indicator of denitrification. Seitzinger and D'Elia (1985) found that the rate of N_2O production could be enhanced in dead coral heads under conditions of acetylene blockage when additional nitrate substrate was supplied. They also found that sediments incubated in NO_3--enriched seawater rapidly consumed this nitrate. Corredor and Capone (1985), using a similar experimental approach, made similar observations, and proposed that close coupling exists between nitrification and denitrification in the reef sediments studied, because measured rates of the two processes were similar and because the nitrate intermediate rarely accumulates. However, neither study experimentally addressed the possibility that assimilatory nitrate reduction consumed nitrate produced by nitrification. Clearly, future studies are needed to compare nitrogen fixation and denitrification rates for a variety of reef environments, so that their relative importance can be assessed.

10.6.3. Ammonification/Ammonium Regeneration

Evidence for ammonification on coral reefs derives from direct observation of the excretion of ammonium by reef grazers, by sediment ammonium fluxes, and from high levels of ammonium encountered in interstitial waters and cryptic habits. Obviously, the ammonium accumulations must be considered indirect evidence for the process, because it is not certain what rates of ammonification account for elevated concentrations, or that the ammonium results from dissimilatory reduction of nitrate to ammonium.

Measured rates of ammonification have been directly determined for some reef invertebrates. For example, Webb et al. (1977) measured the excretion of ammonium by the sea cucumber, *Holothuria atra*, at Enewetak, and found a characteristic power–function relationship between excretion rate and body weight. Coral animal tissue by itself appears to be ammonotelic (Muscatine and D'Elia 1978; Szmant-Froelich and Pilson 1977; Wafar et al. 1985; Muscatine et al. 1979; Kawaguti 1953); however, as discussed above, corals with endosymbiotic algae (i.e., zooxanthellae) do not excrete much ammonia because of algal ammonium uptake and assimilation. Accordingly, the importance of reef corals as N cyclers is probable at the intraorganismal level, if one considers the coral–algal association as a "superorganism."

Recent interest in sediment processes has stimulated measurements of ammonification rates in reef carbonate sands and other benthic environments (Williams et al. 1985a, 1985b; Hines and Lyons 1982; Corredor and Capone 1985; Boon 1986). The sediments near coral reefs represent a difficult environment in which to sample and evaluate the ammonification process. Ammonium standing stocks in reef sediments exceed those in the water column many times and presumably result from regenerative processes. Although concentrations can reach or exceed 40 μM, those in reef sediments are typically lower than in temperate environments (Corredor and Morrell 1985; Patriquin 1972; Hines 1985; Boon 1986; Corredor and Capone 1985; Williams et al. 1985a, 1985b).

Benthic ammonification rates have been measured in tropical calcareous sediments using slurry techniques and have been found to be lower than in temperate systems (Williams et al. 1985b). Hines (1985) presented evidence that much of the regenerated nitrogen does not enter overlying water, but is instead sequestered near the sediment–water interface. Although they did not make direct measurements of ammonium excretion by the different taxa, Moriarty et al. (1985) proposed that protozoans, benthic copepods, holothurians, and filter feeders are major consumers of bacteria, and presumably as such, important regenerators of ammonium. These authors also suggested that the general role of microfauna needs further investigation. They, along with DiSalvo (1973), have published the only estimates of nonforaminiferan protozoans in reef sediments. Such organisms may also play an important role in reef ammonification.

Direct measurements of ammonification by grazing fishes have also been made (Meyer et al. 1983; Meyer and Schultz 1985a, 1985b). While fishes certainly may be responsible for localized increases in ammonium levels that stimulate autotrophic processes, or for affecting nutrient fluxes by altering community structure, no published reports estimate their relative importance in coral reef nutrient budgets.

10.6.4. Nitrification

Nitrification is actually a two step process. The first step is ammonia oxidation, in which ammonia is oxidized to nitrite; the second step is nitrite oxidation, in which nitrite is oxidized to nitrate. Different species of bacteria accomplish each step. The first conclusive evidence of nitrification on a coral reef was obtained by Webb and Wiebe (1975), who: (1) observed the net production of nitrate in the algal pavement area just behind the reef crest at Enewetak; (2) employed an inhibitor, nitrapyrin (N-Serve), to block ammonia oxidation in tide pools; and (3) used a fluorescent antibody technique to verify the presence of one species of nitrifying bacterium, *Nitrobacter agilis*. This experimental study helped explain the observations of Webb et al. (1975), who reported net export of N as nitrate from a windward reef at Enewetak. Webb and Wiebe (1975) inferred that ammonia oxidation and nitrate oxidation are closely coupled because the stable intermediate nitrite did not accumulate.

Corredor and Capone (1985) used the inhibitor chlorate to block nitrite oxidation in unconsolidated reef sediments from Enrique Reef, Puerto Rico. Maximum rates of nitrite accumulation occurred in the upper few centimeters of sediment, with a secondary peak observed between 20 and 25 cm. This observation supports Webb and Wiebe's inference.

10.6.5. Assimilatory Nitrate Reduction

Nitrate can also be removed from the water by uptake and assimilation into plant and bacterial protoplasm. One must differentiate between the steps involved in uptake and assimilation. Nitrate may be taken up and accumulated by some organisms without being reduced and assimilated into the nitrogen metabolism of the organism. Accordingly, simply measuring uptake does not equate with measuring growth. Assimilation is a necessary prerequisite for growth under conditions of nitrogen limitation.

By far the greatest attention has been paid to assimilatory nitrate reduction by corals. The focus on corals has related more to the nutrition of the algal-invertebrate symbiosis than to ecological considerations of corals' roles in ecosystem nutrient cycling. Franzisket (1973; 1974) first observed nitrate uptake by the reef corals *Fungia scutaria* and *Montipora verrucosa*. Subsequent observations have confirmed nitrate uptake by other reef corals and foraminiferans (D'Elia and Webb 1977; Webb and Wiebe 1978), but have shown either no uptake or ambiguous results with other symbiotic associations (Wilkerson and Muscatine 1984; Muscatine and Marian 1982). The ability of symbiotic associations to remove nitrate may vary with host and nutritional state (Wilkerson and Trench 1985).

The technique most often used to determine the kinetic characteristics of NO_{3-} uptake by corals has been to incubate the organism in seawater containing a spike of NO_{3-} and monitor its depletion. Criticisms of this approach are that competing processes for nitrate uptake (e.g., denitrification by bacteria in coral skeleton) may be responsible for observed uptake, or that uptake and accumulation of nitrate in the organisms occur, but that reduction and assimilation may not. However, D'Elia and Webb (1977) have shown that if experimental organisms are chosen with care, specimens free of significant skeletal endosymbionts can be obtained. The corallum (coral skeleton) alone cannot then deplete nitrate from the medium, although intact corals can do so in the light and in the dark. Evidence that assimilation of nitrate occurs is still indirect; isolated zooxanthellae can take up nitrate (D'Elia et al. 1983) and grow on nitrate as a sole N source (Domotor and D'Elia 1984; Wilkerson and Trench 1985). Moreover, Crossland and Barnes (1977) found nitrate and nitrite reduction in zooxanthellae, but not in coral tissue. Far less attention has been paid to uptake and assimilatory nitrate reduction of reef macroalgae, although the uptake capacity and kinetics of macrophytes have been well established (see review by Hanisak 1983).

10.6.6. Assimilatory Nitrite Reduction

Assimilatory nitrite reduction undoubtedly occurs in many, if not most, organisms that assimilate and reduce nitrate to nitrite. Most of this nitrite appears to be retained within organisms, and, although corals with symbiotic zooxanthellae can take up nitrite (D'Elia and Webb 1977), the uptake of nitrite from seawater is unlikely to be an important nitrogen source. Nitrite concentrations in reef waters are typically low, and assimilatory nitrite reduction is not favored kinetically.

10.6.7. Ammonium Immobilization and Assimilation

Ammonium assimilated by reef autotrophs can be derived from reduction of nitrate and nitrite, or taken up directly as ammonium advected, regenerated by heterotrophic processes, or excreted by nitrogen fixers. Which pathways predominate on reefs is uncertain. Uptake and assimilation of ammonium are two separate processes. Uptake, assimilation, and growth of autotrophs may or may not be closely coupled, so growth rates of reef autotrophs cannot be reliably estimated from short-term ammonium uptake kinetics.

The uptake of ammonium by reef corals has been studied intensively. Kawaguti (1953) was the first to observe that reef corals take up ammonium from seawater. Many researchers were skeptical of these observations, owing to methodological considerations. However, subsequent studies have clearly confirmed Kawaguti's observations for corals (Muscatine and D'Elia 1978; D'Elia and Webb 1977; Burris 1983; Muscatine et al. 1979; see also reviews by Muscatine 1980b and Cook 1983) and other zooxanthellae bearing invertebrates (Cates and McLaughlin 1976; Muscatine and Marian 1982; Summons and Osmond 1981; Wilkerson and Muscatine 1984). Ammonium uptake by reef corals may be offset by dissolved organic nitrogen excretion (D'Elia and Webb 1977), but more observations are needed to confirm this. Reef corals with zooxanthellae apparently cycle ammonium internally, although direct measurement of the relative importance of internal cycling vs. the uptake of dissolved and particulate N sources has not been estimated. Coral reef macrophytes, as well as corals (Meyer and Schultz 1985b), may exploit pulses of ammonium derived from excretion by fishes and invertebrates. For example, Nelson (1985) has shown ammonium pulse-stimulated photosynthesis for several species of *Gracilaria* in Guam and Taiwan.

10.7. Phosphorus Cycle

The phosphorus cycle, unlike the biologically controlled nitrogen cycle, is mediated by both biological and chemical processes. Accordingly, autecological considerations of factors that affect phosphorus dynamics must consider both biotic and abiotic processes. Fig. 10.10 shows the major pools and exchanges of compounds important to phosphorus cycling on coral reefs. Each component is discussed separately below. Equivalent emphasis has been given below to both chemical and biological processes because their relative importance has not been adequately evaluated.

10.7.1. Assimilation

Phosphate assimilation by autotrophic reef organisms has not been widely studied, but several studies have focused on corals (D'Elia 1974; D'Elia 1977; Yamazato 1970; Pomeroy and Kuenzler 1969; Pomeroy et al. 1974) and algal mats (Pomeroy et al. 1974). Other important organisms that assimilate dissolved inorganic phosphorus (DIP) obviously include benthic algae, foraminiferans, and bacteria. As with other nutrients, uptake and assimilation of DIP occur in separate steps, and a closely coupled relationship between short-term uptake and organismal growth is unlikely (Atkinson 1983). Corals with zooxanthellae were first observed to behave "autotrophically" by Yonge and Nichols (1931), who found they had the ability to deplete seawater of low concentrations of DIP. Initial studies on DIP uptake were viewed skeptically, given the less precise analytical techniques, the possibility of bacterial interference, and

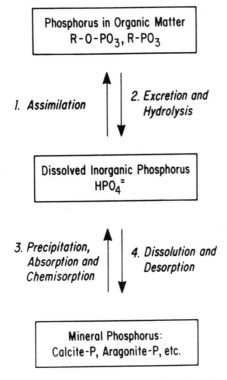

Figure 10.10. Major components of the phosphorus cycle.

the primitive working conditions available to pioneering workers. Subsequent work with improved methodological approaches showed that corals with endosymbiotic zooxanthellae excrete considerably less P than do comparably sized invertebrates without symbionts (Pomeroy and Kuenzler 1969). Moreover, the coral-zooxanthellae symbiotic association could remove net quantities of DIP from seawater even at natural, low concentrations (Yamazato 1970; D'Elia 1977; Pomeroy et al. 1974). D'Elia (1977) found evidence that a bidirectional exchange of DIP occurs: some DIP molecules are released, while others are being taken up. This seems attributable to DIP regeneration by the host concomitant with uptake by the symbiotic alga. As for dissolved inorganic nitrogen (DIN) compounds, the mechanism for uptake appears to be a depletion–diffusion process (D'Elia et al. 1983), in which coral tissues are depleted in DIP by algal uptake, with that DIP being replaced by DIP diffusing in from seawater.

10.7.2. Excretion and Hydrolysis

Pomeroy and Kuenzler (1969) measured DIP excretion rates for coral reef animals at Enewetak. As noted above, corals with zooxanthellae excreted very little DIP and had very long P turnover times. Fishes and molluscs excreted slightly less DIP than predicted by a widely cited statistical model, but in general appeared to display no special mechanisms for retaining P in a low P environment (Johannes 1964). Echinoderms excreted DIP very slowly, but had so little body P content that turnover times fit well with the Johannes (1964) model. Similar observations were made by Webb et al. (1977), who worked with holothurians.

The possible hydrolysis of dissolved organic phosphorus (DOP) may also play a role in supplying P to autotrophs. Alkaline phosphatase, the enzyme responsible for this hydrolysis, has received little attention in the phosphorus cycle of reefs. Dunlap (1985) found diel changes in extracellular alkaline phosphatase in waters flowing over reef flats on the Great Barrier Reef. He did not identify the source of this enzyme activity; however, since he showed that activity also varied along a transect across the reef flat, it may be surmised that the enzyme was excreted by reef organisms or that phosphorus exchange from the reef affected the repression and derepression of enzyme in water advected over the reef.

DIP regeneration in sediments is considered diagenesis. The diagenesis of sediments refers to all physico–chemical, biochemical, and physical processes modifying sediments between deposition and lithification (e.g., Chilingar et al. 1967). Diagenic processes affect nutrient concentrations in sediments and, in turn, nutrient fluxes between the sediments and the water column. Early diagenesis in clastic, siliceous sediments has received considerably more attention than in carbonate sediments; because of the diverse chemical and biological processes involved, phosphorus diagenesis in reef sediments is undoubtedly complex.

DiSalvo (1974) considered the relationship between bacteria and dissolved nutrients in coral reef sediments and found a relationship between soluble phosphorus and amino nitrogen in interstitial waters. He interpreted this as evidence of biological regenerative activity. Similarly, Risk and Müller (1983) and Andrews and Müller (1983), who sampled porewaters in coral heads and in cryptic reef environments, found stoichiometric relationships in nutrient concentrations suggestive of biological processes (Fig. 10.11). Clearly, establishing rates from such stoichiometric models is impossible, as is determining the effect of counterbalancing processes such as assimilation. Reconciliation of possible chemical processes that release phosphate bound to sediment particles of coral reefs will undoubtedly require more research.

10.7.3. Precipitation, Absorption, and Chemisorption

Physicochemical processes such as precipitation, absorption and chemisorption also play a role in determining the partitioning of phosphorus between dissolved and sediment particulate phases. However, the relative roles of these processes are poorly understood. Laboratory experiments have been conducted to determine the chemistry of DIP uptake or "sorption" from seawater to the surface of synthetic mineral substrates. For example, deKanel and Morse (1978)

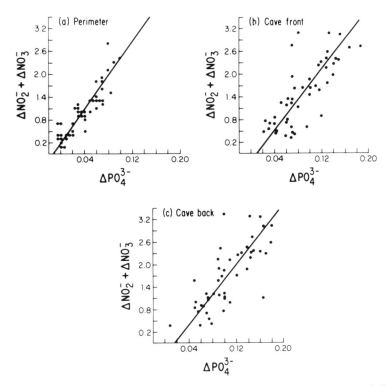

Figure 10.11. Scatter plots of N vs. P nutrient concentrations as evidence of diagenesis in cryptic reef environments. Reprinted with permission from Andrews and Müller (1983), Limnol. and Oceanog.

found that the kinetic expression for Elovichian chemisorption (not the more familiar Langmuir or Freundlich adsorption isotherms) described the sorption of phosphate to aragonite and calcite. However, coral sediments are not homogeneous mixtures of mineral calcite, aragonite and apatite, so it is not surprising that laboratory experiments involving dried natural sediments yield different results: Entsch et al. (1983a) found that a Langmuir sorption isotherm at least partly explained sorption–desorption kinetics.

Other possible explanations exist for the formation of particulate phosphorus phases from DIP. Gaudette and Lyons (1980) compared stoichiometric nutrient regeneration models with observed pore water nutrient samples in shallow water Bermuda environments and found both lower DIP values than expected and higher percentages of organic P than would be found in clastic sediments. In addition to the sorption of DIP to calcium carbonate as discussed above, Gaudette and Lyons (1980) listed the other possible explanations for DIP removal to particulate phases: (1) direct precipitation of authigenic apatite on inorganic mineral phases; (2) direct precipitation of amorphorus calcium phosphate; and (3) apatite replacement of $CaCO_3$. The relative importance of these removal processes to biological ones, such as assimilation by foraminiferans, is unstudied.

10.7.4. Dissolution and Desorption

The dissolution of calcium carbonate and the liberation of bound or sorbed DIP have received extensive attention in calcareous soils (Cosgrove 1977), but relatively little in calcareous marine sediments. Many of the processes are undoubtedly similar. The formation of organic acids could conceivably dissolve P-containing calcareous material in sediments, but this has not been demonstrated, and details of this process are considered controversial by soil scientists (Cosgrove 1977). Tracer studies using the radioisotopes ^{33}P and ^{32}P in natural reef sediments should help to improve our knowledge of the gross exchanges of P between sediment particles and interstitial waters, but generalizations may prove elusive because of the heterogeneity of sediment contents and the multiplicity of processes involved.

10.8. Nutrient-Growth Relationships

Measurement of calcareous accretion rates of present and ancient reefs has long been a preoccupation of geologists, who are concerned with coral reef abilities to grow apace with changes in sea level (Adey 1978). However, few studies have focused on the nutrient–growth relationships of major reef autotrophs. Converting between rates of calcareous and organic growth of corals, for example, is not possible because the coral tissue occupies but a thin veneer on the surface of the calcareous corallum.

A thorough review of organic productivity and calcification has been published by Kinsey (1985). Corals grow vertically, accreting calcimass, but not necessarily biomass. There have been detailed growth measurements of symbiotic zooxanthellae (e.g., Wilkinson et al. 1983), but specific growth rates for coral biomass and that of other reef organisms have rarely been measured. In one of the few studies addressing nutrient–growth relationships, Meyer and Schultz (1985b) have shown that migrating fish schools provide, through their excreta, enough nutrients to corals to increase tissue nutrient content and enhance coral growth rates.

Growth studies of reef organisms can help put autecological nutrient flux studies in perspective. Mathematical models of coral reef nutrient–growth relationships will require such information, and, if ecologists are to develop predictive capabilities for reef response to alteration in nutrient regime, such information will be necessary. The growth of autotrophic organisms can be supported by "new" or "regenerated" nutrients (Dugdale and Goering 1967). New nutrients for a coral reef would include net amounts removed from seawater passing over the reef or nitrogen fixed locally by cyanobacteria. Supplies of new nutrients are necessary for the ecosystem to accumulate biomass, whereas regenerated nutrients can only support the growth of individual organisms at the expense of other organisms. The traditional explanation of high productivity on coral reefs has been that recycling is strong and regeneration of nutrients plays a large role in sustaining productivity. Observations of high gross productivities but low net productivities are consistent with this. However, as this review has shown, evidence for tight cycling of nutrients is not as strong as is generally assumed.

Several possible mechanisms might explain high gross productivity, but low net primary productivity of coral reefs (Table 10.5). Intraorganism and interorganism cycling, as discussed above, clearly exist, but their relative importance to reef function is difficult to evaluate. Another possible explanation is that stoichiometric requirements for nutrients for reef autotrophs are lower than might be expected on the basis of planktonic observations, and primary producers are less likely to be limited by nutrients than by grazing. Atkinson (1981) and Atkinson and Smith (1983) provide some evidence for this.

Table 10.5. Possible Mechanisms to Explain High Gross Primary Productivity and Low Net Primary Productivity of Coral Reefs

Mechanism	Explanation
Intraorganismal cycling	Primary producers retain (cycle internally) essential elements that are in short supply
Interorganismal cycling	Primary producers and consumers retain (cycle externally) essential elements that are in short supply
Low stoichiometric nutrient requirements	Primary producers have relatively low requirements for essential elements. Grazers prevent biomass accumulation, but interorganismal nutrient cycling need not be high to sustain productivity.

The ideas that: (1) the importance of cycling and retention of nutrients are overtouted on coral reefs; and (2) autotrophic organisms on coral reefs are not nutrient limited, each have merit. I believe that coral reefs function more as open systems (Webb et al. 1975) than is widely perceived. The traditional view has been that nutrients cycle internally.

INTERNAL CYCLING
Autotrophs <= = = = = = => Heterotrophs

The more realistic view, in my opinion, is that a higher throughput of nutrients occurs.

INTERNAL CYCLING
EXTERNAL   EXTERNAL
INPUTS -------> Autotrophs <= = = = = = => Heterotrophs -------> LOSSES

Thus, the ratio of external gains and losses of nutrients to internal nutrient recycling is higher than is generally assumed. Future research to evaluate this, to understand trophic pathways by which nutrients pass through the system, and to identify "emergent properties" is an important priority.

10.8.1. Nutrient Limitation

One of the fundamental concerns of ecosystem science is to determine the relationship between the energetics and nutrient dynamics of an ecosystem. Confusion often exists over what a limiting nutrient is (discussed by Smith 1984). This is partly a problem of definition. One component species may not be nutrient limited, but the entire system may be, by virtue of the fact that net community productivity can increase under conditions of enrichment. That is, plant species with less affinity for the nutrient in least supply, but with a higher potential maximum growth rate, may supplant those with the opposite characteristics.

Nitrogen vs. Phosphorus Limitation

In contrast to studies of limnetic, riverine, estuarine, and marine pelagic systems, the literature about coral reefs contains relatively little discussion of whether nitrogen, phosphorus, or neither limits productivity. Smith (1984), who has considered the topic in most detail, has developed nutrient budgets for reef environments and lagoonal reef complexes in atolls. He suggests that nutrients are likely to be nearly conservative in systems with high advective flux, and that neither nutrient limits net ecosystem production. Confined ecosystems, on the other hand, have very slow advective throughput and, in Smith's opinion, are likely to be phosphorus limited on the scale of months. Although Smith et al. (1981) presented evidence that N limits primary productivity in Kaneohe Bay, Hawaii, Smith (1984) suggests that most reef ecosystems are P-limited over the long term, by virtue of their capacities to fix

N at high rates. Nixon et al. (1986) have independently analyzed Smith et al.'s data and found "an association between higher standing crops and primary production of the phytoplankton in this system and higher average concentrations [of DIN]."

Other elements may also limit organic productivity on reefs. Entsch et al. (1983b) hypothesized that iron is a limiting nutrient. Recent observations by Howarth and Cole (1985) suggest that molybdenum may limit the growth of nitrogen fixers in the marine environment. Whether molybdenum availability is an important factor controlling nitrogen fixation on coral reefs is unknown.

Interestingly enough, little consensus exists about the relative roles of different elements as limiting nutrients on reefs. Three recent reviews illustrate this point. According to Wiebe (1985), "a current concensus [sic] is emerging regarding nitrogen and phosphorus limitation on reefs: coral reefs most often *do not appear to be limited* [emphasis mine] by the low concentrations of these nutrients in ambient sea water." According to Gladfelter (1985): "*Nitrogen is usually considered the limiting nutrient* [emphasis mine] ... although phosphorus and other nutrients might be limiting in some reef zones ... ," while according to Kinsey (1985): "A major distinction is made between concentration and supply ... *where a major nutrient limitation is indicated, it is virtually always likely to be phosphorus* [emphasis mine] ..."

Individual organisms in reef environments have been considered vis-à-vis nutrient limitation. Seagrasses in the vicinity of reefs may be limited by N (Patriquin 1972) or P (Short et al. 1985). The control of such limitation is poorly understood. Hatcher and Larkum (1978) showed that standing stocks of epilithic algae on reefs were regulated by grazing or nitrogen, depending on location on the reef and season.

Elemental Ratios as Indicators

One way to evaluate which nutrient limits primary productivity in natural systems is to examine the elemental ratios in primary producers. Unfortunately, in most aquatic pelagic systems, nonliving particulate matter constitutes a substantial enough fraction to limit the utility of this method. However, in the case of reef benthic primary producers, especially macrophytes, one can obtain "uncontaminated" fractions suitable for deriving N/P ratios (Atkinson and Smith 1983). Atkinson and Smith found that C/P ratios in reef plants are higher than are the Redfield ratios. Short et al. (1985) made similar observations on the spermatophyte *Syringodium* and concluded that high N/P ratios in the plants and high N/P ratios of dissolved nutrients available in the sediments for uptake by the plants indicated phosphorus limitation. Once ratios for nutrient replete plants are obtained, nutrient limitation can be evaluated using these ratios.

10.8.2. Responses to Nutrient Enrichment

While nutrient privation might appear detrimental to coral reefs, it may, in fact, be the sine qua non for their existence. Coral reefs may, in a sense, be "disclimax" (stress-controlled climax) communities maintained by the low nutrient levels. Evidence from nutrient enriched reefs suggests that a succession of macroalgae occurs that may outcompete corals. Johannes et al. (1983b) hypothesized that higher latitude reefs have higher nutrient levels, and thus the calcareous organisms there, which are slow growing, are more susceptible to overcovering by faster growing macroalgae. Moreover, since calcification on coral reefs may be reduced by elevated nutrient levels (Simkiss 1964), nutrient enrichment may also profoundly alter geological structure (Kinsey and Davies 1979b). Whether moderate levels of nutrient enrichment ultimately enhance fishery yields has been virtually unstudied and remains unknown.

Perhaps ecosystem science leaves us least capable of evaluating the relative roles of different nutrient elements and chemical forms as growth limiting substances. We lack the ability to predict system responses to perturbations by knowing system behavior under unstressed conditions. Will the addition of nitrogen to a reef system cause a succession to a community less dominated by nitrogen fixers? Will phosphorus addition stimulate gross primary productivity over the long term or will stability be reduced so that the ecosystem is destroyed? What are the sustainable fishery yields of systems with low net primary productivity, and can they be increased with nutrient enrichment? Or can any one nutrient be considered as a growth limiting nutrient for an ecosystem? Such questions are impossible to answer at the present state of the art.

Although a burgeoning literature pertaining to coral reef pollution exists (Johannes 1975), far less attention has been paid to tropical pollution than to that at higher latitudes. Cause and effect relationships are poorly understood, and evidence suggests that tropical marine communities respond differently than their counterparts at higher latitudes (Johannes and Betzer 1975). Adey (1983) has argued that the experimental "microcosm" approach offers a way to test the responses of coral reef systems to nutrient perturbation and to evaluate cause and effect relationships. Odum (1984b) has stressed the general need to employ microcosms (mesocosms, if you prefer) for ecological problems, and Mann (1982a) has provided a good review of some of the advantages and disadvantages associated with them. Johannes et al. (1983a) have stated the need to apply such techniques to coral reef problems. However, although microcosm work has been done with coral reef communities, it has typically dealt with facets other than the relationships between nutrients and ecosystem structure and function.

Several preliminary studies of coral reef response to nutrient enrichment suggest the kinds of response we may expect (e.g., Kinsey and Domm 1974; Kinsey and Davies 1979b; Smith et al. 1981). Kinsey and Domm (1974) noted that photosynthesis, as measured by oxygen evolution, increased in an enclosed patch reef enriched with N and P (Fig. 10.12). Calcification appeared to decrease (Kinsey and Davies 1979b). However, such studies are necessarily preliminary to developing the ability to predict relationships between nutrient inputs and net organic productivity, and are limited to hierarchical levels no greater than those on the scale of the test system. Ironically, the best "nutrient

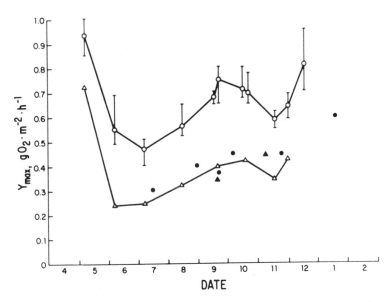

Figure 10.12. Enhancement of net production measured by the O_2 technique in response to experimental enrichment of a reef pool with N and P at a molar ratio of 10:1. Open circles represent experimental pool 1972; open triangles represent control pool 1972–73; solid circles represent experimental pool 1971–72; solid triangles represent control pool 1971–72. Reprinted from Kinsey and Domm (1974), with permission of the Australian Coral Reef Society.

bioassay" experiments with coral reefs have been inadvertent. The anthropogenic sewage enrichment and subsequent mitigation of enrichment of Kaneohe Bay, Hawaii (Johannes 1975; Smith et al. 1981; Laws 1983; Maragos et al. 1985), provided researchers with an ecosystem level experiment of coral reef response to nutrient enrichment (Smith 1979). The Kaneohe Bay experiment has clearly shown that nutrient enrichment enhances organic productivity. However, vast changes occur in the structure of the coral reef ecosystem, and organisms that were typically rare may predominate.

Kaneohe Bay responded to enrichment in three ways (Smith et al. 1981; Laws 1983): (1) the benthic community structure changed radically under conditions of slight enrichment, with macrophytes such as the "bubble alga," *Dictyosphaeria cavernosa*, overgrowing and excluding reef corals; (2) in moderately-enriched areas, phytoplankton productivity increased and subsequently turbidity shaded light from the benthos, resulting in high pelagic and low benthic productivity; and (3) in heavily enriched areas near the sewage outfall, calcification decreased and the deposition of organic material resulting from planktonic productivity caused benthic dominance by deposit and filter feeders. This pattern is likely to be characteristic for any coral reef subjected to anthropogenic enrichment. Naturally enriched systems may exhibit similar characteristics. For example, indirect evidence suggests that high latitude coral reefs are naturally nutrient enriched and are maintained at the first stage discussed above (Johannes et al. 1983b).

The process of nutrient enrichment and its consequences lead one to consider the prospects of mitigation. Here too, ecosystem science as traditionally practiced can leave us inadequately prepared to make predictions. We have very little information on the mitigation of nutrient stress for coral reef systems, primarily because environmental stresses in reef environments are generally increasing, not decreasing. By far the greatest information on mitigation also comes from Kaneohe Bay, Hawaii, where the sewage outfall in the southeastern sector was diverted out to sea in 1977–78, and where the recovery processes were studied by a group at the University of Hawaii (Smith et al. 1981; Maragos et al. 1985). During the period of sewage enrichment, all of the nitrogen and most of the phosphorus delivered to the bay were taken up biologically before water was advected from the bay. The phytoplankton responsible for most of this uptake were consumed by filter feeding organisms or sedimented from the water column. These organisms, in turn, contributed to the destruction of reefs by boring into the substratum, particularly near the outfall. At sites in the central bay, far enough away from the outfall that phytoplankton had removed most, but not all of the nutrients from the water, enough nutrients remained to stimulate the growth of benthic macrophytes.

The response of Kaneohe Bay to relaxation of sewage stress was rapid and striking (Smith et al. 1981; Laws 1983). By the end of the first year after diversion, many of the filter feeding organisms in the southeastern sector of the bay had died. A rapid reduction in phytoplankton standing stocks also occurred. The rate of recovery slowed thereafter, but conditions have nonetheless been improving throughout the bay since. Surprisingly, the high standing stocks of nutrients sequestered in the benthos during the period of sewage enrichment did not sustain high rates of phytoplankton productivity in the water column through mineralization and benthic efflux; this may relate to carbon flux in the sediments.

10.9. Conclusions

The ecosystem science approach has enabled coral reef researchers to establish that coral reefs are the sites of high gross primary productivity, low net community productivity, and unexpectedly low exchange of nutrients with the overlying water column. This holistic approach provides good discipline in thinking of the system's hierarchical organization. It has led to more reductionistic studies, which were necessary to elucidate specific mechanisms. The holistic approach is particularly exemplified by the flow respirometry studies, which have shown that there is less exchange of nutrients between coral reefs and overlying water than might be expected, although there is nonetheless a substantial throughput of N in particular. Coral reefs are undoubtedly the sites of specialized mechanisms for obtaining, cycling, and retaining nutrients, but the holistic approach has not provided adequate understanding of these mechanisms. Research has progressed haphazardly, with inadequate synthesis of previous work. The communication between holists and reductionists has often been ineffectual or antagonistic.

The ecosystem science approach has not enabled researchers to predict with confidence the response of coral reefs to nutrient perturbation, or the relationship between primary productivity and fishery yields. This observation is not unique to coral reefs, but is rather a general failing of ecology.

Most studies of the nutrient dynamics of coral reefs have failed to account for the relationship between nutrient loading and growth of reef organisms. Only recently has any substantial attempt been made to measure growth rates of organisms, and this is primarily for corals and their endosymbionts. Moreover, the remote field locations of most study sites, without permanent scientific staffs and adequate instrumentation, has led to too many short-term studies with very little long-term overview. Major expeditions to coral reef areas are necessarily brief because of funding constraints. Kinsey (1985) has discussed this problem in detail.

There has been a tendency to accept uncritically concepts such as "tight nutrient cycling" on reefs, without attempting to develop an in-depth understanding of the mechanisms involved. While cycling mechanisms undoubtedly exist on coral reefs, other alternatives may explain the productivity of reefs in the face of low nutrient supplies *without* invoking cycling. For example, observations suggest that the stoichiometric requirement of many reef autotrophs is considerably lower than would be expected from planktonic systems where the Redfield ratios may apply.

Considerably more differentiation is needed between "internal" (intraorganismal) and "external" (interorganismal) cycling mechanisms. That is to say, internal cycling within coral–algae "superorganism" associations occurs, but the extent to which this cycling accounts for ecosystem productivity features has not been established. We must better understand trophic relationships and what size classes of organisms are particularly important in determining the gross fluxes of nutrients that are observed from the holistic studies, and we must compare and contrast these cycling mechanisms with other possibilities such as the stoichiometric one.

Much of the holistic work on the structure and function of coral reefs has been done in the Pacific. Windward coral reefs on atolls, in particular, have been the focus of the flow respirometric studies that are the basis for most of our knowledge of the gross nutrient dynamics of coral reefs. While what we have learned from such studies may be broadly applicable to the function of all coral reefs, we must be careful not to overgeneralize. Accordingly, it would be useful to have comparative studies in other localities, and over environmental gradients, to provide contrasting information. More integrated, ecosystem level studies such as the Symbios expedition of the early 1970s are needed. Expeditions like Symbios must include studies of as many hierarchical levels as possible and should involve rigorous hypothesis formulation and synthesis by all scientists involved.

Coral reefs are more variable than was realized a decade ago. Many recent studies have shown the scope of this temporal and spatial variability. Now the problem is determining to what extent the features of this variability are important. We must avoid the mindless accumulation of data to which we can apply little insight. The design of studies to accumulate data on temporal and spatial variability should always, to the extent possible, involve appropriate hypothesis formulation so that sampling can be appropriate for time series analysis.

As Mann (1982) has stated, there are inherent dangers in the reductionistic modeling approaches being used by many on aquatic ecosystems. One danger of employing such an approach is that emergent properties at higher levels are missed by the aggregation of properties at lower levels. We need to understand better holistic and reductionistic properties of systems to know better what we can safely aggregate artificially, and we need to contrast this information with the hierarchical properties of systems.

In summary, we clearly understand certain gross ecosystem level features of coral reefs. We know how many of the parts of the reefs operate, but we have relatively poor understanding of system integration. The goal of ecosystem science in the next decade should be to learn how to study ecosystems in order to understand this integration. The study of coral reefs can further this understanding.

Acknowledgements

I thank C. B. Cook, R. E. Johannes, J. A. Love, J. C. Ogden, and S. V. Smith for making helpful comments on the manuscript, S. Ferguson for help with the bibliography, and T. Heimer for assisting with manuscript preparation. NSF grant No. OCE-86-15699 provided support during the preparation of this paper.

11. Coral Reef Energetics

William J. Wiebe

At first glance, the most impressive features of coral reefs are their great biological diversity and oasislike quality. Hundreds of species of plants and animals compete aggressively for nutrients, food, and space. Order seems nonexistent. The early investigators concentrated on studying the autecology of organisms, particularly corals, describing their distribution, habitats, behavior, and physiology (e.g., Mayor 1924; Yonge 1930). The problem with these studies, and in fact with almost all ecological studies at the time, was that with so many species to examine, there was little hope of identifying their individual roles within the system or the interactions between them. Lindeman suggested what has become another approach to ecological research, ecosystem modeling. The emphasis turned from examination of the behavior, physiology, and distribution of each organism to an examination of community and ecosystem processes. Lindeman (1942) proposed that energy flow could be used to integrate communities into functional units. It is not my purpose here to go into the development of this approach, but rather to point out that the ecosystem modeling approach permitted investigators to ask a new set of questions about entire, functioning systems, based upon their energetics.

Sargent and Austin (1949; 1954) applied this new approach to the study of coral reefs. They utilized a feature found on many reefs, a unidirectional flow of water across their windward sides, to measure upstream and downstream concentrations of dissolved oxygen. By also measuring current speed and water depth, they could calculate the mass transport of oxygen and thus get a net flux across the reef. By making measurements over 24 hr., a production to respiration ratio (P/R) could be calculated; they found a P/R ratio of 1.1. This relatively simple approach opened the way for a series of studies on different zones within coral reefs.

In their study of Enewetak Atoll, H. T. and E. P. Odum (1955) combined the traditional approach of identifying the biota and communities with measurements of in situ productivity and nutrient flows. It is a paper well worth reading today. The approach they used still serves as a model for coral reef ecosystem studies. In a period of 6 weeks, they attempted to identify the trophic structure of the reef communities and their contributions to productivity and nutrient exchange. They found high gross community primary productivity, but like Sargent and Austin (1949; 1954), they measured a P/R ratio of about 1 for the windward reef as a whole. They concluded that ". . . the reef community is, under present ocean levels, a true ecological climax or open steady-state system." (p. 319).

Sixteen years later, R. E. Johannes and L. R. Pomeroy led the Symbios Expedition to the same location at Enewetak, with the specific goal of reexamining the study of Odum and Odum (1955) in light of new methods and ideas about coral reef metabolism. This study (see Johannes et al. 1972 for a summary) confirmed and expanded considerably the results of the Odums, and in addition provided one explanation for the seemingly perplexing question of how coral reefs could flourish when they are surrounded by an oligotrophic ocean. The answer, in fact predicted in 1955 by Odum and Odum, was that high levels of nitrogen fixation occur on the windward reef slope and algal flats (Wiebe et al. 1975). The Symbios Expedition also established the importance, in terms of net productivity, of the algal portions of the reef flats. The entire windward reef had a P/R of ~ 1, the algal portion had a $P/R > 2$, while the coral dominated downstream portion had a ratio of about 0.75 (Johannes et al. 1972). In addition, Smith (1973), using a new method for calculating calcification rates, found that about 4 kg $CaCO_3$ m^{-2} yr^{-1} was produced.

The Symbios Expedition results generated several subsequent metabolic ecosystem studies. The Lizard Island Metabolic Exchanges on Reefs expeditions (LIMER 1975 Expedition Team, 1976) and LIMER II in 1977, at Lizard Island behind the Great Barrier Reef, focused more on the metabolism within reef zones and the contributions of components to the overall system properties (e.g., Crossland and Barnes 1983). Emphasis shifted from flux measurements per se to an examination of processes, particularly microbial processes (e.g., sulfate reduction, nitrification, microbial growth) in water and the benthos. These studies, and more recent ones (e.g., the Microbial Ecology of a Coral Reef expedition (MECOR) in 1984), have concentrated on understanding the mechanisms and, to some extent, identifying key organisms involved in coral reef metabolic processes. In addition, studies on the overall ecosystem metabolism have continued. In this regard, the works of Smith and Kinsey on rates of calcification and primary production have greatly influenced the field (see Kinsey 1983b, 1985; and S. V. Smith 1983 for summaries of these studies).

This brief historical review of coral reef energetics and approaches to its study serves as an introduction to the topic. In the following sections, we will examine current results and future needs for research. Before starting, a word of caution. Coral reef community metabolism studies have largely been confined to single expeditionary visits. We lack both replication and detail. A look through the literature will reveal no more than a few dozen studies conducted at an even smaller number of sites, almost all of which are in the Pacific Ocean (see Kinsey 1985 for a recent review). Our view of coral reef energetics, and indeed many other aspects of coral reef biology, could be greatly distorted by a lack of broad geographical coverage, in addition to the problems inherent in expeditionary research. We clearly need to broaden both the range and intensity of studies.

11.1. Coral Reef Energetics

As Kinsey (1985) has pointed out, what is a "reef" to one worker may not be to another. It is important to define the operationally equivalent parts of reefs. For Pacific reefs, Kinsey (1985) identified eight community types (see Chapter 10); not all are necessarily found at any one reef site. Only six or seven complete reef system studies have been made; most studies have been concentrated on the windward algal–coral reef flats, where water flow rate and direction can be determined with precision. For entire reef systems, the P/R ratios are 1, as shown in Table 11.1. This is true in spite of the approximately threefold variation in the individual production and respiration measurements, and suggests that the entire reef complex forms a tight, self-sustaining ecosystem. But this is not the whole story. As Kinsey (1985) has discussed, most major components within the reef also have P/R ratios of 1 or very close to it. These include reef flat coral–algal zones (Fig. 11.1), shallow lagoonal environments, zones of high activity with hard substratum cover (excluding pavements), and coral outcrops. Thus, individual subsystems also can form tight, self-sustaining units, and, at least in terms of carbon flow, they are relatively independent of each other.

One factor that became obvious during the Symbios Expedition was that although entire windward reef flats showed an overall average P/R ratio of 1, the system could be divided into an autotrophic algal turf zone (Fig. 11.2) and heterotrophic zones of small coral heads (Fig. 11.3) and larger coral heads (Fig. 11.4). Thus, production and consumption may not necessarily occur within the same physical area. This observation also demonstrates how efficient downstream communities are at removing exported material.

Production to respiration ratios have been taken to express the energetic balance of systems. All reports lead to the conclusion that there is "little" carbon left over a 24 hr. period. This holds true across the entire latitudinal gradient over which coral reefs occur (e.g., Smith 1981). For community primary production and respiration there does not appear to be a latitudinal effect. However, as Smith (1983) has stressed, this apparent balance may be misleading, because a small difference in either P or R could lead to large differ-

W. J. Wiebe

Table 11.1. Production and Respiration for Complete Reef Systems

Site	Production	Respiration	P/R	Reference
One Tree Is., Great Barrier Reef	2.3	2.3	1	Kinsey (1977)
Lizard Is., Great Barrier Reef	3.2	3.2	1	Kinsey and Davies (1979a)
Fanning Is., Line Islands	–	–	1	Smith and Pesret (1974)
Canton Is., Phoenix Islands	6.0	5.9	1	Smith and Jokiel (1975)
Christmas Is., Kiribati	–	–	1	Smith et al. (1984)
Takapoto, French Polynesia	4	4	1	Sourmia and Richard (1976)

Figure 11.1. Fore-reef flat of Japtan Island, Enewetak Atoll seen at low tide. W. J. Wiebe photo.

ences in productivity over time, and Kinsey (1977, 1985) has shown that there can be seasonal differences in P/R. This has important implications for the management of reefs and for human use of the biota.

While some uncertainty exists about the details of organic carbon production and consumption, it is clear that inorganic carbon, as $CaCO_3$, is deposited on reefs in large and predictable amounts. A series of studies commencing with the Symbios Expedition has yielded a remarkable set of comparative data (see Kinsey 1985 and Smith 1983 for reviews). For a large number of reefs, including those in the Pacific and Indian Oceans and the Caribbean Sea, three modes of growth can be seen. For lagoons or complete systems, calcification rates are about 0.5 kg $CaCO_3 \cdot m^{-2} \cdot yr^{-1}$; for windward algal reef flats, rates are 4 kg$\cdot m^{-2} \cdot yr^{-1}$; the highest rates, 10 kg$\cdot m^{-2} \cdot yr^{-1}$, occur in both the Caribbean Sea and the Pacific Ocean at discrete locations. This latter rate may be related to areas well below the present day sea level, and thus the reefs are able to grow without vertical restriction. What makes these data remarkable is that the rates are so consistent, in spite of very different conditions and biota. A variety of algae, corals, and foraminifera have been identified as $CaCO_3$ producers on coral reefs, but their quantitative contributions vary little between reef zones and different reefs. While we do not understand what controls calcification

Figure 11.2. Algal turf zone of the "Odum transect," Enewetak Atoll. W. J. Wiebe photo.

rates at the system level, it is clear that the rates are highly predictable, regardless of the location.

The lack of a latitudinal effect is somewhat unexpected, since temperatures on average are lower (Wells 1957) and radiant energy more seasonably variable at higher latitudes. Yet, in the Houtman Abrolhos Islands at 29°S latitude off Western Australia, Smith (1981) found rates of community primary production and calcification as high or higher than those at low latitudes. However, actual reef growth, which is dependent upon corals for structural development, may be less at higher latitudes as a result of more porous (and thus, more easily disintegrated) coral skeletal development (Crossland 1981). Thus, rates of calcification and net community primary production appear to fall within narrow, predictable limits on all whole coral reefs and most reef subsystems that have been investigated, regardless of location. While many of the details of these processes have been worked out at both ecosystem and organism levels, we do not yet have a complete theory to explain what controls these rates.

The energetic considerations of reef metabolism and growth have led many investigators to examine the role of nutrients. D'Elia (Chapter 10) discusses nutrient concentrations and dynamics in detail. Here, I wish only to point out some features that bear on the energetics.

Figure 11.3. Zone of small coral heads of the "Odum transect," Enewetak Island. W. J. Wiebe photo.

As reviewed by Lewis (1977), Crossland and Barnes (1983), and Kinsey (1985), nutrient concentrations in tropical water are generally much lower than in temperate waters. This observation has led investigators to examine how reefs could sustain high productivity, given low nutrient inputs. One apparent answer was provided during the Symbios Expedition at Enewetak Atoll; nitrogen was found to be fixed by cyanophytes on the windward reefs (Johannes et al. 1972; Wiebe et al. 1975; Webb et al. 1975). Nitrate and other nitrogenous compounds were exported to the downstream communities. This has been confirmed at many other reefs (D'Elia and Wiebe, in press).

Pilson and Betzer (1973) found no net flux of phosphorus in the same reef system at Enewetak. More recently, Johannes et al. (1983a) and Atkinson (1983) have shown that phosphorus flux across reefs is dependent on the incoming concentration; if concentrations are high, phosphorus is removed, if very low, it is released. Pomeroy et al. (1974) showed that phosphorus was conservatively recycled by corals, but not by other communities. Thus, the picture emerging is that while nutrient input from the ocean is low, it probably does not significantly limit coral reef metabolism (see Atkinson and Grigg 1984; Grigg et al. 1984; and Smith 1984 for further discussion). In fact, increased nutrient concentrations at higher latitudes have been proposed as one reason coral reef development is limited to low latitudes (Johannes et al. 1983b).

Figure 11.4. Zone of larger coral heads of the "Odum transect," Enewetak Island. W. J. Wiebe photo.

At high nutrient concentrations, macroalgae can outcompete corals for space (Fig. 11.5). Thus, coral reefs may *require* relatively low nutrient concentrations to persist.

Smith (in press) points out that low net ecosystem production, relative to gross production, can be considered a measure of the efficiency of a system to recycle nutrients internally; coral reef *net* community production appears to be only marginally higher than oceanic production. Stated this way, it is clear that there is no paradox concerning low nutrient delivery and high gross primary production. We will return to this topic in section 11.3.

11.2. Constancy and Variability

There have been few seasonal studies of coral reef metabolism. This has resulted from the relative isolation of the reefs, and as a consequence, the expeditionary mode of investigation. It also has resulted from a firmly held view that tropical ecosystems, unlike those in temperate climates, exist in physically constant environments. However, this idea has been challenged since 1957 (Kohn and Helfrich 1957). As summarized by Kinsey (1985), strong seasonal signals have been found for both community primary production and calci-

Figure 11.5. Corals among dense growth of *Sargassum*, Houtman Abrohlos Islands, Western Australia. W. J. Wiebe photo.

fication at all sites where it has been investigated. This is not simply a temperature effect, since the response is seen at approximately the same level at all latitudes. In fact, Kinsey (1985) points out that the greatest winter–summer effect on primary production has been seen at Lizard Island, the reef at the lowest latitude of those studied. It is obvious from these data, as shown in Table 11.2, that *P/R* on a variety of reefs can show seasonal changes. This emphasizes the need for long-term studies and modifies the general concept of an invariant, daily *P/R* ratio of 1.

In addition to seasonality, a number of other less predictable phenomena can alter the energetics and structure of reefs. Hurricanes can severely affect reef systems. In November, 1985, I dived on the reef at Discovery Bay, Jamaica, which was damaged severely in 1981 by Hurricane Allen. The upper 20–30 m of coral, which had covered the outer reef slope, were completely devastated. Corals were replaced by a thick, gelatinous, algal mat covering the dead skeletons. The shallow lagoon behind the reef was accumulating detritus from the reef slope, and the corals and seagrasses were overgrown with a variety of macroalgae. The few living corals left were slowly being smothered.

In this instance, the effect of the hurricane may have been exacerbated by two other conditions. There was a mass mortality of the Caribbean sea urchin, *Diadema antillarium*, in 1983 (Lessios et al. 1984). Furthermore, the reefs are

Table 11.2. Seasonal Responses in Production and Respiration on Coral Reef Flats

Site	Latitude		Production	Respiration	P/R	Reference
Lizard Island	15°S	S	9.7	11.8	0.8	Kinsey (1979)
		W	4.1	3.8	1.1	Kinsey (1979)
Kaneohe Bay	21°N	S	11.0	15.1	0.7	
		W	5.5	6.4	0.9	
Kauai	22°N	S	8.7	7.6	1.1	Kohn and Helfrich (1957)
		W	7.7	7.6	1.0	
One Tree Island	23°S	S	4.1	3.8	1.1	Kinsey and Domm (1974)
		W	1.7	2.4	0.7	
One Tree Island	23°S	S	9.0	7.9	1.1	Kinsey (1977)
		W	3.6	5.3	0.7	
French Frigate Shoals	25°N	S	8.5	4.9	1.8	Atkinson and Grigg (1984)
		W	4.3	2.6	1.7	
Houtman Abrolhos Is.	29°S	S	21.0	19.6	1.1	Smith (1981)
		W	12.1	14.4	0.8	

S = Summer; W = Winter.

intensively fished. Thus, the two major groups of grazing organisms that might control algal biomass (Ogden and Lobel 1978) have been eliminated. Sammarco (1982) examined the role of echinoid grazing in structuring coral communities directly, by manipulation experiments on whole reefs. In the absence of grazing, there was a massive overgrowth of algae, low rates of recruitment of almost all coral species, and a shift in algal taxa. Lewis and Wainwright (1985) have shown by comparison of population densities that herbivory by both fish and echinoids is a primary factor in determining the structure of coral reef communities. Thus, while it has not been directly measured, it is probable that the Discovery Bay reef system has shifted from a low to a high P/R, and that organic matter is accumulating on the reef slope and in the lagoon.

The Crown-of-Thorns starfish, *Acanthaster planci*, has caused periodic catastrophic changes to coral communities. Whether these changes can lead to the destruction or permanent alteration of coral reefs is still being debated (e.g., Cameron and Endean 1982; Rowe and Vail 1984). Certainly, in the short term, these infestations can result in massive mortality to corals (Endean 1973), and even changes in fish abundance and diversity (Sano et al. 1984). There is also some question as to whether these periodic population explosions result from natural causes. Birkland (1982) has shown that infestations are correlated with heavy rainfall; he postulated that increases in nutrients from terrestrial runoff stimulate plankton blooms, and that this condition increases *Acanthaster* larval success. Nutrients from agricultural land, sewage, and land clearing could increase this effect. There are differences of opinion on how, and how rapidly, recovery takes place. Endean (1973) has suggested that the effects are long-term, while Pearson (1981) has shown rapid recolonization by corals. Recently, Wallace et al. (1986) have reported resettlement of *Acropora* larvae on recently denuded tabletops, even while *A. planci* are still feeding. DeVantier et al. (1986) have found that serpulid worms can provide local refuge for coral polyps in the massive coral species, *Porites lutea* and *P. lobata*; these surviving polyps can initiate rapid recolonization of the denuded coral skeleton.

Sustained low temperature affects some coral reefs. Porter et al. (1982) reported that 96% of the shallow water corals in the Dry Tortugas were killed by a 14°C water intrusion. They suggested that these periodic events contribute to major long-term reef growth anomalies, as seen in the geological record.

Exposure of reef surfaces, due to periodic extremely low tides, also alters reef structure and sets limits to the vertical growth of corals (Fig. 11.6). In addition, as described by Banner (1968), the combination of extreme low tides and heavy rainfall can drastically alter the nearsurface biota. During a 10 day period, up to 32 in. of rain fell in the region of Kaneohe Bay, Hawaii, with 17 in. recorded on one day. This removed or killed virtually all of the shallow water biota. The effects of this event are still in evidence today, with corals in the shallow areas much less abundant than they were previously.

This brief section should convince the reader, I hope, that coral reefs should not be viewed as immutable, structurally or metabolically. They are subject

Figure 11.6. Corals exposed at low tide, Houtman Abrolhos Islands, Western Australia. W. J. Wiebe photo.

to seasonal changes and to a variety of disturbances. Coral reefs should be viewed in the same context as other ecosystems, not as special cases.

11.3. Coral Reefs in Relation to Other Ecosystems

The study of ecosystems at the functional level has benefited enormously from the energetics approach. This approach has been particularly fruitful for the study of coral reefs, in large part because of their hydrological characteristics. Community primary production and respiration are now recognized to be tightly coupled at the system and subsystem levels. This approach has also allowed investigators to delineate the nutrient dynamics of these systems.

Coral reefs may be more sensitive to large-scale disturbances than most systems, due to their low abiotic nutrient storage capacity. Many systems, such as mangrove forests, salt marshes, temperate forests, etc., have extensive reserves of nutrients stored in their soils, while, at the other extreme, some systems have virtually all of the nutrients incorporated in the living tissue. Disturbance to this latter ecosystem type, of which coral reefs are an excellent example, results in removal not only of the organisms, but also of the nutrient reservoirs necessary for their recolonization. The new biota must accumulate nutrients transported to them from outside (Pomeroy 1975).

The examination of reefs under stress has taught us a great deal about ecosystem resistance and resilience. For example, the destruction of reef corals by *Acanthaster* was predicted by some to lead to the permanent destruction of entire reefs. However, this has not proven true for any of the sites investigated. Sewage produces a different type of stress. As Smith et al. (1981) reported, the southern end of Kaneohe Bay, Hawaii, was radically altered by sewage; corals were smothered by microalgae; benthic filter feeders bloomed, and nutrient concentrations rose in the sediments. After sewage diversion, however, there was a rapid return of the system toward presewage conditions. Maragos et al. (1985) found that within 6 yr., corals had returned and macroalgal abundance approached presewage levels. However, it is uncertain whether recovery will ever be complete, because of substrate loss through boring activity and heavy terrigenous siltation associated with runoff rather than sewage (S. V. Smith, pers. comm.).

There are some disturbances and combinations of disturbance whose outcome is at present unclear. In the Caribbean, overfishing has reduced fish populations and stimulated the growth of the urchin *Diadema* (Ogden and Lobel 1978). There was massive *Diadema* die-off in the Caribbean (Lessios et al. 1983). This combination of events has reduced algal grazers significantly and resulted in massive benthic algal blooms in many areas. These blooms have smothered the corals and acted as nutrient accumulators, stimulating yet more algal growth. Reports and personal observation from a number of sites in the Caribbean suggest that the reef biota are changing. Certainly research on their progress should be conducted, and the energetic approach should be employed as one facet of such studies. There are also cases in which the results of disturbance are unpredictable. As pointed out by R. E. Johannes (pers. comm.), while *Dictyosphaeria cavernosa* is circumpolar and ubiquitous in its distribution on reefs, and sewage pollution is found in many reef systems, only in Kaneohe Bay, Hawaii, has the alga ever become a problem. We do not yet understand the interactions of various biological and physical factors well enough to explain the bloom of *D. cavernosa* in Hawaii, but not elsewhere.

Coral reefs have most often been viewed as unique ecosystems, somehow "different" from other systems. As Smith (in press) concludes, however, . . . "reefs are not metabolically different from other shoal–water systems, but . . . the lack of strong land-to-sea gradients makes reef mass balances and transports both easy and useful to evaluate." Rather than treating coral reefs as unique systems, they should be viewed as ideal models for examining general ecosystem properties.

11.4. Research Needs

For reasons discussed, systemwide energetic studies need to continue and be expanded, but the emphasis should be less on expeditionary visits and more on long-term and detailed examinations. Smith (1984) and Smith et al. (1984)

have proposed using a stoichiometric analysis of elements at a system level to elucidate the broad modes of biogeochemical cycles. Using this approach, they predicted that significant rates of denitrification, sufficient to . . ."destroy the geological evidence . . ." of nitrogen fixation, should occur on reefs with long water retention times, while smaller, well-flushed systems should export fixed nitrogen. Hatcher (1985) has discussed nitrogen dynamics on reefs in relationship to the physical structure of the system, and the mechanism by which nitrogen is delivered to the reefs. She particularly emphasized the need for proper temporal and spatial scaling in resolving the question of whether a nitrogen limitation exists for any particular reef, and, of course, this emphasis on proper scaling is required for the examination of other variables as well.

Kinsey (1985) emphasized the need for very fine discrimination in both time and space, using high precision, continuous logging equipment. He also suggested that the number of variables monitored should be increased. The need for precision in time and space is particularly acute for coral reef work because very small differences in values are propagated into very large quantities, by virtue of the large amount of water flowing through the system (mass transport). For example, Pilson (pers. comm.) calculated that based on upstream–downstream phosphorus studies at Enewetak Atoll, one could not predict whether the reef would lose all of its phosphorus, or double the amount, in 8 yr. (see Pilson and Betzer 1973 for results). The expeditionary approach has clear methodological limitations. Atkinson (1983) has demonstrated the value of longer-term studies of phosphorus dynamics.

At the organismal level, we have very little energetic information. Which algae contribute to primary production? Which organisms contribute to the calcification rates? While the sedimentary record gives information on carbonate formation (much is of algal and foraminiferal origin), there are no studies of which I am aware, excepting that of Odum and Odum (1955), that combine flow measurements with component contributions. What factor or factors control these rates? They are remarkably predictable, regardless of local biota or location. A number of chapters in this volume (e.g., Reiners, Chapter 5) point out the predictable nature of biogeochemical cycles. What are the forces that set such global and ecosystem limits? How do they control the functioning of diverse biota?

The study of organism interactions remains another problem. Studies of the effect of fish grazing on nitrogen fixation rates (Wilkinson et al. 1984) point out the subtle behavioral interactions that can have systemwide consequences. Mann (Chapter 15) makes an elegant plea for ecological studies at all levels: ecosystem, organism, and molecular genetic. Certainly, such studies are necessary if we are ever to gain an understanding of the mechanisms controlling the phenomena we now measure. I believe that this multilevel approach to ecological problems represents the next step in ecological investigations.

There are two more general types of studies that are critically needed in reef studies and ecology in general. The first is comparative studies. Coral reef energetics research has benefited from the comparative approach, and maybe the expeditionary research contributed to the use of comparisons and the attempt to generalize properties on a systemwide basis. Recent comparisons of Mediterranean ecosystems (Kruger et al. 1983) and the comparative study of estuaries (Anonymous 1984) point out the value of this approach. A much greater emphasis on comparative studies is needed. We have reached a point in ecology where large amounts of data for different sites of an ecosystem type are available, and there are large numbers of studies underway in many systems. These data sets could constitute the basis for beginning comparative studies. In the same way that biochemistry, starting in the 1930s, benefited from a comparative approach, ecological comparisons now can be used to identify similar processes and patterns.

Perhaps the greatest need in ecological research is the establishment of well-funded and monitored research sites for long-term studies. As discussed elsewhere (Wiebe 1984), these would be maintained and investigated for literally hundreds of years. Ecosystems are not being examined over time scales relevant to their development. A limited number of internationally supported, broadly supported ecosystem sites should be established. These should have a resident staff, responsible for monitoring selected environmental variables, and modern facilities for use by the staff and visiting investigators. Most of the investigations should be supported by outside research grants, much as is presently done with research vessels, million volt electron microscopes, and cyclotrons. Such sites would serve as reference systems, where studies at all levels, as proposed by Mann (Chapter 15), could be conducted. These sites should also stimulate comparative studies, just as work on *Escherichia coli* genetics, biochemistry, and physiology has stimulated comparative studies of microorganisms.

Acknowledgements

I gratefully acknowledge the comments and suggestions on the manuscript by R. E. Johannes, S. V. Smith, L. R. Pomeroy, and C. S. Hopkinson. Part of this work was supported by NSF Grant No. BSR-8304928.

12. The Study of Stream Ecosystems: A Functional View

Kenneth W. Cummins

12.1. Historical Roots and Contemporary Theory

To the extent that modern stream ecology constitutes a formal discipline, its roots are found in Europe in the early to mid-20th century (e.g., Steinmann 1907; Thienemann 1925; Carpenter 1928; Wesenberg-Lund 1943; Berg 1948). During the latter part of this period, streams were largely the domain of fishery biologists (e.g., Muttkowski 1929; Needham 1934, 1938). The emancipation of stream ecology from the status of a fishery vehicle and "poor sister" of limnology (e.g., Welch (1952) and Ruttner (1963) devoted less than 10% of their volumes to the subject), occurred with the publication of the treatise *The Ecology of Running Waters* by H. B. Hynes (1970a; 1970b). This was an extremely important publication; not just another lentic limnology text, of which there is more than an adequate supply (Macan (1974) is an exception). Rather, it was a state-of-the-art treatment, with historical perspective, of the discipline of running water investigation. From this view of stream ecology, it is apparent that trophic dynamics and energetics, which were based on such early work as Shelford (1913; 1914) and later Lindeman (1942), formed the conceptional foundation for much of the stream research in the 1960s and 1970s. In the 15 yr. since Hynes's (1970a) landmark publication, the major development in stream ecology has been its integration with such fields as geomorphology, hydrology, and forest ecology, and the recognition of the fundamental importance of spatial and temporal scales in the investigation of streams. These logical marriages were essential in producing a quantum jump in our concepts of running water ecosystem structure and function. Major events in this synthesis were: (1) the watershed perspective (e.g., Fisher and Likens 1973; Cummins 1974; Hynes 1975; Likens 1984), in which streams were no longer viewed merely as conduits exporting products from their terrestrial vegetative biome; and (2) the geomorphic view of streams in a state of dynamic equilibrium in time and space (the quasi-equilibrium of Langbien and Leopold 1964).

It should be made clear at the outset that this chapter is not a general review of running water ecology. For this we already have a number of sources (e.g., Hynes 1970a, 1970b; Macan 1974; Whitton 1975; Lock and Williams 1981; Barnes and Minshall 1983; Fontaine and Bartell 1983; Resh and Rosenberg 1984). Rather, the focus is on selected portions of the inquiry into stream and river ecosystems, in particular the functional organization of their biotic communities.

Quasi-equilibria in the form of channels are the predictable result of interactions between such controlling parameters as parent geology and the derived sediments, magnitude and pattern of flow and temperature regimes, and the soil vegetation system (e.g., Leopold et al. 1964; Cummins et al. 1983). In the River Continuum Concept (Vannote et al. 1980; Minshall et al. 1983, 1985; Cummins et al. 1984), such quasi-equilibria are viewed as predictable physical templates to which stream organisms are adapted. The basic scheme of the concept is that drainage networks of streams and their receiving rivers form a predictable continuum of increasing channel size and attendant biological characteristics (Table 12.1) (Vannote et al. 1980). With increasing channel size, there is a decrease in the direct influence of the immediate terrestrial setting (the riparian zone) on the biological communities in running water, and there is an increase in the importance of inputs from the upstream network as it coalesces into larger and larger rivers. The pattern, which generally holds worldwide, is for headwater streams (approximately orders 0–3, Table 12.1) to be heterotrophic. That is, annual community respiration exceeds gross primary production, with the major inputs to such energy requiring systems coming from the riparian zone (Minshall et al. 1983, 1985). The biological communities, as exemplified by the invertebrates, are dominated by coarse and fine particulate organic matter (CPOM, FPOM) detritivores, or shredders and collectors, respectively (Cummins 1973, 1974, 1975; Cummins and Klug 1979; Merritt and Cummins 1984). Exceptions to this generalization for headwater streams are found at some high latitude or low latitude, high altitude or xeric systems (e.g., Minshall 1978; Fisher et al. 1982; Fisher 1983). Such streams may be at least seasonally autotrophic, where some combination of height, density, and location of the riparian vegetation near the active channel results in reduced shading and direct inputs of particulate and dissolved organic matter (DOM). In addition, of course, disturbance of the watershed and riparian zone can significantly alter the predicted pattern (e.g., Meyer and Tate 1983; Gurtz et al. 1980).

The river continuum concept also predicts that the general condition in midsized rivers (approximately orders 4–6, Table 12.1) will be autotrophic.

Table 12.1. General Characteristics of Stream Ecosystems Along the River Continuum of Stream Order

Stream Order	Descriptor	Average Range in Width at Bankfull (m)	P/R (Gross Primary Production/Community Respiration)		Dominant Organic Matter Resources	Dominant Invertebrate Functional Groups[b]	Fishes
			Annual Ratio	Light			
0[a]	Intermittent streams	0.5–1	<1 or >1; heterotrophic or autotrophic[c]	With or without riparian shading	CPOM or periphyton	Shredders (including specialized life cycles); colonizing collectors or scrapers	None
1–3	Headwater (permanent) streams (brooks, creeks, runs, becks)	0.5–8	<1; heterotrophic	Riparian shading	Riparian CPOM and derived FPOM	Shredders (25–50%); collectors (50–60%); scrapers (<10%)	Intertivorous species
4–6	Midsized rivers	10–50	>1; autotrophic	Open, reduced riparian shading; good penetration due to sparse suspended load	Transported FPOM, periphyton, CPOM from aquatic macrophytes	Shredders (<5%); collectors (50–75%); scrapers (25–50%)	Invertivorous and piscivorous species
7–12	Large rivers	75–500	<1; heterotrophic	Attenuated by heavy particulate and/or colloidal suspended load	Transported FPOM	Collectors (75–90%)	Planktivorous, piscivorous, and invertivorous (bottom feeding species)

[a] The more xeric the region, the more the portion of the drainage network is intermittent.

[b] See Cummins and Klug (1979), Merritt and Cummins (1984), Cummins and Wilzbach (1985). Predators are not included, as they show little variation in dominance and are usually about 10%.

[c] At high and low latitudes, high altitudes, and in xeric regions where riparian shading is minimal, primary producers may dominate during annual periods of flow.

The annual excess of production over respiration is transported along the drainage network. Reduced shading and increasing nutrient concentrations are primarily responsible for the autotrophic status. Algal feeding invertebrates, or scrapers (Cummins and Klug 1979), increase in relative dominance (Minshall et al. 1983). Associated with maximum diversity of food resources in midsized rivers (CPOM, FPOM, vascular hydrophytes, periphyton), invertebrate species richness is also maximized (Minshall et al. 1983, 1985). It appears that this portion of the river continuum concept may not adequately describe the most prevalent aboriginal condition of midsized rivers because of greatly reduced braiding in such channels in postaboriginal times (Cummins et al. 1984; Sedell and Frogett 1984; Minshall et al. 1985), although this was not always the case in larger rivers (Triska 1984). Clearly, extensive braiding in midsized rivers would have generally reduced the autotrophic nature of these channels and extended a greater degree of riparian influence along the drainage net.

The river continuum concept likely requires further modification with regard to large rivers (approximately orders > 6) if it is to reflect the aboriginal condition of such water bodies. In North America, large rivers have been extensively isolated from their flood plains (e.g., Sedell and Frogett 1984; Triska 1984). Further, biological complexity has been reduced because of massive snag (large log debris) removal. Snags constituted the most stable habitat in what are typically very active channels (Marzolf 1978; Cudney and Wallace 1980; Benke et al. 1984; Harmon et al. 1986). While not totally altering the general pattern for an entire stream order, strong localized influences at the point of entry of a lower order channel are important modifiers (Minshall et al. 1985). The magnitude of such influences will be inversely proportional to the difference in size between tributary and receiving channel.

It is a misnomer to characterize a number of the papers recently published (e.g., Winterbourn et al. 1981; Barmuta and Lake 1982; Winterbourn 1982) as "tests" of the river continuum concept. The actual status is that major parts of the generalization hold worldwide (e.g., Barmuta and Lake 1982; Minshall 1985), but there are interesting exceptions (e.g., Winterbourn et al. 1981), many of which reflect various degrees of alteration from the aboriginal condition (Table 12.2). For example, impoundment causes displacement of the general transitions predicted by the concept, creating a "serial discontinuity," as described by Ward and Stanford (1983). The concept predicts maximum species diversity in midsized rivers in conjunction with maximized habitat and organic resource availability and range of thermal variation. There is also an indication of the overlap of terrestrially evolved species which dominate the headwaters with marine derived groups more characteristic of the larger rivers and with greater long-term access to the oceans. The intermediate disturbance hypothesis (Connell 1978) is not easily applied to the entire river continuum concept series (Ward and Stanford 1983), but, just as in marine and terrestrial systems, it likely has application within one order or group of stream orders.

Along the continuum of increased channel size, as described in the river continuum concept, downstream reaches depend to an increasing degree on

Table 12.2. Natural Disturbance Events Altering Stream Channel Geomorphology and Biotic Community Functional Organization Compared to Spatial and Temporal Alterations Attributable to Aboriginal and Postaboriginal Man[a]

Preaboriginal Natural Disturbance	Spatial and Temporal Alterations	
	Aboriginal Man	Postaboriginal Man
Fire	Local: increased burning	Regulation: suppression and controlled burning and increase in accidental fires
Flood/drought	Small-scale effects: tree removal and damming	Regulation: impoundment, dewatering, reduction of peak flows, and augmentation of low flows
Mass movement (mass wasting, debris torrents, etc.)	Minor: increased sediment loading at encampments	Acceleration: greatly increased sediment loading (agricultural, road building for logging, draining)
Wind throw (of trees)	Very minor: localized increases near encampments	Acceleration: greatly increased wind throw of vegetation left exposed on the edges of logged areas
Channel meandering	Very minor: possibly restricted near encampments	Extensive alterations: channelization and dyking isolating rivers from their flood plains and valleys
Organism alterations, extinctions and introductions	Minor: increased dispersal by nomadic tribes	Wide scale reductions, extinctions and introduction of exotics; e.g.: 1. Alteration of major plant biomes and associated alteration of riparian zones 2. Overexploitation of native species 3. Water quality, thermal effects
Input of toxic substances (localized seasonal inputs of local geologically or biologically derived toxins)	Minor: localized organic enrichment near encampments	Dramatic increases: inputs of a tremendous array of natural and manmade toxins

[a] Modified from Cummins et al. (1984), with permission of E. Schweitzerbart'sche Verlagsbuchhandlung.

upstream inefficiency (Vannote et al. 1980; Minshall et al. 1983, 1985). Although streams are efficient at retention of sediments and organic material near their sources of origin (e.g., Young et al. 1978; Speaker et al. 1984), the efficiency is not 100%. The efficiency of retention as a function of slope, current, and bed roughness (including sediments, woody debris, encroaching bank vegetation), as well as geological activity, is embodied in the spiraling concept (e.g., Elwood et al. 1983; Newbold et al. 1981, 1982, 1983). The relationship between material cycling (e.g., of phosphorus, nitrogen, DOM) and the concomitant downstream transport in running water ecosystems makes it important to replace the typical view of closed cycles extensively used in limnology (e.g., Wetzel 1983). Spiraling length is defined as the mean channel length that a material is transported downstream while turning over (cycling) once. Given this definition, the shorter the spiraling length, the greater the retentive efficiency. The approach is particularly important for DOM dynamics. Separate analyses are required for the recalcitrant portion (which has a very slow turnover) and the labile portion (with rapid turnover) because of physical adsorption and microbial uptake.

Developments such as those outlined above have led to the maturation of a highly interdisciplinary science, with conceptual theory as well developed as that of other areas of ecology. Recently there has been formal recognition of the logical extension from trophic dynamics, watershed studies, and the river continuum concept to the inseparable linkage between wetted stream channels and their adjacent riparian vegetation (Minshall et al. 1983, 1985; Cummins et al. 1984; Cummins 1986 a, 1986b; Likens 1984). Hynes (1963) and Ross (1963) drew attention to such land-water interactions as fundamental properties of stream ecosystems. Riparian stream linkage (Fig. 12.1) is so complete

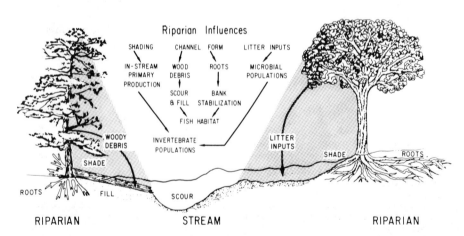

Figure 12.1. A diagrammatic summary of the influences exerted by the riparian zone on the physical characteristics and biological components of the associated stream ecosystem. Reprinted from Cummins (1986), with the permission of the Water Studies Center, Chisholm Institute of Technology, Caulfield, Australia.

that there is little basis for drawing ecosystem boundaries at the position of the stream's edge at any given flow level.

In addition, an extensive hyporheos exists in some streams well beyond the vertical and horizontal extent of typical stream sampling (e.g., Williams and Hynes 1974; Williams 1981; Pugsley and Hynes 1986). Such penetration deep into the bottom sediments or the bank represents an extension of the aquatic system, and is qualitatively different from the association between primarily terrestrial and primarily aquatic organisms that is characteristic of riparian zones.

12.2. Temporal and Spatial Scales

Stream ecosystem studies suffer from a lack of both long-term perspective and a range of spatial scales. The pattern of annual cycles should be viewed over at least decade, if not century, intervals, and study reaches should be placed in the context of watersheds and drainage basins. The magnitude of an event, such as extremes of flow, and the spatial scale associated with the event, should be evaluated within the context of the historical record of flows and the area affected.

Recurrence interval, T_r, which is the average time in years between occurrences of maximum annual flow of a given magnitude, can be used as a basis for comparisons between years (Leopold et al. 1964; Morisawa 1968; Cummins et al. 1983). A specific example can be seen in the detrital dynamics of a small (first order) woodland stream in Michigan (Petersen et al. 1988). The variation (95% CL) of total particulate (0.45 μm) benthic organic detritus, or any of 11 individual particle size categories, was only $\pm 20\%$ of the annual mean value. The study water year (October to October) had a T_r of 3.3 yr. (based on a 30 yr. record), and was preceded by 5 yr. with T_r values ranging from 1.3–2.6 yr. In the following 2 yr., a T_r of 20 yr. was recorded, with a 50% reduction in all categories of benthic detritus. The study indicated that over periods with average recurrence intervals, in the range of 2.4 yr. (Leopold et al. 1964), sedimentary detritus is maintained at a quasisteady state, with inputs approximately equal to processing plus outputs. Although the intensive study was conducted in one 100 M reach, other samples and discharge records suggest that the pattern was applicable at the spatial scale of the entire watershed.

Increased spatial and temporal perspective over the usual reach-oriented seasonal or annual view provides a framework for evaluating disturbance of riparian stream ecosystems. A primary effect of post aboriginal man has been to increase the spatial and decrease the temporal scale of essentially natural events or "disturbances" (Table 12.1; e.g., Cummins et al. 1984; Sedell and Frogett 1984; Triska 1984). Given the appropriate spatial and temporal scales, the probability of, for example, wildfire, and extreme high or low flows somewhere within a given area approaches 1. That is, the types of events are not new, but man has brought about dramatic changes in the distribution and frequency of the disturbances.

12.3. Organic Budgets

The extended temporal scale beyond the annual, and the expanded spatial scale beyond the reach perspective, as discussed above, have major influences on the evaluation of the budget approach to watershed studies. The timing of such investigations can affect the absolute magnitude of the balance between inputs and outputs, although the relative contributions of the components to the budget are likely more constant (Cummins et al. 1983; Petersen et al. 1988). One revelation from organic budget studies, whether of reaches (e.g., Fisher 1977; Mulholland 1981; Petersen et al. 1988), watersheds (e.g., Fisher and Likens 1973; Naiman 1982, 1983; Minshall et al. 1983), or selected organic components (e.g., Gurtz et al. 1980), has been that major shifts in organic storage are under hydraulic control. A smaller but significant portion of the dynamics is attributable to biology: primary production, microbial respiration, and animal assimilation (in the range of 10–25%, Petersen et al. 1988). Part of the perceived dominance of physical processes has resulted from an emphasis on exported (transported) organic matter at high flows (e.g., Likens et al. 1977). If the intent of the budget study is to evaluate the importance of biological processes, the source of the material is important. For example, is the organic matter captured at an export weir of lesser importance to the stream biota for having been excavated from deep anaerobic storage by flood scour or captured from previously dry areas of the upper bank? Is it of greater importance because it is part of the active detrital pool from the wetland channel? Regardless of source, it is important to know if the organic matter is retained in a given reach of channel long enough to allow biological activity to occur on the substrate. The minimum time necessary for utilization of an organic substrate could range from hours for microorganisms to months for invertebrates.

Despite the attention directed to the problems inherent in reach or small watershed budget measurements (Cummins et al. 1983), the limited approach continues to be extensively used. If future studies are to provide new insights, appropriately detailed short-term reach studies (e.g., 24 hr. investigations of DOM dynamics, e.g., Dahm 1984) must be set in the context of long-term (greater than annual) trends and whole watersheds.

12.4. Functional Approaches

A characteristic of stream ecosystem research in the last decade has been the functional view. This approach, whether applied to microbes (cellulose degraders, denitrifiers, etc.), algae (shade adapted photosynthesizers, nitrogen fixers, etc.), invertebrates (leaf litter shredders, fine particle collectors, etc.), or riparian vegetation (root bank stabilizers, large wood source, turnover rates of plant litter once it is entrained in the stream, etc.), emphasizes the system level functional roles played by assemblages of organisms. Although the functional view may not be appropriate for questions addressing some population level phenomena, it permits clustering of genetically and taxonomically diverse entities into groups, or guilds, which share fundamental properties—such as invertebrates having the same morphological-behavioral mechanisms of food acquisition. Thus, the functional approach leads to a worldwide view of riparian stream ecosystems that is much less variable than the taxonomic arrays that inhabit them.

Microbiologists have long taken a functional view of their study organisms, because classical morphological approaches were an inadequate basis for categorizing important physiological and biochemical differences. Morphological, physiological, and genetic "species" usually represent quite different scales of increasing resolution, and the criterion of reproductive isolation used with higher plants and animals is of little use, or at least requires special definition (e.g., Nanney 1963, on Protozoa).

Work on the regional systematics of stream invertebrates has made significant strides (e.g., Merritt and Cummins 1984), but the time when "the species" can serve as the basic unit for many ecological questions lies in the distant future. A functional classification of stream macroinvertebrate associations has been developed over the last 15 yr. (Cummins 1973, 1974, 1975; Cummins and Klug 1979; Minshall et al. 1983; Merritt and Cummins 1984; Cummins and Wilzbach 1985). This evaluation of stream macroinvertebrates according to functional roles grew from a recognition of certain inadequacies of systematic and trophic analyses. Categorization by gut contents revealed that essentially all invertebrate species are omnivores (e.g., Coffman et al. 1971), despite drastically different morphological and behavioral feeding adaptations. In addition, the immature stages of many stream species are extremely difficult to separate taxonomically.

The functional classification depends largely on the food acquisition system; that is, the morphology and the corresponding behavior which act in concert to obtain food. In many instances, the morphology restricts the food which can be efficiently acquired (e.g., filtering collector blackfly larvae cannot shred leaf litter). Although the food acquisition behavior which drives the feeding structures may often be surmised, it harbors the potential for considerable subtlety. Trophic studies (e.g., Coffman et al. 1971) have made it clear that food intake (ingestion) is not synonymous with food preference. Food ingested would be expected to change with habitat, even without changes in food acquisition mechanisms.

The efficiency with which the food acquired is converted to growth is a function of both the assimilation system and the quality of the ingested food (Fig. 12.2). As has been demonstrated for the caddisfly, *Glossosoma nigrior*, samples of the same species collected from habitats differing in food quality show significantly different growth. Undoubtedly, the same acquisition system being applied to different quality food resources causes the assimilation system to operate with varying efficiency (Anderson and Cummins 1979).

The assimilation systems in many species involve complex interactions between the endogeneous invertebrate enzymes and those produced by resident microorganisms or derived from transient forms in the digestive tract (Cum-

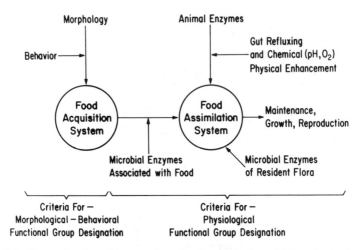

Figure 12.2. Controls of, and interactions between, the acquisition (ingestion) and assimilation (digestion and absorption) systems that are the basis for stream invertebrate functional group designations. Most efficient acquisition is achieved with the best matchup between the morphological–behavioral adaptations and the food resource to be harvested. All food resources acquired are not assimilated with equal efficiency. Maximization of growth occurs with the best matchup between the food acquired and the capabilities for enzymatic (endogenous and exogenous) digestion and absorption.

mins and Klug 1979; Klug and Kotarski 1980; Martin et al. 1980, 1981a, 1981b). The possibility that shredder, and likely many collector, species are dependent to various extents on ingested bacterial enzymes is a most intriguing area of future research.

Invertebrate acquisition systems are categorized on the basis of structures and the behaviors employed to most effectively obtain different classes of food substrates. The general invertebrate functional associations are: (1) coarse particulate organic matter (CPOM, especially leaf litter) and associated microbes (especially aquatic hyphomycete fungi; e.g., Arsuffi and Suberkropp 1985) with shredders (invertebrates with chewing mouth parts); (2) fine particulate organic matter (FPOM) and associated microbes (dominated by bacteria) with collectors (invertebrates that remove particles from the suspended load are filtering collectors; invertebrates that acquire settled particles on or from substrates are gathering collectors); (3) epilithon (periphyton) with scrapers (invertebrates with morpho-behavioral adaptations for removing attached algae and associated detrital–microbial assemblages from surfaces); and (4) prey with predators (invertebrates that are specialized to capture and consume live prey). The evidence at hand suggests that essentially all stream macroinvertebrates are omnivorous, if all growth stages under a variety of habitat and seasonal conditions are considered. Virtually all early instars (or stages) of all species, including predators (e.g., Petersen 1974), ingest FPOM. For example, many predators, such as the megalopteran *Nigronia* (Petersen 1974), ingest significant amounts of FPOM in the first instar. In later instars, the detritus and algae found in predator guts (e.g., Coffman et al. 1971) may represent the digestive tracts of their prey. It is certainly possible that such prey gut contents constitute quite a significant food resource for predators. The FPOM category is qualitatively broad, encompassing at least fine macrophyte fragments, amorphous organic matter (much of it derived from dissolved organic matter), bacteria, fungal fragments and spores, protozoans, rotifers, nematodes, diatoms, green and blue-green algae, microcrustacea, a variety of other microinvertebrates and first instars of some macroinvertebrates (in the size range < 200 μm), as well as various fragments of all the categories. Our interest is in the food acquisition system needed to acquire such a slurry food source, in the particle size range of approximately 0.5 μm-1000 μm, from the water column or the sediments. If a particle of this detrital "porridge" gets beyond a certain size, it will need to be chewed, in addition to, or instead of being filtered from the passing water column or gathered from surfaces or crevices by collectors. One method for dealing with the "ambiguities" that may occur in the functional group approach is to employ the designations of "obligate" (specialist) and "facultative" (generalist) forms (Cummins and Klug 1979), a time tested procedure that has been extremely useful in microbiology, as in facultative and obligate aerobes and anaerobes.

12.5. Some Active Lines of Inquiry

The most telling indicator that stream ecology is a vital, healthy discipline is the burgeoning diversity of stream studies (e.g., summaries in Lock and Williams 1981; Barnes and Minshall 1983; Fontaine and Bartell 1983; Resh and Rosenberg 1984). An important aspect of this diversity in stream studies not discussed here, but extensively treated in the above references, involves research into phenomena at the population level. Of particular note are algal-herbivore relationships (e.g., Gregory 1983; Hart 1985) and predator-prey interactions (e.g., Peckarsky and Dodson 1980). Furthermore, it should be pointed out that in recent years study of the taxonomy (and phylogeny) of stream organisms has advanced as rapidly as any other habitat-defined assemblage (e.g., Patrick and Reimer 1966; Merritt and Cummins 1984). Some selected examples of promising areas of stream research are treated below.

12.5.1. Dissolved and Particulate Organic Matter Interactions

The mechanisms by which organic matter is converted from particulate to dissolved phase (leaching) and the reverse (microbial uptake, coprecipitation with cations, etc.), and the role of these transitions in the biological activity of lotic systems, is an important area of stream research. Identifying the biochemical composition, the sources, and the rates of conversion of the range of labile through recalcitrant portions of these particulate (POM) and dissolved (DOM) pools of organic matter has received considerable attention (e.g., Dahm 1981, 1984, 1987; Kaplan and Bott 1982; Meyer and O'Hop 1983; Meyer and

Tate 1983; Tate and Meyer 1983; Hynes 1983). Because of the rapid conversion of the more labile fractions of DOM to POM (Dahm 1981), much of the earlier data reporting dissolved organic carbon (DOC) concentrations from weekly or monthly grab samples is probably of limited ecological relevance.The proportion of total DOM that is labile varies with season and stream type, but it is always of qualitative and usually of quantitative significance (Dahm 1984). Physical binding (particularly coprecipitation with cations) of the labile portion is on a time course of minutes and microbial uptake occurs within hours. It is this rapid conversion of labile DOM to ultrafine particulate organic matter ($< 50 \mu$m) that likely provides some of the highest quality particulate substrates in the size spectrum of FPOM (< 0.45 m to > 1 mm) for both microorganisms and invertebrates (e.g., Petersen et al. 1987).

Many important questions concerning the role of DOM in stream ecosystem function remain. More detailed analyses of the sources and fates of both the labile and recalcitrant fractions of DOM are needed (e.g., Hynes 1983). The closed circulating chamber technique used by Dahm (1984; Dahm et al. 1987) allows for in situ testing of DOM generated from specific substrates, such as macroalgae and different species of leaf litter, and for the use of tracer labeling methods. Work on specific biochemical constituents of the DOM, particularly of the labile fraction, has progressed little as yet (e.g., Larson 1978).

12.5.2. Food Quality and the Nutritional Biochemistry of Invertebrates

It has been typical in ecology to identify nitrogen, that is protein, as an adequate measure of the quality of organic matter substrate for microbial and invertebrate utilization. A carbon to nitrogen ratio of 17 or less has been generally accepted as indicative of an adequate diet for animals (Russell-Hunter 1970), higher ratios implying that some sort of concentrating mechanism will be required to increase dietary protein (Cummins and Klug 1979). However, because many forms of nitrogen are not readily available to animals, and perhaps more importantly to microbes, C/N may not be a good predictor of substrate quality. For example, Ward and Cummins (1979) showed that microbial respiration and ATP content of detritus, in the particle size range ingested, were far better at predicting growth in the FPOM gathering collector midge *Paratendipes* than was nitrogen. Invertebrates exhibit periods in the growth phase of their life cycles when lipid accumulation exceeds protein elaboration as the major biochemical activity (Cargill 1985a, 1985b). This would be expected in aquatic insects, most of which have non-feeding adult stages. The last instar (holometabolous forms) or last several instars (hemimetabolous) is devoted to lipid storage, accounting for the majority of the weight gain over the entire growth period. These lipid energy stores are mobilized during pupation (e.g., Trichoptera, Megaloptera, aquatic Lepidoptera, Coleoptera and Diptera), in adults for mating, and in females for egg production and oviposition. Feeding strategies of late instars of aquatic insects reflect the need to accumulate lipid reserves. At least some representatives of the shredder functional group can select food on the basis of specific lipids or lipid precursors (Hanson et al. 1983, 1985a; Cargill et al. 1985a, 1985b). Thus, it is clear that nitrogen content, or protein measured directly, does not serve as an adequate indicator of organic substrate quality in many instances. It is worth noting that many researchers abandoned caloric content (a rough index of lipid and carbohydrate content) as a measure of food quality or condition of organism because it did not adequately reflect protein requirements for elaboration of structure and enzymes.

It is now clear that biochemical questions of food quality for stream invertebrates must include measurements of lipids and carbohydrates (possibly indexed to caloric content), as well as protein, together with information on the growth portion of the life history of species of interest. Further, nitrogen is not a suitable index for useable protein, and measures of microbial activity (respiration, substrate uptake) or biomass (e.g., ATP) of detritus may miss a critical feature of quality—namely the presence of key exogenous enzymes.

12.5.3. Anaerobic Habitats

A third example of a research area destined for increased attention in stream ecology is the importance of anaerobic micro- (and macro-) habitats in running waters. These sites, which include the digestive tracts of stream invertebrates (e.g., Cummins and Klug 1979), represent locations of microbial-biochemical activity quite different from the much more extensively studied aerobic portions of lotic habitats. Given that reduced nitrogen as ammonia is usually the preferred form of precursor nitrogen for protein synthesis, anaerobic-aerobic interfaces should constitute sites of intense biological activity. At such interfaces, efficient aerobic organisms can incorporate anaerobically-generated reduced nitrogen, which is otherwise quickly oxidized in aerobic environments. Thus, although anaerobic microhabitats may be spatially restricted, they would be expected to constitute extremely critical regions that are functionally important to overall stream ecosystem metabolism.

The digestive tracts of stream invertebrates constitute widely distributed and strictly controlled anaerobic microhabitats (e.g., Martin et al. 1980). Anaerobic stream habitats in the sediments and in organic detritus accumulations are undoubtedly important sources for continued colonization of invertebrate digestive tracts with key anaerobic microorganisms.

12.5.4. Riparian Influences

Focus on the reciprocal interactive nature of the land-water interface (i.e., the riparian-stream ecosystem) is a logical consequence of the river continuum concept. The influences of stream-side riparian vegetation (Fig. 12.1) on the biology of stream organisms has been recognized (e.g., Vannote et al. 1980; Minshall 1983, 1985), and the possibility that riparian setting can override geomorphic features in many cases has been proposed (e.g., Cummins et al. 1984). The relative dominances of stream invertebrate functional feeding groups reflect this linkage between the riparian setting and the organization of stream communities (Cummins et al. 1984).

As an example of such linkage, Fig. 12.3 represents the relationship between litter input and shredder and collector invertebrate response. Any riparian vegetation system can be categorized according to the temperature-specific rates at which the plant litter derived from these stream-side zones is processed by the biota of the receiving channel (e.g., Petersen and Cummins 1974; Webster and Benfield 1986). Processing is defined here as the conversion of the litter, once it is entrained in the channel, to CO_2, organism biomass, smaller particulate (FPOM) and dissolved (DOM) organic matter. Categorization into processing classes (fast, medium, slow; Table 12.3) allows a riparian zone to be evaluated, along with information on input and retention, as to its impact on the in-stream biology of the receiving channel.

The input schedule of a particular litter-processing class is generally related to the timing of plant tissue abcission of species belonging to that class. Measurements of percent plant cover in the riparian zone and percent retention (Speaker et al. 1984) of each litter-processing class in the stream channel provide the background for analyzing riparian influence on shredder and collector invertebrates. The use of temperature-time (degree days) allows for the comparison of processing rates between seasons and between streams. Analyzing a given riparian zone on the basis of percent cover of fast, medium, and slow processing classes can be related to the timing of input and the relative percent of retention of each general category in the stream channel. If processing time for a given litter class is taken to start with the first retention during the year of that litter class, the processing curve (left-hand solid curve, Fig. 12.3) would be the maximum processing interval for litter belonging to that category. Similarly, if processing is taken to begin with the last retention of litter in the processing class, the maximum processing interval would be described. A mean processing interval—lying between these earliest and latest processing curves for a given class—can be taken to start at the time of maximum retention of litter in the category (middle processing curve in Fig. 12.3). The maximum shredder response to a given litter class has been observed when approximately 50% of the post-leaching weight loss (processing) of the litter has occurred (e.g., Petersen and Cummins 1974; Swift et al. 1987). Maximum response is defined as maximum shredder biomass per unit of litter mass. The major interval of shredder utilization of a given litter class (activity) is encompassed by the 50% processed points between the curves for minimum and maximum processing intervals (Fig. 12.3). Once the time of maximum retention has been determined for a given litter class, analogues of natural litter accumulations (Cummins et al. 1980), termed leaf packs (Petersen and Cummins 1974; Merritt et al. 1979; Hanson et al. 1985), are made of plant litter representative of the litter-processing class in question. Such packs can be used to determine the biomass values at which shredders and collectors are maximized. The magnitude of such values determined for each of the three litter-processing classes, and the time of year at which they occur, can be used to compare streams, assuming that maximum values represent the most efficient couplings between riparian litter and detrital-feeding shredders and collectors.

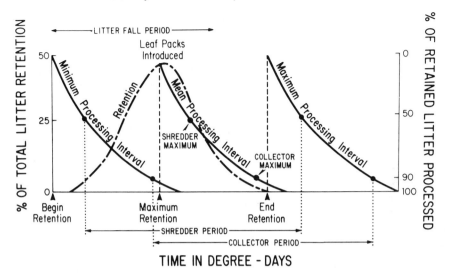

Figure 12.3. A generalized conceptual model of the relationship between litter inputs to a stream reach and the response of shredder and collector invertebrates. The generalized case is for a particular litter processing class (i.e., fast, medium, or slow turnover of the litter once it is entrained in the stream). Following the litter fall period for a given processing class, a retention curve (perforated line) can be described from an initial (begin) through a maximum to the final (end) point in time designated as cumulative temperature (degree days). The scale of percent litter retained for this curve is on the left ordinate. A curve (solid line curve at left) of litter processing (i.e., the conversion to FPOM, CO_2, and organism biomass that begins with the first litter retained (begin)) describes the minimum degree day processing interval that would be observed for the particular litter category. Similarly, the maximum processing interval would begin at the point when the last litter in the category was retained (end) in the stream channel. A mean processing interval can be taken to begin at the point of maximum retention of the litter belonging to the given litter processing class. Shredder invertebrates are maximized at approximately the 50% processed point of the processing interval (scale on the right ordinate) and the litter-associated collectors at approximately the 90% processed point. Shredder and collector maxima are taken as the points of maximum invertebrate biomass per unit of litter mass. The major periods of shredder and collector activity are delineated by the degree day interval between the 50–90% processed points of the first litter of a given class that is retained, and the 50–90% processed points of the last litter of the class that is retained. Leaf packs, i.e., artificial accumulations prepared as analogs to natural litter accumulations, made of plant tissue of a representative species in the litter processing class, are introduced at the time of maximum retention for the litter category. Loss of mass and invertebrate colonization representative of the category are thus measured in a standard fashion.

12.6. Conclusion

The study of stream ecosystems has made many important strides and a wide range of exciting problems remain to be investigated. The future is a bright one as the field is rich in interdisciplinary interaction and has both developed

Table 12.3. Categories of North American Riparian Plant Litter Based on
Approximate Ranges of In-stream Processing Rates[a]

Processing Categories	Approximate Processing % per Day	Approximate Rates % per Degree Day	Common Eastern Deciduous Forest Examples
Fast	>1.5	>0.15	Basswood, species of ash, elm, alder, most species of maples, most herbaceous plants
Medium	<1.5 >0.10	<0.15 >0.10	Species of willow, hickory, shrubby dogwoods, viburnum
Slow	<0.10	<0.10	Beech, aspen, oaks, rhododendron, conifers, most grasses, sedges, and ferns

[a] Based on data from Peterson and Cummins (1974), Hanson et al. (1984), and Webster and Benfield (1986).

and incorporated from other areas integrative concepts in ecology. Important advances will undoubtedly accrue from further integration of studies conducted at the population level with those focused on the ecosystem level.

Acknowledgements

I thank Dr. Margaret A. Wilzbach for many helpful suggestions in preparing this paper and two anonymous reviewers for their constructive criticisms. Research supported by the U. S. Department of Energy Grant No. DE-FG05-85ER60301.

13. Energy Flow and Trophic Structure

Stuart L. Pimm

As Odum (1971a) succinctly put it,

> The transfer of food energy from the source in plants through a series of organisms, with repeated eating and being eaten, is referred to as a *food chain*. At each transfer a large proportion, 80 to 90 percent, of the potential energy is lost as heat. Therefore the number of 'links' in a sequence is limited, usually four or five ... The number of consumers that can be supported by a given primary production output very much depends on the length of the food chain; each link in the chain decreases the available energy by about one order of magnitude.

In their most general form, Odum's remarks are about energy flow and the trophic structure that modifies, and is modified by, that energy flow. In this chapter, I review some of the ideas about energy flow and the particular aspect of trophic structure—the length of food chains—referred to by Odum. We can easily rephrase as a set of predictions the idea that the amount of energy flow limits food chain lengths. First, if we increase the energy entering the food chain (the primary production), then we should get an increase in food chain length. Suppose the efficiency of transfer from one level to the next were 10%, then an increase in one order of magnitude of primary productivity should add a trophic level. Higher efficiencies will lead to greater changes in food chain length with increases in productivity; lower efficiencies, the reverse. Second, it follows that if we alter the efficiency of energy transfer between levels, we should alter food chain lengths accordingly.

In the first part of this chapter, I will attempt to show that a naive energy flow hypothesis is inadequate to explain the patterns of food chain lengths we observe in nature. Certainly, the species at the end of the food chain must be limited by their food supplies; it follows from how we define "the end of the food chain." But a good correlation between the overall level of energy flow and food chain length does not follow.

In the second part of the chapter, I will argue that food chain lengths are set by a combination of two factors—how violently ecosystems are disturbed and how quickly their constituent species can recover from those disturbances. The first factor may be thought of as the variance of energy flow, over time and space. The second factor begs our asking: what determines the rate of recovery? I will argue that the flux of energy and nutrient flows through a system and the openness of nutrient cycles are essential features in determining this rate of recovery. Food chain length modifies these rates, of course, but it is not the only contributor. I will conclude that spatial and temporal variations of energy and nutrient cycles are the essential ingredients in any comprehensive theory of trophic structure.

13.1. What, if Anything, is a Trophic Level?

The definition of the length of a food chain is the number of trophic levels the ecosystem contains. But how do we define a trophic level? Starting with "one" for plants, we add levels when the plant is eaten by an herbivore, when the herbivore is eaten by a carnivore, and so on. This is fine at the level of an introductory ecology text, but every ecologist knows that reality is more complex than this. As one reviewer wrote, should we "officially declare the notion (of a trophic level) dead and move on?" If we could not count trophic levels, I would agree. But I will consider the number of trophic levels of nearly 200 food chains in this chapter, so I am clearly counting something. The criticisms of "trophic level" fall into several groups.

1. Not all species can be assigned a unique trophic level. For example, what happens if a species feeds on both herbivores and plants? Studies of food webs typically record the mean or modal number of all the possible pathways of feeding interactions from the base of the food chain to the top predators. Recent studies (e.g., Cohen et al. 1986) look at the distribution of these pathways. Most statistics in ecology have a variance; food chain lengths are no exception.
2. Some species feed at different trophic levels as they develop. Fishes are good examples of species that feed their way up food chains as they grow. One convention for handling this variation would be to take the trophic position of the adult as the trophic level of the species.
3. What about parasites? We could add "one" to the end of the food chain, since very few reported food chains include parasites. Parasite–host interactions are dynamically rather different from predator–prey interactions, so the usual convention is to not count the parasites.

4. Matter constantly cycles through the food chain, so where does the food chain end? Dynamically, the interaction between two trophic levels is very different from that between living organisms and the detritus it becomes after death. The former interaction will depend on the densities of both the predator (recipient) and the prey (donor). The latter interaction is totally donor controlled. Matter cycles through the ecosystem and may pass through a given species more than once. The number of these cycles, or the residence time of an atom within the ecosystem, is an interesting parameter. Such parameters tell us one thing; the length of food chains tells us about something else—the number of dynamically coupled interactions. Because of this argument, I also count detritus as trophic level "one," and detritivores as level "two," etc.

Trophic levels are countable. This does not mean that the number of trophic levels is the only parameter we should know. Nor are counts of trophic levels always adequate. Many small organisms (bacteria, for example) have trophic roles in the ecosystem that we do not understand. And, some of the conventions I have just discussed are arbitrary. But these difficulties pervade all science. What is important is that we can use trophic levels, as I have defined them, as a means of comparing ecosystems.

13.2. Energy Flow and Food Chain Lengths

To evaluate a simple energy limited model of food chain lengths requires only comparable data on energy flow and food chain lengths for a variety of ecosystems. Were such data available! In their absence, we must resort to a number of less direct approaches.

1. As Odum (1969) anticipated, the International Biological Program assembled many studies on primary productivity. Some of these studies show that there is sufficient variation in primary productivity and ecological efficiency for us to expect substantial variation in the amount of energy flowing through food chains. Yet such variation is not obvious.
2. In section 13.2.2, I will try to correlate energy flow with food chain length for some systems where both have been measured simultaneously, though by different groups of scientists.
3. In section 13.2.3 I will review a small number of studies where the same scientists have looked at food webs in the same ways in widely different places. There are distinct patterns of food chain length variation, but they do not appear to correlate with what can be measured or inferred about the level of energy flow.
4. I will conclude this part of the chapter with a discussion of an experiment which directly tests the effects of energy input on trophic levels in section 13.2.4.

13.2.1. Gross Patterns of Food Chain Length, Primary Productivity, and Ecological Efficiencies

Food Chain Lengths

Two studies of food chain lengths are summarized in Table 13.1. An obvious limitation of extracting data on food chains from the literature is that they may be incomplete. Frequently predators are missing from published food webs, and, equally commonly, the species at the bottom of the food webs are not plants, but herbivores or even carnivores, whose prey are not reported. Cohen et al. (1986) analyzed 113 food webs, and from their data I have extracted the modal food chain lengths. In addition, I have presented elsewhere (Pimm 1982) a compilation of food chain lengths for a smaller data set, from which I separately recorded webs with small invertebrates as top predators. Bearing in mind the incomplete nature of some food webs, it seems clear that most food chains are three or four trophic levels long, and the percentage of chains with five or six trophic levels is small. There are clearly some systems that do have long food chains, and Coleman's studies of soil food chains (Chapter 8) are a good example. Nevertheless, for the entirety of marine, freshwater, and aboveground terrestrial systems, we are faced with surprisingly little variation.

Primary Productivity

Primary productivity is known to vary widely. The data in Table 13.2 on the ranges of productivity are illustrative and by no means comprehensive. These data from various sources were usually reported with the hope of finding consistencies between different studies, and this will tend to decrease the range of values reported. The ranges of value seem to increase as better data become available. Note, for example, the difference between the values for oceanic ecosystems reported by Ryther (1969) and those of later studies. But, whatever the details, the ranges of values are large. Certainly, if the average level of

Table 13.1. Distribution of Food Chain Lengths from Cohen et al. (1986) and Pimm (1982). Those from Cohen et al. (1) are the single most common length over all the top predators in each web. There is no attempt to exclude data where the basal species are animals (i.e., plants or detritus are omitted from the published web) or from which the top predator is unlikely to be the end of the food chain. In the studies compiled by Pimm (2), each top predator is counted separately, and webs were excluded if plants or detritus were not at the base of each chain. Webs for which small invertebrates were not the end of the food chain (and so are less likely to be incomplete) are shown in the third line of the table (3).

| | Trophic Levels | | | | | | |
	1	2	3	4	5	6	Source
Frequency	2	45	37	22	5	3	(1)
	0	25	44	25	7	2	(2)
	0	2	23	24	5	2	(3)

Table 13.2. Studies to Illustrate the Range of Primary Productivities in Aquatic Ecosystems

Productivity	Kind of System	Source
Marine		
mg C m^{-3} day^{-1}		
0.1–100	Various marine planktonic	Koblenz-Mishke et al. (1970)
200	Coastal upwellings	Barber and Smith (1981)
1000	Coastal upwelling	Barber and Chavez (1986)
mg C m^{-2} yr^{-1} (i.e., productivity integrated over water column and over year)		
50	Open oceans	Ryther (1969)
82	Sargasso Sea	Platt and Harrison (1985)
50	Sargasso Sea	Jenkins and Goldman (1985)
690–720	NW Africa, Peru Coastal upwellings	Barber and Smith (1981)
100	Coastal zones	Ryther (1969)
200–300	Three shelf–sea ecosystems from Bering Sea, Oregon, and New York	Walsh (1981)
> 1000	Coastal zone, Peru	Walsh (1981)
150	Upwellings	Ryther (1969)
Freshwater Lakes		
mg C^{-2} yr^{-1} (i.e., productivity integrated over water column and over year)		
5–6400	Various lakes	Likens (1975)
9	Lake Myvatn, Iceland	Johanssen (1979)
22	Ovre Heimdalsvatn, Norway	Larsson et al. (1978)
430	Lake Suwa, Japan	Mori and Yamamoto (1975)
450	Lake George, Uganda	Burgis et al. (1973)

Estimates based on daily rates.

energy flow sets the limits to food chain lengths, these data provide plenty of opportunity to observe variations in food chain length.

Trophic Efficiencies

The amount of energy transferred from one trophic level to the next depends on a variety of factors:

1. First, different systems accumulate or export very different amounts of energy. In Odum's (1971) terms, we now know that community production to respiration ratios vary from near unity (no net accumulation) to large values. (These large values may only be temporary, but they could provide an interesting test of how energy flow affects food chain length.)
2. The amount of energy passing through the grazing food chain also varies. Some systems, like the forests discussed by Reiners (Chapter 5), pass very little energy through the grazing chain; most of the energy in plant material passes via litter into a decomposer-based system.
3. Finally, there are very different efficiencies of converting energy (*A*) into production (*P*): the production efficiency (*P/A*). Since much of assimilated energy goes towards respiration, the range of production efficiencies reflects

these costs. *P/A* ranges from typically 1% in endothermic birds and mammals, through 10% in long-lived ectotherms like fish, to as much as 40% for herbivorous invertebrates and perhaps nearly 60% for carnivorous invertebrates (McNeill and Lawton 1970; Humphries 1979). These differences have the potential to cause considerable variability in the amount of energy flowing through food chains.

Secondary Productivity

Of course, it is possible that highly productive systems may be dominated by energetically profligate endotherms, and barren systems only by efficient insects. Or perhaps in productive systems, either energy accumulates, or more of it passes into the detritus-based food chain. Such effects would reduce the expected variation in food chain lengths imposed by energetic constraints. As I will also show, I think these compensatory mechanisms are unlikely. More probably, the effect of combining variations in productivities, efficiencies, and pathways is to increase, rather than decrease, the potential for variation in food chain lengths.

A simple way to look at this is to document the ranges of secondary productivity. If energetically efficient animals predominate in unproductive environments and vice versa, then the ranges of secondary productivity will be much smaller than those for primary productivity.

Odum (1971a) provides a compilation of secondary productivities in fish populations, where, for obvious economic reasons, there are good data. For herbivores, he shows studies that range from 22–3024 kcals m^{-2} yr^{-1}, and for carnivores, from 0.3–450 kcals m^{-2} yr^{-1}. Despite the different units from those I have used in Table 13.2, the pattern is clear. There is at least as much variation in the productivity of the second and third trophic levels as there is in primary productivity. I would not want to claim statistical significance for this observation; ranges are difficult to estimate statistically. It appears, however, that there is more variation in the productivity of progressively higher trophic levels.

In sum, there is considerable potential for variation in the amount of energy entering any given trophic level. Indeed, this variation may well become greater, the higher the trophic level. Yet the variation in food chain lengths observed in the real world is very small. Does what little variation there is in food chain lengths correlate with differences in energy flow?

13.2.2. Primary Productivity and Trophic Structure

Given measurements of energy flow, we can compile data on food chain lengths from a general description of the ecosystem's natural history.

Some Terrestrial Systems

In grassland and tundra ecosystems, primary productivities range from 35 g C m^{-2} y^{-1} (Bliss 1977) through temperate grasslands with values of 36–234 g

C m^{-2} yr^{-1} (French 1979a) to tropical grasslands with productivities as high as 690 g C^{-2} yr^{-1} (Lamotte 1975). The trophic structures of these systems are not markedly different, despite the twentyfold variation in energy flow (Pimm 1982). All these systems have predators, and there is a suggestion of a fourth trophic level in all the systems. I say "suggestion" because the food chains are not always completely described in these studies.

There could be compensatory changes in the amount of energy flowing through the grazing (vs. detritus) food chains, or in the groups of organisms constituting the food chains. These changes on energetic grounds alone might make it unlikely that we would see variations in food chain lengths, since compensatory changes would reduce the variability of energy reaching higher trophic levels. Yet factors affecting the flow of energy to higher trophic levels might, more realistically, tend to amplify rather than decrease the differences in food chain lengths that we might expect. The predators at the barren tundra sites are endothermic vertebrates (foxes, falcons, and the like), whereas the major tropical grassland predators are ectothermic vertebrates (lizards and snakes). This latter group is an order of magnitude more efficient at transferring energy, and there could be a twentyfold difference in the amount of energy entering the systems. Other things being equal, there ought to be 200 times more energy to be had from feeding on the tropical grassland's snakes than on the predators in the tundra.

Nor do grasslands accumulate more energy than tundra (and so pass proportionately less energy down the grazing food chain). In fact, the opposite seems to be the case. Tundras have large reserves of dead plant matter, grasslands have much less (Post et al. 1982).

Freshwater Ecosystems

Patterns similar to these emerge from studies of freshwater systems. Lake Myvatn in Iceland and Lake Ovre Heimdalsvatn in Norway have very low productivity (Table 13.2), yet both have a fourth trophic level (Johanssen 1979; Larsson et al. 1978). Some temperate and tropical lakes have productivities nearly 60 times that of Lake Myvatn (Table 13.2), yet have no more trophic levels.

Marine Systems

Ryther (1969) provides a discussion of marine productivity and trophic structure exactly comparable to what I have attempted for terrestrial and freshwater systems. He has assembled both measurements of productivity and descriptions of food chain lengths. He contrasts oceanic, coastal zone, and upwelling ecosystems.

The productivity of the open ocean is relatively low (Table 13.2). At the base of the food chain are oceanic nanoplankton that are too small to be effectively filtered by most of the common zooplankton crustacea. Intermediate between the nanoplankton and the carnivorous zooplankton is a group of herbivores, the microzooplankton, which includes protozoans and larval nauplii of microcrustaceans. Feeding on these tiny animals are many carnivorous zooplankton. In turn, these animals may be fed upon by chaetognaths. The food chain described so far involves three or four trophic levels from the very tiny phytoplankton to animals about 1 cm long. How many trophic levels from these organisms to the familiar tunas, dolphins, and squids is not clear. For these predators feed on fishes, and, if the fishes feed on the carnivorous zooplankton, we have a food chain of five trophic levels: nanoplankton, microplankton, carnivorous plankton, fish, and top predators.

In the coastal zone, primary productivities may be about twice that of the open ocean. The phytoplankton of the coastal zone are large enough to be filtered directly by common crustacean zooplankton such as copepods and euphausids. Ryther concludes that food chains here—at least those which people exploit—are three trophic levels long.

Upwellings are highly productive ecosystems (Chapter 9) and Ryther suggests an average productivity of 150 g C m^{-2} yr^{-1}. More recent studies (Table 13.2) suggest that the productivity may be much greater than this. It is in upwellings that food chains are shortest. The phytoplankton are of large size and many species are colonial in habit, forming colonies as large as several cm in diameter. Such aggregates of plant material can be readily eaten by quite large fishes. Many of the clupeoid fishes (which include anchovies and sardines) have specially modified gill rakers for removing the larger species of phytoplankton from the water. Thus, it would appear that humanity is exploiting a system with two trophic levels.

Finally, Ryther goes on to consider the efficiencies of these systems. They tend to become more efficient, rather than less, as the food chains become shorter. Ryther is considering food chains involving humans as predators—on tuna in the open ocean or on sardines in upwellings. But people are not obviously different from other predators in selecting fishes: people still hunt them, albeit with some sophisticated technology. So, the relationship of the length of the marine food chains to productivity is of considerable interest to us. Quite simply, food chains are longer in the *least* productive marine habitats.

13.2.3. Some Detailed Studies of Food Webs in Different Systems

The construction of a food web from field observations requires two key pieces of information—the list of species present, and which feed on which. For most cases, one or the other of these sets of information will prove disabling, or at least restricting, in any attempt to produce a reliable food web. Several problems underlie this tendency. The incompleteness of most species lists, the inability to determine the diet of each species, and the generally open nature of the habitats concerned all contribute. Phytotelmata—the class of habitats made up of plant-held waters—provide us with systems that can minimize many of these problems. These aquatic situations usually contain a detritus-based animal community which is highly endemic to the particular situation, highly circumscribed physically, and much replicated within the wider ecosystem in which it occurs. Five basic classes of phytotelmata occur in living plants: water bodies which occur: (1) in the axils of leaves, petals, and bracts of trees; (2) in the internodal spaces of bamboos; (3) as "tanks" within bromeliads; (4) in pitcher plants; and (5) in tree holes (Kitching and Pimm 1985). The range and frequency of occurrence of such habitats can be judged from Fish (1983), who records 150 plants from 28 families from which insect containing communities have been recorded. Yet Fish excludes, for practical reasons, bamboos (an estimated 80 genera), bromeliads (about 40 genera), and trees from his list. Phytotelmata are worldwide in distribution, though not surprisingly, they are more numerous and diverse in the wet tropics.

Pitcher Plant Faunas

The pitcher plants represent three families, the Nepenthaceae, Sarraceniacaeae, and Cephalotaceae. The last two families contain only four genera. Sarraceniaceae has three genera and is Pan-American, but disjunct in its distribution. Cephalotaceae contains a single genus restricted to Western Australia. The Nepenthaceae, in contrast, contains some 70 species scattered throughout the Indo-Pacific region. Beaver (1985) summarizes their distributions. The genus has some 28 species known from Kalimantan (formerly Borneo), 10 from the Philippines, 19 from Malaysia, down to single species known from Sri Lanka, the Seychelles, Australia, and New Caledonia. There are two species on Madagascar.

Beaver (1985) has documented the organisms which live in the various species of *Nepenthes*, as well as reviewing a scattered literature. Beaver (1985) summarizes five webs from his own work in Malaysia and that of other workers on the faunas from the pitchers of Sri Lanka, Madagascar, and the Seychelles. All the webs have the same energetic base: the insects that are captured by the pitchers. At the next trophic level are saprophitic organisms that feed either on the dead, undigested material or on the older detritus (and possibly the complex of microorganisms associated with it). Some webs contain another trophic level: predators on the saprophages.

After examining these webs, Kitching and Pimm (1985) agree with Beaver (1985) that the distributional pattern of the host family seems paramount in explaining the features of the trophic structure. The webs from the center of *Nepenthes* distribution in Malaysia are notably more complex than the others in having both more predators and saprophages. The webs from outlying areas—Australia, Madagascar, and the Seychelles—have fewer saprophages and lack the third trophic level.

We do not know the level of energy input to any of these systems. Yet the similarity of pitcher structure, the habitats in which the pitchers occur, and the constraints imposed in how they capture insects, suggest that they do not differ markedly in the amount of energy they capture.

Tree Hole Faunas

Kitching and Pimm (1985) compiled four webs for tree holes. The most dramatic pattern in these webs reflects gross climatic change roughly on a latitudinal basis. The richest web in both species and trophic levels is that from a subtropical rain forest in Australia (3–4 trophic levels); the sparsest is from England (2 levels). The leaf fall of each of these locations is within a few percent of the other—roughly 1 kg dry wt m^{-2} yr^{-1}.

While Kitching and Pimm had no webs from tropical forests, we know that some tropical forest tree holes contain such predators as odonate nymphs, including several genera which appear to be specialists in this situation (Corbet 1983). The presence of these odonates, along with the increased diversity of frogs and other groups with predatory larvae that use tree holes, leads us to suspect that the trophic structure of tropical tree holes would be even more complex. Thus, Kitching's and Pimm's compilation only hints at the possible variation in trophic structures in food webs. Yet large differences in the energy entering these systems is not very likely.

In sum, it seems unlikely to me that the natural variation in food chain lengths could correlate with different levels of energy input. The pitchers are broadly comparable in both structure and the habitats in which they occur, over their entire range. Similarly, correlation of food chain lengths and level of energy input also seems unlikely for tree holes.

13.2.4. An Experiment on Energy Flow and Trophic Structure

Pimm and Kitching (1987) experimentally induced variation in food chain lengths in water filled tree holes in subtropical southeast Queensland. These tree holes commonly contain three trophic levels, and, as already noted, comparable tree holes in temperate forests typically have only two. What can we do to these communities that will lead to shorter food chains?

The three levels in these communities are: (1) basis detritus which enters the habitats allochthonously; (2) saprophages, including scirtid beetles, culicid, chironomid, and psychodid fly larvae, ostracods and algophagid mites; and (3) predators, such as arrenurid mites, the tanypodine chironomid, *Anatopynia pennipes*, and tadpoles of the leptodactylid frog, *Lechriodus fletcheri* (Kitching and Pimm 1985; Kitching and Callaghan 1981).

This community will also develop in analogs of the habitat, water-filled pots, placed around the bases of host trees in the rain forests in which natural holes frequently occur. Pimm and Kitching (1987) performed experiments using such tree hole analogs to test the effects of levels of energy input on the lengths of the resulting food chains. Three levels of leaf litter input were maintained over a 24-week period at three study sites, with three replicates per site, in the Green Mountains Region of southeast Queensland. The levels of leaf litter input approximated the average level (approximately, 1 kg m^{-2} y^{-1}) in a nearby rain forest site (Plowman 1979), twice the average, and half the average. Litter was added every 3 weeks to 1-liter pots from which natural leaf input had been excluded by coarse netting. This did not interfere with the ingress of ovipositing insects or frogs. Pots were sampled destructively at 3-week intervals.

Who eats what? The tadpoles caused a significant reduction in the number of mosquitoes. Placing tadpoles in similar pots in the laboratory also showed how quickly they could reduce mosquito numbers. In contrast, the tadpoles caused a slight increase in the numbers of saprophagous chironomid larvae. The predatory chironomid, *A. pennipes*, reduced the numbers of saprophagous chironomids. Simply, tadpoles eat mosquitoes and the predatory chironomids eat saprophagous chironomids. Both saprophagous chironomids and mosquitoes were present at our first sampling, i.e., after 3 weeks, and increased thereafter to initial peaks at 15 weeks. For neither group was there any relationship between numbers and the level of energy input. The larvae of *A. pennipes* began to appear in the samples after 9 weeks, but were rare until about 18 weeks, and were still increasing dramatically at the end of 24 weeks, and even at 30 weeks in the limited samples we took at the end of the experiment. *A. pennipes* numbers increased slightly, but significantly, with the level of energy input. The increase in numbers with time, however, was highly significant and closely approximated exponential growth with different rates for the different levels of energy input.

Could this increase be simply due to a seasonal increase in the numbers in the surrounding habitat? Pimm and Kitching (1987) thought not. In contrast with this experimental result, *A. pennipes* is present in most natural tree holes throughout the year, with little discernible variation in numbers. Kitching (1987) found that of 11 natural tree holes sampled monthly for a year, *A. pennipes* was present in eight of the holes at least 10 months of the year. Abundances—averaged across the holes—showed no discernible trend from September through May (during which our experiments were conducted). There was a slight and nonsignificant increase in numbers during the winter months, June through August.

The density of the tadpoles of the top predator, *L. fletcheri*, also showed a pattern related to the level of energy input. But it was the *opposite* to that expected by the energy constraint hypotheses—density decreased with increasing levels of energy input.

Kitching and Pimm showed that one predator, the tadpole, avoided rather than preferred the most productive of the habitats they presented to it. The numbers of *A. pennipes* did increase slightly with increasing energy input. More importantly, the food chains are shorter early in the successional sequence. *A. pennipes* did not colonize the pots until its prey were not only abundant but also sufficiently large. When it did colonize, it significantly lowered the numbers of its prey. We argue that in our treatments, shortening the supply seems a plausible explanation for the variation in the food chain lengths. And we have seen the effect of an experimental disturbance, which shortened the food chain in tree hole analogs. Moreover, it is certainly plausible to suggest that the upwelling systems discussed by Ryther (1969) may be more variable in time and space than the open ocean systems with their lower productivity.

Pitcher Plants and Tree Holes

It is reasonable to assume that *Nepenthes* pitchers on plants at the extremes of the distribution of the genus represent more difficult habitats for colonizing insects than do those closer to the center of the distribution of the genus. Those plants on the periphery of the distribution encounter their edaphic limits and, accordingly, present a spatially dispersed and often isolated environment for the infauna. Or, if the genus is assumed to still be in the process of radiation, peripheral isolates may represent the more novel situations in evolutionary terms. In either case, environmental uncertainty interacting with the dynamics of the webs could produce the observed patterns of shorter food chains in the peripheral areas. Similarly, the variation in tree hole webs observed along a gross climatic axis may be explained as representing variations in environmental predictability—which, arguably, may be greater in more equable warmer forests than elsewhere.

By introducing pristine habitat units into an environment which contains an abundance of natural tree holes and, presumably, a well-stocked pool of potential ovipositing females, Kitching and Pimm (1987) simulated an environmental disturbance, yet it is hardly a major one. They noted that larval *A. pennipes* are widespread and common throughout the year in natural holes. But is this disturbance comparable to the disturbances tree holes experience in temperate forests that could lead to their lacking a third trophic level, whereas tropical forests lack the disturbances and have the third level? There is no absolute way of answering this question. We must look to natural systems for confirmation that the experiments involve a small yet influential perturbation, and that the significant factor causing the difference in food chain lengths between tropical and temperate tree holes is the severity of the disturbances. In those systems which have short food chains, over what spatial scales are species densities severely reduced by perturbations, such as unusually low or high temperatures, or flooding? I suspect that, for natural holes in a temperate forest, winters (particularly severe ones) may influence species densities on a far broader geographical scale than in our experiments. While in a mild winter most holes may retain some unfrozen water, in severe winters, very few holes may escape freezing. The experimental disturbance is probably small compared with that of a temperate winter where trophic activities and colonizations are suspended by the cold temperatures, on a continental scale.

The critical questions, therefore, are whether temperate systems experience more severe or more frequent disturbances than tropical systems and, if they do, why are predators unable to develop adaptations which permit them to survive these difficult periods? While it is useful to have a knowledge of the driving climatic variables, the essential comparison involves the animals themselves. We cannot easily answer why species fail to make adaptations; we are always forced into admitting they cannot because they do not. What we can do is to ask whether the animal communities themselves show greater spatial and temporal variability in temperate systems. We need comparable data on tree hole faunas through time in various locations. We would also need a knowledge of the spatial scale of variation in communities: how close to each other do holes need to be before they strongly affect each other's dynamics? We need this knowledge of spatial correlation in order to estimate how quickly the communities would recover given the severity of disturbances and the areas over which they occur. It is clear that it will be difficult to measure the severity of disturbance, though Kitching's (1987) studies of spatial and temporal changes in tree hole communities show how this might be done.

13.3. What Determines Resilience?

A second set of predictions about food chain lengths comes from considering what affects the rate at which species can recover. How this rate interacts with the severity of disturbance to affect population persistence will be considered later. First, we need to consider what modifies these rates of recovery. Resilience measures the speed at which a system recovers after a disturbance; $1/T_r$, where T_r is the recovery time, provides a suitable measure of resilience. In an important paper, DeAngelis (1980) shows that system resilience can be related to two fundamental structural concepts: (1) the energy flow through the system per unit of standing crop, and (2) the recycling index that measures the mean number of cycles a unit of matter makes in the system before leaving it. A simple index which combines both of these, the transit time of a unit of matter through the system, provides a rigorous measurement of resilience. DeAngelis reviews three sets of studies:

1. In an attempt to understand underlying similarities and differences among ecosystem types, O'Neill (1976) used data from six diverse ecosystems: tundra, tropical forest, deciduous forest, salt marsh, spring, and pond. O'Neill set parameters for a nonlinear energy flow model for each of the six ecosystems. The six models had the same compartmental structure and were subjected to a standard perturbation. O'Neill found that recovery times decreased as energy input per unit of standing crop increased. The tundra system had the longest recovery time. The pond, with a low standing crop and high biomass turnover had the shortest recovery time. O'Neill related his results to a suggestion by Odum and Pinkerton (1955). They defined *power capacity* as the quantity of energy processed per unit of living tissue and hypothesized that greater power would result in a greater capacity to counteract change. In other words, more power, greater resilience.
2. By using simple Lotka–Volterra models, Pimm and Lawton (1977) examined resilience as a function of the number of trophic levels. We found that, as the number of trophic levels increased, so the average recovery time also increased. This result, DeAngelis argued, is entirely compatible with O'Neill's ideas. The flux of energy or biomass through the system has an important effect on resilience. The higher the flux, the more quickly the effects of the perturbation are swept from the system and so the higher the resilience. Long food chains may be expected to decrease this flux. Of course, there are likely to be other factors which will affect the flux and these also will account for system to system differences in resilience.
3. Finally, DeAngelis discussed the recycling of nutrients. Some nutrients may be held very tightly by a system and recycled many times before being lost,

others may be washed out of the system much more rapidly. Jordan et al. (1972) argued that models of nonessential elements tend to be more resilient than models of essential elements. They suggested, as an explanation, that essential nutrients, such as calcium or phosphorus, tend to be tightly cycled. A perturbation to these elements tends to dampen only slowly. Minerals that are not essential (cesium, for example) are lost at a high rate from the system and the disturbance disappears quickly. A similar argument was made by Pomeroy (1970), who pointed out that coral reefs and rain forests are examples of systems with tight nutrient cycles. When systems of this type are disturbed, recovery may be slow, because there is little throughflow of nutrients.

The basic factor determining the resilience of nutrient cycling models seems to be the degree of recycling, while the factor determining the recovery of food web or energy flow models (with no feedback) is the energy or biomass flux per unit standing crop. DeAngelis points out that these factors are really similar. They both relate to the rate at which a given unit of material or energy is carried through the system from the point where it enters to the point where it leaves. He then goes on to show that these intuitive ideas can be mathematically formulated. He shows that it is possible to define a single index, transit time, that incorporates both power capacity and recycling. Transit time is the expected time that a unit of energy or matter remains in the system; i.e., the time from input to output. Transit time is strongly related to recovery time: the longer transit time, the longer recovery time.

13.3.1. Resilience, Disturbance, and Species Persistence

How do these proposed changes in resilience translate into variability in a population subject to random disturbances? Resilient populations (i.e., those that return quickly to equilibrium following a perturbation) are able to maintain relatively steady trajectories, because they spend more time close to some average value (e.g., equilibrium). A stiff population (i.e., one with long return time), in contrast, will tend to be further from equilibrium at the time of the next perturbation, and will thus reach more extreme abundances than a resilient population. Short return times should correlate with low population variability.

Unfortunately, this simple idea is not the only possible relationship between population variability and resilience. A second alternative scenario suggests that a long return time renders a population less responsive to changes in environmental quality (it "sits it out" until more favorable conditions prevail), and so population variability is relatively low. A population with a short return time tends to track the variations in the carrying capacity, and is thus relatively variable. In addition, a population which is too resilient may first overshoot, and then as a consequence, undershoot its equilibrium level. The more resilient the population, the greater the population variability; indeed, this is the mechanism that can yield increasingly chaotic dynamics, and an increased chance of extinction, with increasing resilience (May 1974).

These alternatives need not be mutually exclusive, since the relation between the environment and the dynamics of a population will not necessarily be the same for different species. Moreover, different time scales of variability also may affect which alternative is obtained. Consider a population that has a relatively short return time (high resilience). It may have a relatively high variability over the short term (alternative 2), because it overcompensates for short-term changes in the environment. At the same time, it may have a relatively low component of variability due to long-term trends (alternative 1). Conversely, a population with a long return time may show a large trend component of variability (alternative 1) (since, at the time of the next perturbation, it has recovered only slightly from the previous one), but a small short-term component (alternative 2).

Lastly, the explanations for the second alternative become less appealing than those for the first alternative, if populations spend most of their time recovering from rare but large perturbations to their abundances. This is the case for some terrestrial bird populations (Pimm 1984), and we are currently investigating whether it is a general ecological phenomenon. The bird populations, as a consequence, show a negative relationship between resilience and population variability. The idea that this might be a general phenomenon comes from considering the nature of environmental variation.

The amplitude of environmental variation increases as the period (P) of the variation increases; that is, large perturbations occur relatively infrequently. It has been suggested that in marine systems this amplitude is proportional to P (Steele and Henderson 1984).

In short, it is likely that long recovery times decrease species persistence, even though the opposite effect is possible. There is a clear need for a better understanding of environmental variability and its interaction with resilience, but long recovery times—and the long food chains that can cause them—seem to be unlikely in a constantly varying world.

13.4. Summary and Discussion

Food chain lengths, once thought to be determined by the overall level of energy flow through an ecosystem, do not appear to correlate well with energy flow. What few data we have suggest that the severity of disturbance is a better correlate: it is how often communities experience energy shortages, or are deprived of the inability to exploit energy, that counts. The length of food chains is determined by a balance between factors which shorten the chains (disturbances) and the species' ability to recover from these disturbances.

Where these ideas become complicated is also where they fall short. I am led to conclude that what matters is the spatial and temporal variation in population density and resilience. These factors are familiar in ecosystem ecology, but they are not ones we understand very well. What we lack are empirical data on the spatial and temporal variation of populations, communities, and patterns of energy flow. Steele's (1985) observation that there might be a fundamental difference between the temporal variability of marine and terrestrial ecosystems is both fascinating and indicative of our ignorance; we have very little knowledge of variability over either time or space. There is not even much theoretical exploration of these ideas. And while we have scattered results on the rates of energy flow and nutrient recycling, there are precious few attempts to compile these data and ask comparative questions. Yet, for a complete understanding of food chain lengths and, I suspect, many other aspects of trophic structure, we will have to document ecosystem variability and resilience.

14. Scale, Synthesis, and Ecosystem Dynamics

H. H. Shugart and D. L. Urban

Systems analysis in ecology has tended to condense and aggregate complex systems, a tendency that has a tradition going back to the classic work of Lindeman (1942). This orientation has been reinforced by a hearty interest in systems analysis techniques that use a small number of differential equations to represent the compartments into which an ecosystem is divided (for examples see Patten 1971, 1972, 1975, 1976; Shugart and O'Neill 1979). The use of such models in the synthesis of element cycles is such that compartment models and "box and arrow" diagrams dominate most documentations of ecosystem element cycles. However, several of the preceding chapters cite the need to incorporate mechanisms and processes in ecological models to a greater degree to further our understanding of ecosystem dynamics.

Despite the importance of compartmental models to understanding element cycles in particular, and to ecosystem science in general, these relatively sparse representations of ecosystem components are not necessary to ecosystem science. Nor does an interest in understanding ecological processes lead to the morass of reductionism which is often seen as the antithesis of ecosystem science. It is appropriate to consider the original definition of the term, ecosystem, as developed by Tansley (1935), and the context in which this word was coined. Tansley was one of a large number of prestigious ecologists who wrote in a *Festschrift* dedicated to Henry Chandler Cowles, who, like our honoree in this volume, Eugene P. Odum, was a visionary ecologist able to inspire his colleagues with his breadth of interest and his development of "big picture" concepts (ecological succession). Tansley began his paper by noting that he would clarify vegetational theory, particularly with respect to statements by Phillips (1934, 1935) and F. E. Clements (1916). Tansley was critical of the dogmatic positions taken by Phillips on the organismal theory of vegetation ("Clements appears as the major prophet and Phillips as the chief apostle, with the true apostolic fervor in abundant measure."). Tansley recognized the importance of processes and dynamics to the development of ecology as a science. To this end, he defined a new term, the ecosystem (Tansley 1935):

> The more fundamental conception is, as it seems to me, the whole *system* (in the sense of physics), including not only the organism–complex, but also the whole complex of physical factors forming what we call the environment—the habitat factors in the widest sense. Though the organisms may claim our primary interest, when we are trying to think fundamentally we cannot separate them from their special environment, with which they form one physical system.
>
> It is the systems so formed which, from the point of view of the ecologist, are the basic units of nature on the earth . . . These *ecosystems*, as we may call them, are of the most various kinds and sizes. They form one category of the multitudinous physical systems of the universe, which range from the universe as a whole down to the atom. The whole method of science, as Levy (1932) has most convincingly pointed out, is to isolate systems mentally for the purposes of study, so that the series of *isolates* we make become the actual objects of our study, whether the isolate be a solar system, a planet, a climatic region, a plant or animal community, an individual organism, an organic molecule or an atom. Actually the systems we isolate mentally are not only included as parts of larger ones, but they also overlap, interlock and interact with one another. The isolation is partly artificial, but it is the only possible way in which we can proceed.

The ideas put forth by Tansley, now more than 50 years ago, have been recently reiterated in what is now called "hierarchy theory" (Allen and Starr 1982; O'Neill et al. 1986). One of the principal advantages of this theory is the reconciliation of two approaches to the study of ecosystems. The first approach, which proceeds "from the bottom up," attempts to assemble large-scale phenomena from smaller-scale components; the emphasis is on larger-scale implications. An example from forest dynamics is the use of computer models simulating the processes that affect the growth of a single tree as a basis for projecting the long-term dynamics of forest landscapes (Shugart 1984). A contrasting "top–down" approach would involve analyzing the patterns seen in a forested landscape to infer the processes that generated the pattern; i.e., working out finer-scale details. There is a tendency to think that one or the other of these approaches should typify the modus operandi of the ecosystem scientist, particularly in ecosystem dynamics.

Hierarchy theory, which is useful for structuring mathematical models to simulate multiscale phenomena, may also help reconcile the "bottom–up" and "top–down" approaches to studying ecosystems. A hierarchy is a partially ordered set in which the elements are ranked by asymmetric relationships among themselves. Any number of criteria might be used to order a hierarchy.

For example, the higher levels in the ordering might: (1) be larger than, (2) behave more slowly than, (3) contain, or (4) control lower hierarchical levels (Allen et al. 1984). The following example, in which classic control theory is applied to simple models of element cycles, illustrates how such an ordering might occur.

14.1. Behavior of Simple Element Cycling Models in the Frequency Domain

Child and Shugart (1972) developed an analysis in the frequency domain of a simple model of magnesium cycling in a Panamanian rain forest. This work was continued in a later project (Shugart et al. 1976) involving the analysis of a more complex calcium model for a temperate deciduous forest. These papers discussed the difficulties in analyzing the control aspects of negative feedback cycles, if such cycles were installed on the output of an ecosystem element cycle (as represented by the model).

While these analyses were strictly theoretical, the feedback systems would be relatively easy to imagine. If one were measuring the response of any ecosystem component to a given input (e.g., the output from an industrial plant) in order to assess the effect on the ecosystem, the analysis would determine if a feedback system that reduced the external input in proportion to measured increases in the element content of the compartment was stable (and vice versa—increased inputs if the element content decreased). The inertia in responses at the measuring point could produce system instabilities, as could the amount of gain (how much the input was changed with respect to a given measured deviation in input). Monitoring ecosystem response as a basis for making decisions regarding the magnitudes of inputs to ecosystems is a standard practice in environmental policy at all levels.

One way to express the effectiveness of such control systems is by using compartmental models (linear models with a conservation of mass assumption) to compare the response of the total system (the ecosystem plus the feedback cycle) over a range of frequencies with and without the control system in place. The resulting curves (formed by plotting the ratio of the variation in the measured ecosystem component with feedback to the variation without feedback) are called deviation curves. A set of deviation curves for the performance of several different feedback cycles, each monitoring a different component in a terrestrial calcium cycle, is shown in Fig. 14.1. These negative feedback cycles use information on the variation of a monitored ecosystem component from a desired set point to alter the input. The cases shown are among the approximately 20% of the 116 monitoring and negative feedback control systems investigated by Shugart et al. (1976). These cases are interesting because they exemplify the behavior of linear systems in the frequency domain. The deviation curves shown (Figure 14.1) feature three regions of interest.

1. The negative feedback control systems successfully reduce the effects of periodic low frequency inputs, as shown by the index of the performance

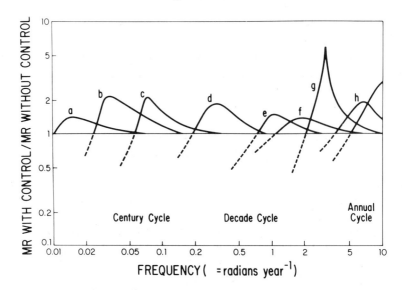

Figure 14.1. Deviation curves for negative feedback systems that amplify input variations in 1–100 yr. periodicities. Deviation curves are plots of the ratio of (MR with feedback/MR without feedback) vs. frequency (in radians per year). MR is the component amplitude/input amplitude. Note that both axes are logarithmic. The feedback control systems shown are:

a. monitoring Ca content in the boles of canopy trees, and altering the input with a gain of −10.
b. monitoring Ca content in the central roots of canopy trees, altering the input with a gain of −100.
c. canopy central root, gain = −1000.
d. standing dead wood, gain = −1000.
e. litter layer woody material, gain = −100.
f. decomposed litter, gain = −10.
g. decomposed litter, gain = −100.
h. small dead roots, gain = −100.
i. leafy litter, gain = −100.

From Shugart et al. 1976.

of the control system having values of less than one on the left of a deviation curve. This effectiveness results from the tendency of the system (as represented by the model) to "track" low frequency inputs. Therefore, at low frequencies, this monitoring point provides a reasonable estimate of the current input to the system.

2. There is an intermediate range of frequencies at which the system dynamics (as manifested at the particular monitoring point) are a major part of the signal that is measured. In the cases shown, the control systems take on positive values in this intermediate range of frequencies, indicating that the control systems actually amplify variations that are periodic at these fre-

quencies. These are very undesirable attributes for any control system. These cases occur because internal system inertia causes the measured responses to give misleading information as to the nature of the input. The increases in variation of system response with the control system in place can be as much as six times greater than without the control system. This is evidenced by the positive values for the deviation curves (Fig. 14.1).

3. At the highest frequencies, the system essentially attenuates the input signal, and the deviation curves become asymptotic around 1. The deviation curves illustrate the possibility that monitoring schemes that do not consider internal system dynamics can perform poorly, even in relatively sluggish and mathematically well-behaved systems such as those represented by element-cycle compartmental models. This analysis also demonstrates a categorization of dynamics that is central to hierarchy theory.

14.2. Hierarchical Concepts and the Ecosystem

The hierarchical concept in the preceding section is related to the three zones in the deviation curves that we have discussed. The dynamics of large complex systems (such as those that Tansley seemed to imply in his "ecosystem" definition) are of three types. With a given time (and space) scale as a reference point, one can think of the system as being composed of:

Mechanisms. These parts of the system have rapid responses, and behaviors which are often used as a basis for understanding the system.

Constraints. These parts of the system respond slowly with respect to the time scale of interest, and are essentially nondynamic within this time frame. Strictly, constraints represent environmental factors or boundary conditions that govern the behavior of the mechanisms.

Phenomena. These system behaviors are dynamic in the time frame of interest. The behaviors of the components have their origins in the interactions of the mechanisms, as modified by the system constraints.

To clarify this terminology by an example, consider a forested ecosystem at a time scale of approximately 100 yr. To predict the increase in biomass of this system, typical state variables of interest would be the biomass of bole wood, the leaf area, or the average tree diameter (all of which could be quite dynamic at this time scale). The mechanisms associated with these dynamics might involve the annual tree energy budget and include such factors as whole tree physiology, root respiration, or annual productivity. Rapidly responding parts of the system, such as the leaf stomatal resistance or the translocation of photosynthate within the tree, function at too fine a scale to be considered. Slow (but nonetheless, dynamic) processes, such as pedogenesis of the soil and the evolutionary responses of the trees, would be virtually constant at this time scale and would be constraints on the range of expression of the system behavior.

14.3. The Importance of Scale to the Ecosystem Ecologist

This may seem rather curious terminology to ecologists not oriented toward models, but one attempt to unscramble data collected at the wrong sample interval with respect to the phenomenon of interest will inspire a keen attention to space and time scales in any scientist. In building interdisciplinary research teams, close attention to matching scales of time and space may not always guarantee success—but to not do so will inevitably lead to failure. Attention to scale may be the most important contribution of hierarchy theory to ecosystem ecology.

Consideration of scale necessarily originates in hierarchy theory. A number of authors in other fields have attempted to categorize phenomena of interest in a time and/or space scale domain. A good example is the diagram of Delcourt et al. (1983) of the space and time domain of several different phenomena (Fig. 14.2) involved in the dynamics of forests. In this example, one sees the scale of exogenous factors ("environmental disturbance regimes"), the biotic responses of ecological systems, and the domain of vegetational patterns. If the panels in the diagrams (Fig. 14.2) were stacked as if they were playing cards and a pin were driven through them, the processes in the vicinity of the pin hole would represent the phenomena of interest. Processes to the right and above the pin hole would be constraints; to the left and below would be the mechanisms.

The point of these comments is to provide perspective on the importance of reductionism to the ecosystem scientist. Ecosystem scientists often treat as mechanisms the same phenomena that other biologists treat as constraints. Just as in an apartment building, where one man's floor is another man's ceiling, one scientist's mechanism is another's constraint. These discussions can quite properly be directed to E. P. Odum, for in him exists an intellect that slides across organizational levels—with mechanisms transformed to constraints and vice versa—in a manner that is unique and exemplary. Usually the ability to look down for mechanisms and up for constraints is not found in one individual. The experimental ecosystem ecologist often considers his system of interest on a large scale and looks for mechanisms that cause system-level behavior. At the same time, the ecosystem modeler may be attempting to combine what is understood about ecological processes with the larger system behavior. If synthesis and ecosystem observation work in different directions on the organizational scale, this does not imply ecosystem scientists do not know how to proceed. To the contrary, it is a sign of vigor in our subdiscipline.

14.4. Hierarchy and Forest Ecosystem Dynamics

Various criteria can be used to order a hierarchy. The higher levels in a hierarchy may be larger than, behave more slowly than, contain, or control lower hierarchical levels. In the case of terrestrial landscapes, all four of these criteria

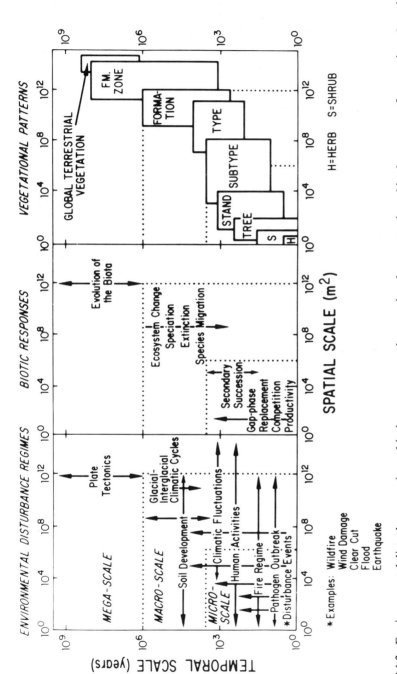

Figure 14.2. Environmental disturbance regimes, biotic responses and vegetational patterns viewed in the context of space–time domains in which the scale for each process or pattern reflects the sampling intervals required to observe it. The time scale for the vegetational patterns is the time interval required to record their dynamics. The vegetational units are graphed as a nested series of vegetational patterns. Reprinted with permission from Quaternary Science Reviews, Vol. 1. Delcourt, H. R., P. A. Delcourt, and T. Webb, III. Dynamic plant ecology: the spectrum of vegetational change in space and time. © 1983 Pergamon Press, Inc.

generally apply. The fact that landscapes are spatially nested (higher levels contain all the lower levels) provides for an especially rich conceptual and analytical framework (Allen and Hoekstra 1984).

In an idealized hierachical system, the first level comprises a number of small interacting components (Fig. 14.3). Subsets of these components that interact much more among themselves than they do with components further removed can be aggregated to form the components of the next higher level. These in turn can be aggregated into a higher level component, again based on their degree of interaction. The general rule is that within aggregate interactions are more frequent and intense than among aggregate interactions. The among aggregate interactions at one level become the within aggregate interactions at the next level. In this way, the dynamics of the system propagate mechanistically upscale. In this particular example the system is rate structured: the ordering priniciple within the hierarchy is in the patterns of the rates of interactions. Also in this example, the higher levels are larger because they are composed of the lower level components.

A forested landscape can be thought of as a hierarchical system (Table 14.1). The forest gap, the scene of gap–phase regeneration, makes a good reference level. Gaps have a spatial scale defined by the influence of a canopy tree and a natural frequency that reflects the life span of the dominant species (Shugart and West 1979, 1981; Shugart 1984). Higher levels of a forest hierarchy can be aggregated from gaps, based on the degree of interaction. Important interactions involved in these higher level aggregations in a forested landscape might include seed dispersal, pest outbreaks, or fire propagation. Because landscapes consist of spatially nested subunits, interactions are correlated by similarity of such attributes as species composition, soil type, etc., according to the interactions of interest. In this way, stands can be aggregated from similar, interacting gaps; stands interact among watersheds; and landscapes are composed of interacting watersheds.

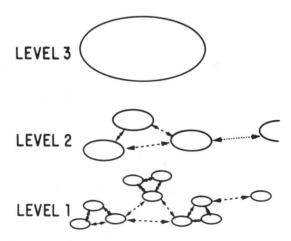

LEVEL 3

LEVEL 2

LEVEL 1

Figure 14.3. Schematic diagram of a generalized hierarchical system.

Table 14.1. Representation of a Forested Landscape as a Hierarchy

Level	Boundary Definition	Scale
Landscape	Physiographic provinces; changes in land use or disturbance regime	10,000s ha
Watershed	Local drainage basins; topographic divides	100s–1000s ha
Stand	Topographic positions; disturbances patches	1s–10s ha
Gap	Large tree's influence	0.01–0.1 ha

For example, consider the biomass dynamics of a relatively small area in an intact forest. Watt (1925) recognized that the dynamics of a forest at the ca. 1/10 ha scale were cyclical (actually an irregular repeating curve with a shape like a saw blade). Watt (1947) reiterated the importance of this small-scale disequilibrium in ecosystems in general in a presidential address to the British Ecological Society that provided examples from several different ecosystems including heathlands, grasslands, and forests. Because the period of this sawtoothed curve is more than 100 yr. in forested systems, the actual pattern of biomass response of a small area in a forest is rarely measured. The curves are patched together from observations of several gaps presumed to be in different stages of the cycle (as they were in Watt's (1925) doctoral work) . The shape of the biomass curve from a forest succession model (Fig. 14.4) illustrates the dominant features of the cycle. The death of a large tree causes the biomass curve to drop abruptly. Following this canopy mortality event, the smaller trees under the previous canopy dominant begin to grow and there is a flush of seedlings (because of enhanced seedling survival on the forest floor). This burst of growth causes the biomass of the gap to increase rapidly after the minimum that followed the canopy tree mortality event. The cohort of trees released by the canopy opening begins to grow, and the trees compete with one another for resources (light, water, nutrients). This competition and associated thinning cause the biomass response to slow to some degree. Eventually, one tree comes to dominate the stand and this tree grows until its death, which restarts the recovery cycle. Because the length of time that a tree lives has a stochastic component, the periodicity between cycles is variable.

This relatively complex curve can be aggregated to obtain an indication of the expected biomass dynamics of a forested landscape. A landscape disturbed by a large exogenous event that killed the canopy trees of several gaps should produce the following biomass response. Starting at a minimum after the mortality event, all of the areas that comprised the landscape would be synchronized and in the recovery phase of the cycle described above. The aggregated biomass curve would increase and would reach a maximum at the time in which each of the component plots would have one or two dominant trees. If

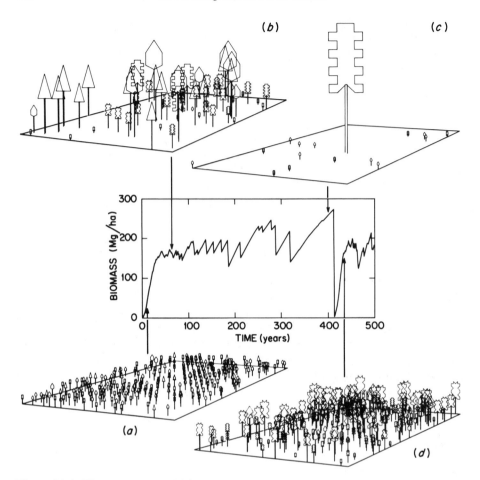

Figure 14.4. The structure and biomass dynamics of a forest gap as simulated by the FORET model for different parts of the gap–scale biomass curve (Mg/ha of trees versus time). a. Structure of regenerating forests. b. Canopy competition in an even-aged patch. c. Dominance of the plot by one canopy tree. d. Regeneration following the death of the dominant tree. The horizontal scales are twice the vertical scales. From Shugart 1984.

the mortality of these trees were not synchronized by another exogenous event, large tree mortality on the subareas would be desynchronized, the biomass curve would drop from its maximum, and the long-term aggregate biomass dynamics would settle about one equilibrium value. This larger-scale view of forest ecosystem dynamics is referred to as the "shifting mosaic steady state" concept of ecosystem dynamics by Bormann and Likens (1979a, 1979b) and the "quasi-equilibrium" concept by Shugart and West (1981) and Shugart (1984).

These two forest biomass curves, gap and landscape, differ radically from each other. The biomass response expected from a forest depends largely on the spatial scale used. In fact, the kinetics at either scale of a forest can be much richer than the examples shown here (see Shugart 1984, Chapters 5–6), but the point here is that scale is important in understanding and communicating ecosystem dynamics.

14.5. Conclusion

An important future challenge in ecosystem ecology is to develop the ability to transcend scale in ecological systems. This is not a matter of endorsing the efficacy of one procedure in science over another. Given the "top–down" and "bottom–up" approaches to science, our problem as ecosystem scientists is not in choosing one or the other. Two larger problems facing us today involve the following relationships.

1. The relation between processes at small scales and the ecosystem, (e.g., understanding how the small-scale responses of organisms to stresses such as air pollution relate to ecosystem response);
2. The relation between processes at large scales and the ecosystem (e.g., understanding the role of ecosystems in global element cycles and in the earth's heat budget).

These challenging problems both require a thorough understanding of the importance of scale in the dynamics of ecosystems. Considering scale as a prerequisite to progress in ecosystem studies may seem trivial to scientists who are involved primarily in studies with a narrow range of space or time. In fact, many subdisciplines in ecology have not ventured across certain scale boundaries, in part because the scientists in the discipline are not comfortable (in an intellectual sense) moving to different scales. For example, the motivation behind a great deal of physiological research (pollution effects, toxicological effects) is to understand a tissue level response, so this knowledge can be scaled to an ecosystem. Demonstrations that such scaling up can be done successfully are not abundant.

Acknowledgements

Research supported by the National Science Foundation's Ecosystem Studies Program (Grant No. BSR-85-10099) to the University of Virginia.

15. Towards Predictive Models for Coastal Marine Ecosystems

Kenneth H. Mann

Of the various themes suggested for discussion in this volume, I have chosen to concentrate on the relationship between ecosystem ecology and population ecology. There are scientists devoted to championing the cause of one or the other of these approaches, and there are those who, baffled by the complexity of ecosystems, wonder whether it is sensible to continue that line of work when the general principles seem so elusive. My position is that ecosystems obviously exist, and those who are interested in them should be encouraged to study them. The outcome of their work may lack the fine scale of resolution of those who choose to study smaller, more narrowly defined problems, but conversely, ecosystems will not be understood in their entirety by making a large number of studies of subsystems. Capra (1982), a subatomic physicist who has come to the same conclusion about physical systems, states:

> The systems view looks at the world in terms of relationships and integration ... Instead of concentrating on basic building blocks or basic substances, the systems approach emphasizes basic principles of organization ... What is preserved in a wilderness area is not individual trees or organisms but the complex web of relationships between them ... Systemic properties are destroyed when a system is dissected, either physically or theoretically, into isolated elements. Although we can discern individual parts in any system, the nature of the whole is always different from the mere sum of its parts. Another important aspect of systems is their intrinsically dynamic nature. Their forms are not rigid structures but are flexible yet stable manifestations of underlying processes.

The debate about "reductionism vs. holism" is an old one, and people have been inclined to take sides in it according to their philosophical preferences. In tracing the history of the debate about the relative merits of population ecology and ecosystem ecology as tools for the management of fish stocks in coastal marine habitats, we will see that a shift towards ecosystem science has come not from philosophical considerations but from practical necessity. In one of our examples, the view is taken that ecosystems exhibit properties that are self-generated; that fluctuations in populations having a certain periodicity are not necessarily traceable to external forcing functions of the same periodicity, but may be generated internally within the ecosystem. In other situations, the self-organizing properties of systems may lead to the existence of two or more stable points separated by areas of instability. Those accustomed to making ecosystem models by joining together a larger number of predetermined submodels will find the idea of self-organization difficult to accept. Certainly, much more evidence is needed to substantiate the idea, but it deserves serious consideration, for it seems to fit many situations in nature.

Between the extremes of those who study only the dynamics of a named species, paying scant attention to the environment of that species, and those who study ecosystem fluxes, caring little about the species that mediate those fluxes, there is a middle ground occupied by those who are concerned with the dynamics of one or more named species, but who try to take into account the whole suite of complex interactions with other organisms, and the physical and chemical environment. There is clearly a continuum between population ecology and ecosystem ecology, but there is a tendency for research projects to be clustered towards the two ends of the scale, creating the impression of a divergence, both conceptually and in people's assessment of what is useful.

15.1. Fish Population Biology and Ecosystem Properties

Nowhere has the clash of opinions about the relative merits of ecosystem science and population biology been more evident than in the arena of fisheries management. The traditional approach has been through very detailed studies of populations of commercial importance. Data on distribution, abundance, growth rates, life history, and recruitment have rightly been seen as the basic information required for the effective management of an exploited fish stock. However, it has gradually become apparent to fisheries managers that population data on the stock of interest are not enough to permit prediction of its future state. An indication of a change in thinking can be seen in the report of a symposium entitled "Multispecies Approaches to Fisheries Management Advice" (Mercer 1982).

> Traditional approaches to fisheries management, both in terms of stock assessment and management policy, have tended to treat species as if they existed independently of other species and were harvested independently of other species. The multispecies *or ecosystem* approach to resource assessment and development is accepted in principle ... (my italics).

The topics discussed in that symposium ranged from simple two species interactions, such as the predation of cod on capelin off Newfoundland, to considerations of the functioning of whole ecosystems, such as George's Bank or the Gulf of St. Lawrence. In general, the tone of the symposium was exploratory rather than definitive. Nevertheless, in the last few years, evidence that commercial stocks are strongly influenced by ecosystem properties has been accumulating rapidly. At this stage of our understanding, the evidence is often of a correlational nature, demonstrating coupling between changes in the physical environment and changes in the fish stocks. A full explanation of the coupling mechanism awaits further information on events at the ecosystem level. Other studies show clear evidence for interactions between the fish stocks and biological components of the system.

15.1.1. The Peruvian Upwelling System

Perhaps the most talked of example in recent years is the consequence of the 1982—83 El Niño (Barber and Chavez 1983 and Chapter 9). The Pacific Ocean in tropical latitudes adjacent to the coast of South America is normally capped by a layer of remarkably cool water, around 16–17°C. It used to be thought that this was due simply to the local upwelling of cool water from great depth, but it is now seen that the explanation is much more complex and involves events on the scale of the whole Pacific basin.

In normal conditions, the cool surface layer is relatively thin, about 40 m deep, and below it are nutrient-rich waters that are readily upwelled to the surface by winds blowing along the coast. As a result, this area is normally very productive of plankton, fishes (primarily anchovy or sardine), and fish-eating birds. An El Niño event occurs when this structure breaks down. The surface layer becomes warmer and thicker and the coastal winds upwell water that is no longer rich in nutrients. Primary production falls off sharply and this is reflected in successive links of the food web, so that there is poor reproductive success of fishes, birds, and sea mammals.

In the 1982-83 El Niño, the first report of a biological response was the reproductive failure of seabirds on Christmas Island (2°N, 157°W). By 1983, it had been reported from southern Peru that all the Galapagos fur seal pups born in 1982 had died; probably of starvation. By November 1982, the catches of hake off Peru had fallen catastrophically, apparently because the fishes had moved to deeper, cooler water. Jack mackerel, which normally occupy the offshore boundary of upwelling areas, were driven by the anomalous conditions to a narrow band of coastal waters where they were easy prey for large predatory fishes (tuna, bonito, dorado) that are tolerant of warm water. Anchovy, as they had done during previous El Niño events (Fig. 9.2), showed a marked decrease in abundance, with dead anchovies floating at the surface.

15.1.2. The North Pacific

Evidence of anomalous warming of surface waters was seen all the way north to California and beyond, and Fiedler (1984) has reviewed the consequences for northern anchovy along that coast. In general, these fishes exhibit enhanced population growth when temperature is higher, but in the second year of each El Niño event, the growth of individuals is depressed, probably by reduced productivity of food organisms.

Pacific mackerel, on the other hand, seem to do well in Californian waters during El Niño years. At these times, sea levels along the California coast are anomalously high, but as Sinclair et al. (1985) have shown (Fig. 15.1), there is a clear positive correlation between high sea levels and survival of Pacific mackerel. The explanation is not clear, for plankton productivity is reduced during the El Niño events. The authors speculated that, in normal years, large numbers of mackerel larvae are transported away by the California current and lost to the stock, while in El Niño years this effect is reduced.

Similar arguments have been brought forward to explain variability in recruitment of Pacific hake along the west coast of the USA. Strong offshore Eckman transport, associated with upwelling and with enhanced plankton production, is negatively correlated with larval survival. It is suggested that in years of strong offshore transports, many larvae are advected offshore and fail to return to the coastal nursery areas.

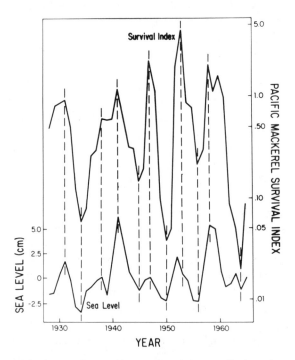

Figure 15.1. Survival index for Pacific mackerel (upper line) and sea level along the California coast. High sea levels indicate El Niño events. From Sinclair et al., reprinted by permission of Can. J. Fish. Aquat. Sci. (1985).

15.1.3. The North Atlantic

Leggett et al. (1984) studied the early life history of capelin (*Mallotus villosus*), which spawn on the beaches of Newfoundland. Larval emergence is keyed to the infrequent occurrence of onshore winds. As the waves break high on the beach, the larvae emerge from the gravel and enter a relatively warm water mass that is characterized by high food abundance and low predator densities. In the converse situation, when the prevailing summer wind blows offshore, cold water upwells and is characterized by high predator densities and low food abundance. Reduced wave action and low water temperatures prevent the larval fishes from emerging during offshore winds. When the interval between onshore winds exceeds that required for complete yolk sac absorption, larval condition deteriorates rapidly and subsequent survival and recruitment is poor. It is important to note that this situation does not reduce to a simple cause and effect relationship between wind and larval survival. The mechanisms of the effect involve the distribution of food and predators in the coastal waters, and is clearly an aspect of ecosystem functioning.

15.1.4. Recruitment Variability

Among fisheries managers, the debate about the relative importance of population parameters and ecosystem (or environmental) parameters often centers around the question, "What is the cause of year to year variation in recruitment to each fish stock?" Ricker (1954) distinguished between density dependent mortality (proportional to the density of the spawning stock) and density independent mortality caused by factors in the "environment" of the stock (or the ecosystem in which it lives). His equation was

$$R = aBe^{-bB}$$

where R = recruitment, B = spawning stock, a is the coefficient of density independent mortality, and b is the coefficient of density dependent mortality. This yields dome-shaped curves in which the density-dependent effect gives rise to reduced recruitment at high stock densities. Csirke (1980) found that by adding an index representing the degree of aggregation or dispersion of the stock (in response to environmental conditions), the modified Ricker equation could account for 80% of year to year variance in recruitment of Peruvian anchovy. Further analysis (Ware and Tsukayama 1981; Dickie and Valdivia 1981) showed that of the 80% explained by Csirke's (1980) equation, more than half was attributable to environmental factors.

While it is perhaps not surprising that, in a system as dynamic as the Peruvian upwelling, environmental changes appear to be the chief determinants of recruitment variability, it is interesting to note that similar results have been obtained elsewhere. Welch (1984) analyzed time series of 16 heavily fished marine stocks and found that he could separate out a variability in the stock recruitment relationship that was too short-term to be density dependent, and a component of variability that showed clear density dependent characteristics. The short-term environment (i.e., ecosystem) signal was the dominant cause of the variability in almost every case. The chief exception was Pacific halibut, where the density dependent component was strong enough to drive population oscillations.

15.2. Theories of Marine Food Webs

15.2.1. Trophic Dynamics

Charles Elton (1927), in his classic text, *Animal Ecology*, set the scene and stated the challenge in the following passage:

> When we are dealing with a simple food chain it is clear enough that each animal to some extent controls the numbers of the one below it. The arrangement we have called the pyramid of numbers is a necessary consequence of the relative sizes of the animals in a community. The smaller species increase faster than the large ones, so they produce a sufficient margin upon which the latter subsist. These in turn increase faster than the larger animals which prey upon them, and which they help to support; and so on until a stage is reached with no carnivorous enemy at all. *Ultimately it may be possible to work out the dynamics of this system in terms of the amount of organic matter produced and consumed and wasted in a given time, but at present we lack the accurate data for such a calculation, and must be content with a general survey of the process.* (my italics).

He really said it all. A more succinct statement of ecosystem dynamics would be hard to achieve. Almost 40 yr. later, in 1965, he was invited to write a preface to a paperback edition of the 1927 book. In it he made an interesting comment on the relationship of population dynamics to what we would now call ecosystem ecology:

> Enormous emphasis has been given to different theories about how population limits are determined and fluctuations regulated; and especially to the relative importance of inter-relations between animals (as in competition and the action of enemies or parasites) and that of other things like climate. Again, there are theories based on relations between species, and others on relations within species. *It begins to look as if ecology needs to develop, not swarms of mutually antagonistic theories about the regulation of populations, but a new kind of comparative ecology that would bring them all into perspective. For unfortunately, Nature cannot be understood by pretending that it is simple.* (my italics).

Lindeman's (1942) trophic–dynamic model may be seen as a response to Elton's original challenge, but as several people have pointed out (Murdoch 1966; Rigler 1975a; Cousins 1980; Platt 1985), the concept of a trophic level embodied in it has proved to be a major difficulty. Organisms do not conveniently fit into trophic levels and if the biomass of a trophic level cannot be specified, no predictions of the model can be falsified. To take an example with which I am familiar, the two most abundant fish species in the River

Thames, roach (*Rutilus rutilus*) and bleak (*Alburnus alburnus*), were found to be feeding on periphyton, plant and animal detritus, zooplankton, insects, filter feeding invertebrates, browsing and grazing invertebrates, and carnivorous invertebrates (Fig. 15.2, Mann et al. 1972). To which trophic level should they be assigned? Cummins et al. (1966) assigned the biomass of a particular species among several trophic levels after analysis of gut contents, but as Rigler (1975a) pointed out, if one wished to manipulate primary production and test its effect on herbivore biomass, it cannot be done, because the biomass in the herbivore trophic level is an abstract calculation, not a concrete entity.

Meanwhile, other workers had been baffled by the sheer complexity of natural systems. Beginning with Hardy's (1924) complex diagram of the food relationships of herring, it has become abundantly clear that a species by species modeling of complete food webs is unlikely to lead to enhanced predictive capability. In the early 1970s, there was a return to emphasis on size as a basis for grouping organisms in ecosystem models. H. T. Odum (1971), in his book *Environment, Power and Society*, worked out a hypothetical example of a planktonic food chain in which size of organisms increased logarithmically from phytoplankton to top carnivore, while the energy required to maintain a gram of living matter decreased in the same direction. Dividing the estimated energy flow in the food chain by the energy required to maintain a gram of living matter enabled him to deduce that the biomass of each size class should lie in the same range of 20–50 g dry wt m^{-2}.

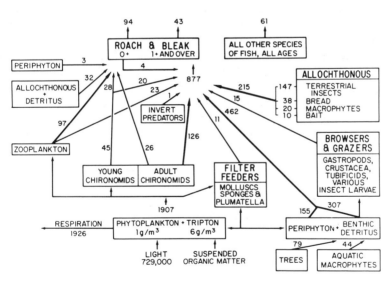

Figure 15.2. Food web in a freshwater section of the River Thames, as determined from energy budgets and examinations of gut contents of fish and invertebrates. Note that the cyprinid fish, roach (*Rutilus rutilus*) and bleak (*Alburnus alburnus*), consume periphyton algae, herbivores and carnivores, as well as detritus. Arrows indicate trophic transfers, in kcal m^{-2}yr^{-1}. From Mann et al. (1972).

15.2.2. Planktonic Systems

Meanwhile, Sheldon et al. (1972, 1973) had independently begun to develop size dependent models of planktonic systems. Using electronic particle counters, they began to investigate the size spectrum of particles in the plankton and noted that the volumes of particles in each logarithmic size class, over the range measured (1–10,000 μ), were roughly equal within an order of magnitude. Using arguments remarkably similar to those of Elton (1927), they argued that the higher rate of increase of the smaller particles would enable them to supply the food requirements of larger organisms. They even speculated that the trend would continue beyond the scale of their measurements, so that the biomass spectrum would be approximately flat over the size range from bacteria to whales.

In subsequent reviews, Cousins (1980) and Platt (1985) have pointed out that this was an implied rediscovery of the Eltonian pyramid for ". . . if there is a pyramid of numbers, and if it is smoothly sloping, it is always possible to find a simple logarithmic transformation that will map the pyramid into a rectangle" (Platt 1985).

Another line of thought developed the same concept. In a fisheries symposium on environmental conditions in the Northwest Atlantic, Dickie (1972) drew attention to the importance of body size in determining energy transfers in ecosystems. He mentioned particularly the size dependence of population generation time, and of production to biomass ratio. He remarked, " . . . these findings indicate we may be able to overcome the obstacle of unsatisfactory definitions of successive trophic levels . . ."

The way was open for an approach to ecosystem modeling based on the size dependence of metabolism, growth, life expectancy, and various other important properties of organisms. Platt and Denman (1977, 1978) developed a continuous steady state theory for the abundance of organisms in pelagic ecosystems as a function of body weight, incorporating the ideas of Fenchel (1974) and Banse (1979) on the size dependence of metabolism, growth, and rates of population increase. They used as their framework a logarithmic size spectrum in which the characteristic weight of each size class is double that of its smaller neighbor and one half that of its larger neighbor. In this arrangement, the width of a particular size class, Δw, is close to the characteristic weight, w.

Then, if a normalized biomass density function, $\rho(w)$, is defined as

$$\rho(w) = b(w)/\Delta w$$

where $b(w)$ is the biomass in the size class, this approximates as

$$\rho(w) = b(w)/w.$$

When the flow of biomass energy through the pelagic ecosystem was followed as it migrated from particles of small size to those of large size, calculating the turnover of material within each size class from the weight dependent rates of reproduction and respiration, it was found that the biomass spectrum had the form

$$b(w) \approx w^{-0.22}.$$

The practical difficulties associated with applying this type of model in biological oceanography have been discussed by Platt (1985). One of them is the assumption that all material flows from the smaller to the larger size classes. An obvious exception is the production of dissolved organic matter by phytoplankton of mean size about 25 μm, which is rapidly taken up by bacteria of about 1 μm. Clearly the model has to be modified to accommodate this process, which is not insignificant. The great advantage of the model is that it lends itself to verification using automated particle counting and data acquisition systems. For example, Harrison (1986) has used a size dependent model to show that in Canadian Eastern Arctic waters in August and September, most of the phytoplankton biomass and production was by particles larger than 35 μm (probably diatoms), while most of the community respiration was by much smaller organisms, and over 50% was by organisms of less than 1 μm.

Platt et al. (1984) showed that the microscope derived empirical biomass spectrum for the central gyre of the North Pacific published by Beers et al. (1982) was an excellent fit to their size dependent model, the slope being -0.23 ± 0.3 for the 0–20 m layer, and -0.20 ± 0.02 for the 100–120 m layer. They then used their model to calculate the total community respiration for particles of all sizes down to about 0.5 μm and found it to be about 720 mg C m^{-2} d^{-1} for the total water column. In the steady state situation, this is an approximation to the level of gross primary production in the central gyre of the North Pacific, and this figure has been used as a contribution to the debate about the widely differing estimates of the true level of primary production in the oceans.

Until recently, the distinction between living and dead particles, or between autotrophic and heterotrophic particles, was difficult to make without visual inspection, but the adaptation of the techniques of flow cytometry to biological oceanography have now made it possible to both count and sort particles on the basis of their fluorescence signature. It is therefore possible to frame hypotheses and test them operationally with the modern particle counters. Results of ecosystem level studies using this technique have not yet been published, but the value of the technique is well demonstrated at the organismal level by the work of Shumway et al. (1985), in which differential utilization of various phytoplankton species by filter-feeding molluscs was demonstrated, using flow cytometry to analyze changes in the proportion of cell types in the ambient water, the pseudofeces, and the feces of the molluscs.

15.2.3. Benthic Systems

The idea of using particle spectra has now been adapted to benthic studies (Fenchel 1969; Gerlach et al. 1985; Schwinghamer 1981, 1983, 1985; Warwick

1984). It has been found that the biomass spectrum is not flat, but has two well-defined and reproducible minima, at about 8–16 μm and 500–1000 μm equivalent spherical diameter (ESD). These correspond to the earlier notion of separating bacteria, meiofauna, and macrofauna. It has been suggested that the peaks represent optimization for quite different modes of life, with bacteria living on the surfaces of particles, meiofauna in the interstices between them, and macrofauna having a sedentary life in which they are adapted to the bulk properties of the sediment. Organisms of less than 500 μm ESD tend to be capable of producing several generations per year, to develop directly without planktonic larvae, to reach an asymptotic adult size, and to be nonsessile. Organisms larger than 500 μm ESD tend to be longer-lived (one or a few generations per year), to have planktonic larval dispersal, to continue growth throughout life, and to be sessile or sedentary. Hence it appears that meiofauna and macrofauna may constitute two separate evolutionary trends, each with a coherent set of biological characteristics. The troughs in the biomass spectra show that there is a characteristic size range associated with each mode of life, and there are very few organisms in the size classes representing a transition from one to the other. To the extent that they have this biological understanding, the benthic aquatic ecologists are in a better position to build size dependent models than are their planktonic colleagues.

15.2.4. Terrestrial Ecosystems and Coastal Macrophyte Systems

An obvious difficulty in trying to apply the planktonic size dependent model to terrestrial systems is the fact that primary producers on land are often very large particles indeed, and the flow of energy and materials is from large to small. Cousins (1980; 1985) has proposed a "trophic continuum" in which autotrophs exist in a whole range of sizes of particle (e.g., seeds, leaves, wood), each of which may be consumed by a herbivore of the appropriate size. Each type of autotroph and heterotroph generates detritus of various particle sizes, but whereas living organisms tend to grow, or to be eaten by larger organisms, detritus particles always get smaller (by fragmentation, erosion, and loss to detritivores). Hence, in parallel with the size-dependent flux from small to large in the living component of a system, there is a flux from large to small in the nonliving organic components (Fig. 15.3). I am not aware of any serious attempt to model a terrestrial ecosystem on these principles, but at least it is possible to see that there are no insuperable conceptual problems.

Coastal marine systems dominated by macrophytes such as mangroves, seagrasses, or seaweeds would fit this type of model. Primary production in the form of leaves or fragments of seaweed detritus could be fed into the model at the appropriate place in the size spectrum, and the fate of this material could be followed through the network.

15.2.5. Allometry as a General Principle

The ecological importance of body size, emphasized by Elton (1927), has its parallel in the physiological literature (Huxley 1932; Brody 1945; Kleiber 1961).

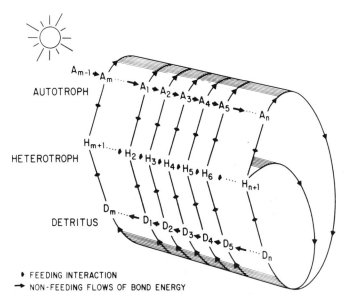

Figure 15.3. The Cousins model of the "trophic continuum," showing increase in particle size from left to right along the autotroph axis, transfer at each size class to an appropriate size class of heterotroph, but decrease in particle size with progressive fragmentation along the detritus axis. Broad arrows are trophic transfers, other arrows are nonfeeding flows. From S. H. Cousins, A trophic continuum derived from plant structure, animal size, and a detritus cascade. J. Theor. Biol., Vol. 82. © 1980 Academic Press Inc. (London) Ltd.

The common theme is that many physiological functions relate to weight, w, by the expression

$$y = aw^b.$$

This has become known as the allometric equation. The physiological property is related to weight not only by the coefficient a, but by the exponent b, the "other (allo-) measure of allometry" (Calder 1985). When data fitting this equation are plotted on log–log axes, the coefficient a is the intercept and b is the slope. Fenchel (1974), Banse (1979, 1982), Lindstedt and Calder (1981), Peters (1983) and Calder (1985) have elucidated a remarkable set of generalizations about organisms and populations that show great promise for incorporation into size dependent models. Basal or standard metabolic rate of mammals, R, has been shown to be approximated by

$$R = aw^{0.76},$$

from which it follows that the weight-specific metabolic rate is given by

$$R_s = aw^{-0.24},$$

showing that the tissues of larger mammals dissipate energy at a lower rate than those of smaller animals.

It is found that among the larger animals the life span ($t_{ls,\ max}$) follows the size relationships,

$$t_{ls,\ max} = cw^{0.20},$$

from which it follows that the energy consumed in a lifetime is given by

$$E_{total} \propto w^{-0.24} \times w^{0.20} \approx w^0,$$

so that mammals of all sizes use about the same amount of energy per unit of weight in their lifetimes. The small animals use it fast, the large ones slowly.

When we turn to groups other than mammals, it appears that their metabolism is related to weight by the same allometric laws, with approximately the same value for the exponent b. Differences between groups are reflected in differing levels of the coefficient a. For example, Fenchel (1974) suggested that unicellular, heterothermic, and homeothermic animals have characteristic values of a approximately in the ratio 1:8:225.

From the point of view of ecosystem models, population production is of great interest. It is a complex quantity made up of growth production and reproduction, and at first appears unlikely to conform to simple rules. Fenchel (1974) found that the "intrinsic rate of natural increase", r_m, was well described by the allometric equation,

$$r_m = aw^{-0.27},$$

clearly very similar to the equation for weight specific respiration. Since that time numerous analyses have confirmed that population production follows the same allometric rules as population respiration, so that within groups of metabolically similar animals, population production is proportional to population respiration. Since production, P, and respiration, R, are scaled in the same way to body size, it follows that production efficiency, $P/(P+R)$, is independent of body size. Also, it would be expected that the population production to biomass ratio, P/B, would be scaled in the same way as R/w and P/w; i.e., proportional to $w^{-0.25}$. Dickie et al. (1987) have reviewed the recent data and found that, taking the total range from unicells to mammals, this generalization is true. However, the data of both Humphreys (1978, 1981) and Banse and Mosher (1980) reveal that if animals are grouped in quasitaxonomic categories having a limited range of body size, regressions with steeper slopes of the order of -0.4 or -0.5 appear within the larger pattern. Dickie et al. (1987) have assembled evidence to suggest that this second order scaling is a function of spatial and temporal distributions of predators and prey in the natural environment.

It now appears that the groundwork necessary to convert biomass spectra into dynamic ecosystem models has been laid out free of the ambiguities imposed by generalizations about trophic levels. The use of size as the basis of classification opens the door to the powerful generalizations discussed in this section. On the other hand, some of the species–specific properties which have important influences on ecosystem structure (see Section 15.3) cannot be approached through size dependent models.

15.2.6. A Practical Application

To bring the discussion to practical questions of coastal zone ecology, we may note a recent example of the use of a particle–size spectrum model to make a rough estimate of fish stock sizes in unexploited areas. Moloney and Field (1985) began by testing the applicability of the biomass spectral model put forward by Sheldon et al. (1977) (Fig. 15.4) to two exploited regions of the southern African coast: the Benguela region and the Namibian region (Fig. 15.5). The theory is that if the shapes of all organisms are expressed as equivalent spherical diameters (ESD) and plotted in logarithmic intervals on a universal grade scale (Sheldon 1969), the particle concentrations (in parts per million by volume) in each size class are approximately constant over the whole range of sizes. If we take the specific gravity of marine organisms as 1.0, volume in ml is readily converted to biomass. Taking the southern Benguela region as an example, Moloney and Field (1985) estimated that the ESD of phytoplankton filled 21 size classes on the Sheldon scale, while fish covered 3.3 size classes when caught in a 32 mm net, and eight size classes when caught in a 12.7 mm net. If biomasses in all size classes are the same, the stock vulnerable to a 32 mm net, S_{32}, can be calculated as:

$$S_{32} = \text{Phytoplankton biomass}/21 \times 3.3.$$

Phytoplankton biomass was estimated from satellite imagery and ship-based

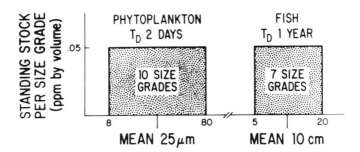

Figure 15.4. Diagram illustrating the "Sheldon Spectrum," applied to phytoplankton and fish. In the example given, phytoplankton occupy 10 size classes and have a doubling time of 2 d. Fish occupy seven size classes and have a doubling time of 1 yr. Note that the size classes are based on "equivalent spherical diameter." Actual lengths of fish would be 3–4 times greater. Making the assumption of a flat biomass spectrum, information on phytoplankton biomass can be used to estimate fish biomass. From Sheldon et al. (1977). Reprinted with permission from J. Fish. Res. Bd. Canada.

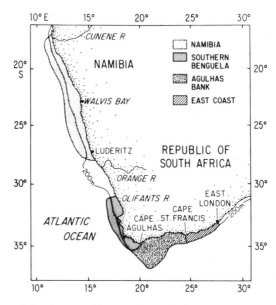

Figure 15.5. Map of Southern Africa to show the Southern Benguela Region and the Namibia region, for which the Sheldon Spectrum model (Fig. 5) was shown to give a reasonable prediction of fish stocks. The model was then used to make predictions for the Agulhas Bank and East Coast Regions (see Table 15.1). From C. L. Moloney, and J. G. Field (1985). South African J. Mar. Sci.

measurements to 50 m depth as 6.687 million tons wet weight, yielding an estimate of 1.050 million tonnes of pelagic fish vulnerable to a 32 mm net, or 2.57 million tonnes vulnerable to a 12.7 mm net. Of these, 50–75 % were estimated to be species of commercial interest, the remaining biomass being rejected from the catches. Similar calculations were made for the Namibia region (Table 15.1). Details of how comparisons were made between these stock estimates and the commercial catches need not concern us here; suffice it to say that the model appeared to give reasonable estimates (Table 15.1). Phytoplankton data from two unexploited areas were then used to provide estimates of fish stocks in those areas, and some suggestions were made about probable yields.

In assessing the utility of this model, we may note that the estimates of fish stocks are no better than the estimates of mean phytoplankton biomass. Before the advent of satellite imagery, such data were extremely difficult to obtain owing to the problems of assessing spatial and temporal variability. Even with good satellite coverage, many assumptions are necessary to extrapolate from what the remote sensor sees at the water surface to total biomass in the water column.

Secondly, the flatness of the biomass spectrum is only a rough approximation with many assumptions built into it that have not yet been investigated. These include the refinement introduced by Platt and Denman (1977, 1978) sug-

Table 15.1. Calculations of the Biomass of Pilchard and Horse Mackerel in Four Regions of Southern Africa, Using the Sheldon Spectrum Model[a]

Region	Biomass	Production	Fish vulnerable to a 32 mm mesh net		
			Estimated Yield	Maximum Catch	Mean Catch
Southern Benguela	525–790	525–790	130–395	497	288 (± 122)
Namibia	1320–1985	1320–1985	330–990	1548	443 (± 362)
Agulhas Bank	290–430	290–430	70–215	—	—
East Coast	65–100	65–100	16–50	—	—

Region	Biomass	Production	Fish vulnerable to a 12.7 mm mesh net		
			Estimated Yield	Maximum Catch	Mean Catch
Southern Benguela	1270–1910	1910–2865	475–1430	509	396 (± 47)
Namibia	3205–4810	4810–7215	1200–3610	1337	600 (± 362)
Agulhas Bank	700–1050	1050–1575	260–790	—	—
East Coast	160–240	240–360	60–180	—	—

[a] It was assumed that pilchard and horse mackerel constitute 50–70% of the total fish biomass. Annual production (column 3) was derived from P/B considerations, and yield was estimated as 25–50% of annual production. These figures were then compared with maximum and mean catches from the two exploited regions (columns 5 and 6). All values in thousands of tons. From Moloney and Field (1985).

gesting that the spectrum is not flat, but has a slope of -0.22; also, the question of whether an area of coastal water such as that studied by Moloney and Field (1985) is a sufficiently closed system for the biomass spectrum to be in equilibrium. From the point of view of a fishery manager, figures about total biomass of fish are of limited utility if the species composition cannot be specified. Nevertheless, the indications of rough agreement between the predictions of the model and the realities of the catch data give indications that the approach has promise, and that here may be a tenuous bridge between the population biologist and the ecosystem ecologist.

15.3. Moving Between Levels of Organization

The day to day problems of a fishery manager illustrate well the futility of trying to emphasize population biology at the expense of ecosystem ecology, or vice versa. Undue emphasis on the population dynamics of the species of interest leads to the situation where the present state of the stock and the biology of every stage from egg to senescence may be very well known, but an unexpectedly good or bad year class will come along because of seemingly unpredictable events in the environment; i.e., at the ecosystem level. Conversely, ecosystem models that do not take into account the characteristics of particular species may well miss biological interactions that have major significance at the ecosystem level. This is well illustrated by events in the rocky subtidal system of Nova Scotia (reviewed in Mann 1982b, 1985).

15.3.1. Kelp Bed Ecosystems

The kelps *Laminaria* and *Agarum* are the dominant benthic primary producers in the euphotic zone of nearshore waters, but the sea urchin, *Strongylocentrotus droebachiensis*, when it reaches a threshold of abundance, is capable of grazing all kelp from the rock surfaces over areas amounting to many square kilometers, and of preventing recolonization by any macroalgae for periods of about a decade. Such drastic changes alter the whole character of the ecosystem. Miller et al. (1971) published an energy flow model of the ecosystem which assumed some kind of steady state, with the production of the macroalgae greatly exceeding the needs of the herbivores and consequently a massive export of detritus to neighboring coastal waters. When, a few years later, the kelp beds had been destroyed by urchin overgrazing, there was no longer a surplus of primary production and export. The model predicted that the sea urchin population would soon collapse from starvation. This did not occur, and it was necessary to move to the microbial level to demonstrate that the urchins had a gut microflora capable of fixing gaseous nitrogen (Guerinot et al. 1977; Guerinot 1979), and of synthesizing essential amino acids (Fong and Mann 1980). With this kind of physiological support, the urchins persisted at high population densities, effectively preventing the regeneration of the kelps. During this period, catches of lobsters, and presumably their population biomass, declined drastically in the kelp–urchin ecosystem. Factors in this decline were the loss of productivity in invertebrate food webs leading to lobsters, and increased mortality of young lobsters by predation, caused by loss of algal cover (Johns and Mann 1987).

Moving again to the ecosystem level, the waters of the Nova Scotia coast had above average surface temperatures during the years 1980–82, and a pathogenic amoeba with a temperature threshold for its development in sea urchins was able to decimate the sea urchin populations, permitting the reestablishment of kelp beds and the resumption of detrital export to adjacent waters (Miller and Colodey 1983; Scheibling and Stephenson 1984).

Many analogous examples have been documented from intertidal ecosystems. Manipulations of populations of limpets, littorines, chitons, sea urchins, starfish, or fishes lead to profound changes in the communities of which they are part, and hence to processes at the ecosystem level (Paine 1977; Mann 1985).

15.3.2. Salt Marsh Ecosystems

On sedimenting shorelines in temperate latitudes, large areas of the upper intertidal are often dominated by *Spartina* marshes. The study of these areas from an ecosystem point of view was pioneered at the University of Georgia and its Sapelo Island field station (Odum and Smalley 1959; Teal 1962; Odum and de la Cruz 1967; Odum and Fanning 1973; Odum 1974). One of the conspicuous features of such marshes is that the *Spartina* is taller and more productive along the edges of the creeks, and shorter and less productive as one moves away from the creek banks. E. P. Odum (1974) drew attention to the energy subsidy provided by tidal flow, and quoting from Odum and Fanning (1973), showed that creek bank *Spartina* produces annually about 4000 $g \cdot m^{-2}$ dry matter, while grasses further from the creek, receiving gentle tidal irrigation, average 2300 $g \cdot m^{-2}$ annual production. Areas of high marsh that are inundated only on spring tides average only 750 $g \cdot m^{-2}$.

At this time, the mechanism by which the tides influenced primary production was not clear. Obviously, most flowering plants are not well adapted to inundation with salt water, and some ecologists sought the key to *Spartina* performance in salinity adaptations. When *Spartina* was grown under controlled conditions at a range of salinities, optimum growth occurred at 5–10‰, and this was true for both low marsh and high marsh plants. However, when the light intensity was varied, it was found that at high irradiance the plants could compensate for the stress of high salinity and photosynthesize at a rate only 10% lower than the controls (Longstreth and Strain 1977). It was found that NaCl is sequestered into the cell vacuoles, and this is balanced by the free amino acid proline and the quaternary ammonium compound glycinebetaine in the cytoplasm. This provides the internal osmotic pressure necessary for the plants to function in a saline medium (Cavalieri and Huang 1981). In light of this mechanism, explanations of the variation in *Spartina* productivity were sought in terms of salinity differences. For example, in warm dry climates evaporation of salt water in areas remote from creek banks leads to salt accumulation and increased salinity stress (Nestler 1977). However, attempts to alleviate this salinity stress by watering plots remote from creek banks did not lead to marked changes in plant productivity.

Others sought to explain differences in productivity by differing nutrient regimes. Valiela and Teal (1974) fertilized areas of salt marsh and demonstrated that *Spartina* is normally nitrogen-limited. Others (Patriquin 1978; Patriquin and McLung 1978) had shown that organisms associated with the roots of *Spartina* were capable of extensive nitrogen fixation, although the rate of fixation is inversely proportional to nitrogen content of the soil pore water. It can now be seen that salinity and nitrogen requirements interact to produce the observed variation in primary production. Plants in areas remote from creek banks experience greater salinity stress, and to combat it they need to store nitrogen compounds in the cytoplasm. Since these areas are inundated infrequently, the plants do not derive much nitrogen from the tidal waters, but survive by supporting nitrogen-fixing organisms at considerable metabolic cost to themselves. This drain of metabolic products to support the nitrogen fixers and combat the salinity stress is at the expense of net growth, and explains why primary production decreases with distance from creek bank. An understanding of a salt marsh as an ecosystem requires a detailed understanding of the biology of one genus of plant, and any suggestion that one level of organization is more important than another is inappropriate.

In ecology, the true generalist, the modern equivalent of the Renaissance man, is one who can comfortably move between different levels of organization, from cell biology through organismal and population biology to ecosystem properties. Only in this way can a concept like "tidal energy subsidy" be used to explain the high level of general primary productivity and its spatial variability in a salt marsh ecosystem.

15.4. Some Attempts at Radical Reassessment

In the world of fisheries, enormous advantages would be conferred by an ability to make predictive models of the marine or freshwater fish-producing ecosystems. In a review, "Mathematical Models in Biological Oceanography," Platt et al. (1981) concluded that the common practice of constructing ecosystem simulation models by means of coupled differential equations had led to only limited success in the making of predictions about the consequences of perturbing those systems. They therefore proposed that attempts be made to radically reexamine the properties of ecosystems, in the hope that new and more fruitful approaches to prediction of ecosystem behavior would be found. Five lines of inquiry were proposed: thermodynamics, statistical mechanics, input-output analysis, information theory, and ataxonomic aggregations. The discussion of size-dependent models given earlier is an example of the pursuit of ataxonomic aggregations as units of a model. Further development of those five lines of inquiry can be found in Ulanowicz and Platt (1985).

15.4.1. The Capacity for Self-Organization

From Ulanowicz and Platt (1985), I wish to select the contribution that is perhaps the most radical of all. Its title is "Ecology, Thermodynamics, and Self-Organization: Towards a New Understanding of Complexity" (Allen 1985). The theme of the paper is a difficult one for those of us who tend to think in terms of mechanistic models. It is that complex systems (of which ecosystems are examples) may possess a capacity for self-organization. The idea has its roots in the early work of Prigogine (summarized in Nicolis and Prigogine 1977).

Early attempts to apply the ideas of classical thermodynamics to ecology used the thermodynamics of closed systems. In a closed system there is an inexorable trend towards uniformity or maximum disorder. However, natural systems are open, and usually have strong inputs and outputs, and a strong internal coupling between the elements. All this has the effect of maintaining the systems far from equilibrium. They often maintain an appearance of stability, but it is a very dynamic stability, accompanied by incessant fluxes of energy and materials. The radical idea which we owe to Prigogine is that the structures which we see in a system are in some sense a product of the fluxes themselves. He coined the term "dissipative structures" to emphasize that the structures are formed in response to the fluxes, are maintained and developed by them and in the process cause them to be degraded and dissipated. Prigogine developed his ideas in relation to certain chemical systems, but they have been more widely applied to living systems by others (for review see Jantsch 1980).

A well-known chemical example is the Belousov–Zhabotinski reaction. It is based on the oxidation of malonic or bromalonic acid by potassium bromate, catalyzed by ceric sulphate dissolved in ceric acid. Under appropriate conditions of flow of the reacting substances, the system goes into oscillations with sharp increases and decreases in the concentrations of the reacting substances, which can be observed as pulses of color change. In a shallow container such as a Petri dish, these oscillations form concentration waves in intricate patterns (Fig. 15.6). The important point to note is that these spatial structures are formed within the system, not imposed from outside, and obtain the energy for their formation by dissipating the chemical energy of the substances flowing through the system.

In Allen's (1985) exposition of the effect of dissipative structures in natural systems, he introduces us to the idea of bifurcations. In Figure 15.7, the left side of the diagram represents thermodynamic equilibrium. Allen explains that, at distance 1 from equilibrium, the system can be characterized by a single value of x with small fluctuations about it. However, as the system is pushed further from equilibrium, it encounters a bifurcation point, and any increase in distance from equilibrium forces it into a situation in which it must adopt one of two stable states, and may under certain circumstances flip from one to the other. At distance 2 from equilibrium, the diagram shows that there are four possible stable states. Changes from one state to another may involve quite drastic alterations in system structure, yet they may be triggered by combinations of system variables that are themselves relatively small.

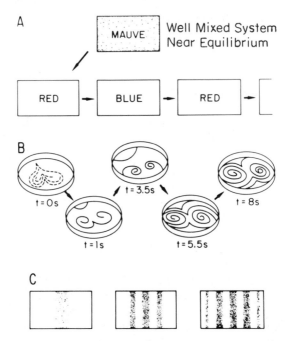

Figure 15.6. Illustrations of the Belousov–Zhabotinsky reaction: A, in a well-mixed system near equilibrium; B, examples of the many patterns that can emerge spontaneously in an unstirred dish; and C, more linear, symmetrical patterns that can occur in unstirred systems moving further from equilibrium. From Allen (1985). Reprinted with permission from Can. Bull. Fish. Aquat. Sci.

The idea of bifurcations and multiple stable states is not, of course, new. The mathematics of limit cycles, multiple stable states, and bifurcations had been explored by May more than a decade ago (e. g., May 1974, 1976; May and Oster 1976), but most would judge these papers to be closer to population ecology than to ecosystem science. In fisheries science, an important contribution was made by Peterman (1977), who showed that bifurcations in the properties of salmonid populations can be caused by a combination of depensatory mortality (mortality rate that increases as the stock becomes smaller) and commercial exploitation. The importance of Allen's contribution is that it sets these population phenomena in the context of ecosystem thermodynamics, and forces us to confront the possibility that the ecosystem fluxes generate these structures in a way that can be justly termed self-organization.

15.4.2. The English Channel

The ecosystem example chosen by Allen (1985) was the western English Channel. A mechanistic simulation model had been published by Brylinsky (1972), working with Patten (Figure 15.8). At the time, it constituted a valuable synthesis of the classical data of Harvey (1950) on the functioning of this eco-

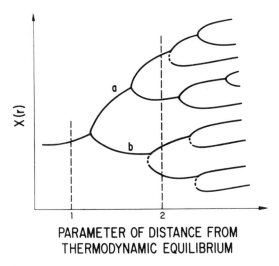

PARAMETER OF DISTANCE FROM
THERMODYNAMIC EQUILIBRIUM

Figure 15.7. Diagram representing the bifurcating tree of possibilities that emerges as a system moves further from equilibrium. Vertical axis: $X(r)$ = value of system property X at point r in the system. From Allen (1985). Reprinted with permission from Can. Bull. Fish. Aquat. Sci.

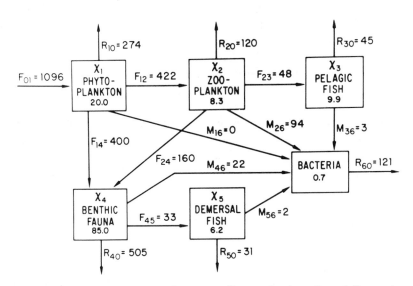

Figure 15.8. Diagram of mean biomasses (boxes, kcal·m^{-2}) and fluxes (arrows, kcal·m^{-2}·yr^{-1}) for the English Channel, calculated by Brylinsky (1972) based on the data of Harvey (1950). Reprinted by permission from B. C. Patten, Ed., *Systems Analysis and Simulation in Ecology, Vol. II.* © Academic Press, Harcourt, Brace, Jovanovich.

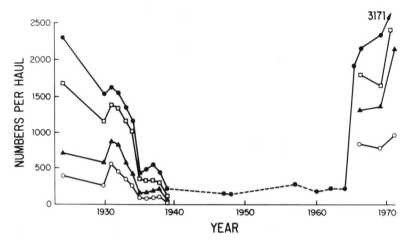

Figure 15.9. The Russell Cycle: changes in the numbers of flatfish, gadoid and *Callionymus* larvae, are shown from 1924 to 1972. Based on F. S. Russell (1973), "A summary of the observations on the occurrence of planktonic stages of fish off Plymouth 1924–1972." Open circles indicate gadoids; triangles indicate flatfish; open squares indicate *Callionymus;* solid circles represent the total. Reprinted with permission of Cambridge Univ. Press. © Marine Biological Association of the U.K. (as reviewed in Cushing 1981).

system. Allen (1985) asked, in effect, whether the structures in the model could be regarded as fixed and unchanging, or whether there was a possibility of the system adopting a totally different configuration. In hindsight, we can see that more than one configuration does exist, and it has been described as the "Russell Cycle" (although some would have reservations about applying the term "cycle" to a system that has been observed to go from one state to another, and back again, once in recorded history). About 1925, the herring stocks in the western Channel began to decline and were progressively replaced by pilchard. The winter phosphate values declined sharply, macroplankton biomass decreased, and the chaetognath *Sagitta elegans* was replaced by *Sagitta setosus*. Flatfish, gadoid and larvae of the dragonet fish *Callionymus* fell to very low levels. Between 1965 and 1970, the situation reversed itself (Fig. 15.9). As Cushing (1981) expressed it: "Each niche sampled in the ecosystem was changed, and the implication is that whatever caste of players is on stage at any time, there is an alternative set of understudies waiting in the wings."

Much has been written about the possible causes of the Russell Cycle in the English Channel. It appears to be correlated with changes in the strength of the North Atlantic current and a northward shift of warm-water species (Longhurst 1984), but it is clear that a model that includes the possibility for self-organization in response to changes in fluxes is more appropriate than a mechanistic box model that has no latitude for such change.

15.4.3. Kelp-Urchin Systems

The large-scale oscillation between a kelp–dominated system and an urchin–dominated system on the coast of Nova Scotia appears to be an analogous situation (Mann 1982). Our present understanding of the mechanism causing the oscillation is that there is a bifurcation point in the system when sea urchin population densities are of the order of 20–30 m^{-2}. If factors tending to reduce that population density (e.g., predators, disease) are greater than factors tending to increase it (e.g., good recruitment), the urchins continue to feed primarily on detritivores, living in crevices and feeding on detrital material that breaks off the kelp plants. If, on the other hand, the balance between recruitment and mortality leads to population densities higher than about 30 m^{-2}, the urchins are unable to maintain themselves by passive detritivory. Instead, they aggregate around individual kelp plants and consume them in a destructive manner. In this way, whole kelp communities can be destroyed over very large areas, leaving a quite different system in which the only macro-algae are corallines, and a sea urchin population at a density of 30–100 m^{-2} prevents the regeneration of fleshy algae for periods of the order of a decade. In our experience, the only perturbation that restores the kelp–dominated ecosystem is the spread of a fatal disease through the urchin population (Miller and Colodey 1983; Scheibling and Stephenson 1984).

This situation is not unique to Nova Scotia. Leighton (1971) reported widespread destruction of kelp beds at Point Loma, California, by the grazing of *Strongylocentrotus franciscanus* and *S. purpuratus*, and Foreman (1977) documented destructive grazing by *S. droebachiensis* in western Canada. The occurrence of the urchin–dominated state with absence of fleshy algae has been described from areas as far apart as Norway and Japan (Hagen 1983), and return of the kelp after removal of algae has been observed in Britain (Jones and Kain 1967). Hence, multiple stable states appear to be a widespread characteristic of subtidal macrophyte–based systems, and the possibility that the switch from one to the other is generated internally by ecosystem fluxes in the presence of forces holding the system far from equilibrium presents a plausible and exciting line of inquiry.

15.4.4. The Scotian Shelf Ground Fishery

One of the key properties of self-organizing systems is that they can be thrown into major structural changes by small random fluctuations in system variables. Allen and McGlade (1986) demonstrated this very clearly in a model of the Nova Scotian ground fishery. They developed Volterra–Lotka equations for interactions between fishing effort and stock sizes of haddock in a multispecies, multifleet spatial model, allowing for changing economic factors such as the feedback from quantities of fish landed to the market price of the fish. The system of equations was shown to produce a catch which went through damped oscillations towards a stable stationary state over a period of about 70 years (Fig. 15.10). Then, the birth rate of the fish was allowed to fluctuate randomly

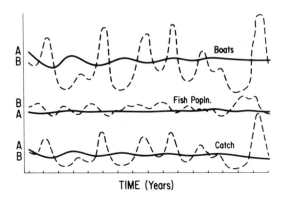

TIME (Years)

Figure 15.10. Output of the model of Allen and McGlade (1986).

A. (Heavy lines) Output of the model when birth rate of fish was held stable. Damped oscillations reaching a stable stationary state after about 70 yr.

B. (Broken lines) Output of the model when the birth rate of the fish was allowed to fluctuate randomly from 5–145% of the long-term mean. Reprinted with permission from Can. J. Fish. Aquat. Sci.

about the same average value b, so that in good years it was $1.45b$, in medium years it was $0.75b$, and in poor years $0.05b$. These were rather conservative fluctuations compared with what happens in nature. A rerun of the model showed that these random fluctuations led to major fluctuations in the catch and in the number of boats that found it economic to fish. The periodicity of the fluctuations was approximately that of the original deterministic Volterra–Lotka model. The structure of the fishery, in terms of species abundance and fishing effort, underwent internally-generated fluctuations analogous to the fluctuations in concentrations of reactants in the Belousov–Zhabotinski reaction.

Allen and McGlade (1986) developed their model further to explore the consequences of various strategies of communication among fishermen and various management strategies. The details need not concern us here. The point of bringing forward this example is to show that while the idea of self-organization in ecosystems is at present a very radical and unproven concept, it does seem to fit a number of observations on fluctuations in coastal ecosystems, and provide a framework for modelling that was entirely lacking in earlier, mechanistic models.

15.6. Conclusions

As we have seen, the scientific debate about which is the more important, population ecology or ecosystem ecology, has been reflected in the theory and practice of fishery managers. Until the 1970s, species were treated as if they existed independently of other species and were harvested independently of other species (Mercer 1982), and almost all attention was paid to either the individual (its growth rate, fecundity, etc.) or the population (its stock size, recruitment rate, fishing mortality, natural mortality, etc.). There was a tacit assumption that the most important parameters were density–dependent and that manipulation of stock density by fishing would give good control of population size.

In recent years, it has become more and more apparent that factors outside the population, which fishery scientists call environmental factors, but which might with equal justification be called ecosystem properties, have a major effect on the commercial species. The dramatic examples, in which stocks crash to levels where it is uneconomic to fish them, are the areas that have proved most challenging and most baffling, but there is now reason to suppose that a great deal of the year-to-year variation in stocks is understandable only in terms of the changing properties of marine ecosystems.

This perception led to a steady, sustained effort to find useful ecosystem models. When the techniques of computer simulation were first borrowed from engineering, they were used to build box models that had as their state variables taxonomic groups or groups classified according to biological function. Some success was obtained using this approach, but predictive capability was limited. In the search for alternatives, the use of organism size as an "alternative taxonomy" was advocated. The size dependence of metabolism of an individual according to the allometric equation $R = aw^b$, and the allometric relationships of population respiration, population production, and production to biomass ratios, has provided a powerful basis for size–dependent modelling which has already yielded useful insights and offers the prospect of further development. Such models have been used to make rough estimates of fish biomass and yield in areas that have not yet been exploited by fishermen.

Underlying almost all modelling efforts by those concerned with management questions have been traditional ideas of cause and effect. If a fish stock showed fluctuations with a periodicity of, say, 8 years, it has been assumed that somewhere there is to be found a driving variable with the same periodicity. Recent developments of systems theory suggest that this approach may be too simplistic. As the model of Allen and McGlade (1986) shows, a relatively long-term oscillation may be triggered by a short-term random fluctuation in a system variable. One way of describing this effect is to say that open systems have the capacity for self-organization. They may also exhibit bifurcations in properties, so that there are two or more stable states separated by areas of instability, but when one looks for the underlying cause of the multiple stable states, one is driven to saying that they are produced by the fluxes in the system, which interact to produce the observed structure. They are products of the system's capacity for self-organization, not patterns imposed from outside.

In the system with which I have been concerned for many years (Mann 1982b), there has been a cycle lasting 15–20 years, with alternation between a kelp–dominated system and an urchin–dominated system. We have been searching for an external driving variable (fishing pressure on lobsters, or temperature cycles in the sea), but are now forced to conclude that in a system held far from equilibrium by predator removal, the pattern may be generated as it were spontaneously, within the system.

It was Eugene Odum's writings that first stimulated me to take an ecosystem view of freshwater and marine habitats, and none of the foregoing text would have been possible without the imaginative insights contained in his studies of Eniwetok Atoll (Odum and Odum 1955) or in the successive editions of his textbook. It is a pleasure to acknowledge my indebtedness as I attempt to take stock of ecosystem processes as they affect coastal fisheries.

Acknowledgements

Many of the ideas in this essay are borrowed freely from the publications of colleagues at the Marine Ecology Laboratory, Bedford Institute of Oceanography, whose work is supported by the Canadian Department of Fisheries and Oceans. I thank the following for reviews of the manuscript in various drafts, or for stimulating discussions: Lloyd Dickie, Ken Frank, Glen Harrison, Steve Kerr, Alan Longhurst, Jacqueline McGlade, Trevor Platt, and Bill Silvert.

16. Problems and Challenges in Ecosystem Analysis

Lawrence R. Pomeroy and
James J. Alberts

Although the ecologists who contributed chapters to this volume represent several subdisciplines between which there has been little recent communication, we see several themes running through the book. Some of these are bits of classical ecological theory or dogma that persist, although we can perceive changes in their acceptance, use, and interpretation. Where we see general agreement, these serve as indicators of the state of ecosystem studies. More often, we see ferment and change. Some are new syntheses or at least new viewpoints of ecological research.

16.1. Stoichiometry in Ecosystem Analysis

Liebig (1855) emphasized the significance of the observation that each species of plant sequesters essential elements in its tissues in a unique ratio. Although Liebig's interest was in the production of field crops and forest trees, ecologists embraced the concept and applied it to hierarchical levels of ecosystem organization that Liebig never intended. Ecologists' fixation on limiting factors and on *the limiting factor* can be traced to the influence of Liebig. We now see a realization that the simple concept of a limiting factor—which even Liebig never intended to apply to an ecosystem—does apply only in relatively rare instances. Limits are many. They may be synergistic or sequential in effect. They are no longer the major focus of ecosystem studies they once were, although they are still a subject of interest.

Long after Liebig, an additional and related concept was introduced into marine ecology and oceanography which was, like Liebig's "law ," seized upon and misused. In 1934 Alfred Redfield observed that the atomic proportions of nitrogen and phosphorus in both plankton and seawater in the North Atlantic Ocean were very nearly 15:1 in a considerable number of samples, and that a very similar relation seemed to hold for samples taken from the Pacific and Indian Oceans. This ratio, now known as the Redfield ratio, took on almost magical qualities and was used to explain many things. Deviations from the ratio were taken as evidence for nutrient limitation by N or P, depending on which element failed to meet the norm. This sort of reasoning was extrapolated from the ocean basins, where the original Redfield data were taken, to continental shelves, estuaries, and even lakes and rivers. Although the basis for the constant ratio in ocean waters and seston was, and is, unknown, there was no reason to expect that it could be extrapolated to other environments, and indeed, it cannot. Although one still sees this sort of mindless invocation of the Redfield ratio in the literature from time to time, there is now much more awareness of the variability of certain processes in the cycles of nitrogen and phosphorus, such as nitrogen fixation and denitrification, and the resulting deviations from the 15:1 ratio (or 16:1 in later versions). We further recognize that in the presence of excess available N and P, a skewed ratio may be telling us something, but it is not telling us that one of those elements is limiting the production of phytoplankton.

What, then, is the "correct" ratio of N/P in any given system? Liebig gave us the answer, and the terrestrial ecologists have followed his directions. One must do analyses on the proportions of the essential elements in the tissues of the organisms of interest, usually those dominating the biomass, in any system under study. The stoichiometry will be relatively constant within a species or even larger taxa, but will vary substantially between major taxa (Reiners 1986). The elemental analyses are easy to do for the living plants and animals, but they may be extremely difficult to do for microorganisms and nonliving litter. Physical separation of living and nonliving components of litter in an absolute way is not ordinarily feasible. Terrestrial litter may contain too little living biomass to significantly change the atomic ratios, but this should not be assumed to be true. Bacteria in particular have a high percentage of nitrogen, phosphorus, and other essential elements, relative to litter, so a small proportion of bacterial biomass might bias the elemental analysis. Similarly in the ocean, samples of "plankton" almost invariably contain a large component of litter (Perry 1976). Probably most of the material taken with a fine net is nonliving. A filtered sample from seawater is mostly detritus. Indeed, the original Redfield samples suffered from this same deficiency. This is another limit to the utility, or at any rate the precision, of elemental ratios—we cannot always separate the components of greatest interest.

This is in no way intended to suggest that ratios and absolute amounts of the essential elements are not significant to ecosystem studies. They are very significant, and there is much incentive to get them right. In many ecosystems one or more essential elements are limiting in some sense. They will limit the flora to those species capable of sequestering essential elements efficiently from the dilute or refractory sources available to them. This is certainly true of oceanic phytoplankton as well as the trees of tropical rain forests (Smetacek and Pollehne 1986; Vogt et al. 1986a). In both cases, energy has to be expended to sequester nitrogen, phosphorus, and other essential elements. In both cases, virtually all of the nitrogen and phosphorus are in the biomass at any instant, although the ocean contains the lowest biomass per m^2, and the rain forest the highest. The clue to the nutrient limitation is the residence times. The residence time of phosphate in the euphotic zone of some Canadian lakes is on the order of minutes, while in the euphotic zone of the open ocean it is on the order of one to several hours, and in estuaries it is on the order of several days (Rigler 1956; Pomeroy 1960; Perry 1976). The residence time of phosphorus in forest floor litter in tropical rain forests is on the order of a year, while in temperate forests it is on the order of 10 yr., and in boreal forests more than 100 yr. (Vogt et al. 1986a). So there are both terrestrial and marine ecosystems in which the recycling of essential elements places limits on both structure and function at the ecosystem level. Understanding the mechanisms of recycling and sequestering nitrogen and phosphorus is, however, largely a physiological question at a more reductionist level. D'Elia (Chapter 10) has pointed out how important this kind of information is to ecosystem studies. Understanding how marine phytoplankton and tropical tree roots take up phosphorus is essential to informed, effective measurements and experiments at the ecosystem level.

As elements cycle through organisms and ecosystems, heavy isotopes are segregated from lighter ones. This has been known for a long time, and $\partial^{13}C$ has been a useful, albeit controversial, index of certain carbon sources for many years. However, danger lurks in the uninformed usage of single isotope ratios. Dogma, experimental difficulties, and funding constraints have historically dictated the use of one ratio, usually the $\partial^{13}C$ of the whole parent plant, for end members of mixing models used to predict the amount of carbon contributed to a specific pool from various sources, i.e., marine plankton or marsh plants. Recent estuarine studies have shown that certain classes of carbon compounds, e.g., lignin, or even different growth forms within a species, may have significantly different $\partial^{13}C$ ratios than a single representative of the whole plant (Ember et al. 1987). Hence, knowledge of the decomposition pathway of the parent material and the relative release rates of the labile and refractory carbon pools may be required to interpret adequately isotopic mixing models.

Considerable progress has been made through the careful use of single isotopic ratios. Now the realization has come that if one ratio is good, several are better, and in situations of limited complexity, isotopic ratios promise to be increasingly useful. Peterson et al. (1985) have shown the value of this approach in analysis of estuarine food chains. There is also an awareness that other biomarkers such as lignin oxidation products (Ertel and Hedges 1985) and carbohydrate patterns (Cowie and Hedges 1984) may be coupled with more conventional isotopic ratios to form a suite of measurements which promise to provide an extremely powerful tool in following the cycling of elements at the ecosystem level. Thus, new analytical techniques and more careful consideration of hypotheses are leading researchers to new insights at both the trophic and systems levels of understanding.

16.2. Trophic Levels

The trophic level concept receives diverse treatment by the authors in this volume, as it does in the recent literature generally. It is interesting that the specialists on ecosystems, who were the ones originally to embrace the concept, are now less enamored of it than are theoreticians and population ecologists. Empiricists now tend to see it as an oversimplification, one that creates confusion by offering false security and unreal regularity. Theoreticians see it as a workable generalization, though they accept that there may be exceptions. Moreover, the trophic level concept is seen as useful and valid by those who work with macroorganisms and often seen as not useful and invalid by those who work with microorganisms. From the ecosystem point of view, we might hope to consider all organisms, and concepts that are applicable to all. Both Pimm (Chapter 13) and Wiegert (Chapter 3) embrace the Lindeman trophic level concept, although Wiegert has revised it to attempt to correct for "split-level" organisms (Wiegert and Owen 1971). At the other extreme, Mann (Chapter 15) finds problems with the concept, citing the trophic continuum concept of Cousins (1980) as a theoretical approach that is compatible with an ecosystem concept in which most of the metabolism is microbial and much of the food nonliving. Cousins stresses the significance of the size of a food packet, rather than whether it is living or nonliving. He views attempts of theoreticians and modelers to retain the trophic level concept as contortions which do not wholly succeed, and he proposes three criteria for ecosystem models:

1. No discrete herbivore and detritivore systems,
2. The categories "all detritus" and "all plant material" are not single reference points,
3. Trophic transfers are not equivalent.

This is a bother to theoreticians who have invested time and thought in the trophic level concept, and it may require a somewhat different modeling approach. However, the major difference is the need to introduce a number of microbial compartments (that are not necessarily "levels") into models and to link them appropriately with the classic food chain of macroorganisms. A number of models do the former more or less well (Pace et al. 1984; Fasham 1985), but perhaps none do the latter. This is partly because we lack empirical information about the details of the microbial food web in soils and waters. However, it might be helpful in focusing our empirical efforts to have some hypotheses to test, based on modeling activities. At any rate, for those of us who have difficulty fitting our round beasts into square trophic levels, the Cousins concept offers an alternative rather than a simple negation. The one conclusion we can reach with certainty at this writing is that there is controversy about the trophic level concept, and changes in our paradigm can be expected.

16.3. Trophic Efficiency

A companion piece to the trophic level concept is the more recent idea of ecological efficiency, and specifically the generalization that 90% of the organic energy at one trophic level is lost in the transfer to the next. Aside from the obvious need to believe that there are trophic levels in order to proceed, a considerable body of data are now available on assimilation efficiency of everything from bacteria to whales. The generalization that emerges is that there is not a generalization. Efficiency of individual organisms and populations changes through their life histories, as the result of both physiological and environmental changes. Some organisms, some of the time, appear to have remarkably high assimilation efficiency. Bacteria utilizing high quality dissolved substrates are said to be more than 50% efficient (but see below). Juvenile metazoan animals often have assimilation efficiencies in the range 15–25%. On the other hand, many adult metazoans assimilate much less than 10% of the food ingested, and bacteria utilizing refractory or nutritionally unbalanced substrates commonly assimilate 10–20% of it, respiring the remainder. Thus, another lovely generalization, dear to theoreticians, falls.

As Wiegert points out in Chapter 3, Battley (1987) attempts to sustain a generalization of his that assimilation efficiency (in yeasts, at least) cannot exceed 35%. Battley does not accept "growth yield" data of the microbiologists on the grounds that all absorbed substances may not be assimilated. While this objection seems unlikely, at least in the case of bacteria that go through many generations during an experiment, someone must devote the same care Battley has with his yeast experiments to investigate the efficiency of other microorganisms, especially bacteria.

16.4. Stability and Diversity

Stability and diversity are very general terms that have been given varied meaning by various ecologists, contributing to the diversity but not the stability of concepts and theories relating to the stability of ecosystems. How stable an ecosystem seems to be depends upon the hierarchical level at which the ecosystem is viewed. If we view the whole ecosystem in terms of total metabolic functions, such as photosynthesis, respiration, and flux of essential chemical elements, most ecosystems exhibit great stability, even when perturbed. If we view the ecosystem in terms of component communities or species populations, the system is less stable, especially if perturbed or if viewed over long time spans relative to the human lifetime. The amount of stability that exists, and the continuity of not only genomes but of ecosystems, is remarkable and certainly

worthy of the attention it has been paid. Many questions relating to stability
have been raised in this volume, relating to all of the hierarchical levels of the
biosphere. These are important issues, and our message is only to remind the
reader that under the rubric stability there are many issues at many levels of
organization. No single, simple statement will suffice to cover the subject.

In that regard, a message that comes through clearly from several chapters
is the need for investigators at the ecosystem level to assimilate and utilize
knowledge of population ecology just as they have long utilized chemistry or
physiology. Experience suggests that as a general rule scientists utilize as given
the body of knowledge of scientific disciplines below theirs in the hierarchy
of nature (Chapter 1) and ignore or dismiss as trivial the higher levels. If this
trend holds true, the burden of assimilating all lower levels, including that of
species populations, falls upon those who study ecosystems. The explosion of
knowledge and theory at the population level in the past two decades provides
a wealth of potentially valuable information on which to build a better understanding of ecosystem function. Though, as pointed out in Chapter 1, ecosystem level processes cannot be understood simply by summing up the
processes at lower levels, logical though it may seem.

16.5. Hierarchy

The concept of hierarchical systems is not new, but its significance for systems
analysis in general and ecosystem analysis in particular has been the subject
of much recent literature (e.g., Allen and Starr 1982; O'Neill et al. 1986). The
importance of this viewpoint for ecological theory can hardly be overempha-
sized. To make sense out of the organisms, forces, and events in an ecosystem,
the investigator must perceive structure, order, and relationships. Hierarchy
theory is a sound basis for that, and it permeates this volume. It is one thing
on which investigators of ecosystems agree, and which most of them utilize
in their thinking, modeling, and analysis. If some population ecologists fail to
see its significance, that is their problem.

A problem that remains for all of us, however, is the need to study in a comprehensive way what Odum in his first ecology textbook described as the entire
layer cake: everything that goes into the ecosystem at all hierarchical levels,
seen in various perspectives by cutting through the structure in various ways
(Odum 1971). Because of the complexity of what they study, ecosystem ecologists recognize the need to compartmentalize and distinguish hierarchical levels. Generally, each level is studied by specialists who tend to see the information
on lower levels as given and the information on higher levels as garbage. The
challenge is that the impediments are largely sociological, not scientific.

16.6. Ecosystems as Units of Study

From a hierarchical perspective, studies of the entire biosphere are at the
highest level of integration. They are largely the domain of the geochemist or
the biogeochemist, because most of the data on biological processes of the

biosphere are in the form of chemical changes. Although the situation may be improving, there is remarkably little interaction between biogeochemists and ecosystem ecologists. While one would expect that persons studying the whole biosphere would feel a need to understand how ecosystems function, this does not seem to be a compelling consideration. An example of how global processes affect ecosystems (the impact of the ENSO cycle on regional oceanic ecosystems) is presented by Barber in Chapter 9. We also find an interesting parallel between Chapters 2 and 9. In the former, Rich points out that it is not the flow of energy but the redox potential that drives ecosystem function biologically. In the latter, Barber points out that it is the temperature gradient of the planet, and the resultant pressure gradients, that drive the physical circulation and set up the oscillations of cycles of atmospheric and oceanic climate. In both cases, we see again the major impacts of small changes in key parameters.

In contrast to these global processes, we see throughout the volume repeated expressions by ecosystem ecologists of the need to understand processes at the level of species populations or local interbreeding populations, thus encompassing evolution and population dynamics. We are also finding in the recent literature a number of attempts by population ecologists to incorporate energetics as a variable. While much of that seems naive to those of us who have been quantifying energetics for years, it is no more so than some of our current attempts to assimilate population level processes into ecosystem concepts. Overall, this is a healthy trend, one that must be encouraged in spite of the pressures to specialize.

16.7. The Time Warp

Ecologists are aware that their view of ecosystem processes is warped by the length of the human lifetime, if not by the length of the research grant. Many strategies for extending the time line and linearizing our interpretation of it are discussed in this volume. For the very long-term changes, of course, we must rely on mining data from lake beds and glaciers. For the shorter term, and looking toward the future, we see repeated requests for the establishment of long-term study sites. This may be to some degree a naive goal, because it implies that we know what future generations will want us to record. Our experience with the few long-term studies which exist suggests that there is an element of luck, but it helps to have someone unusually prescient to decide what data to collect. As we gain experience, as our methods of automatic data collection improve, and as we think about this challenging problem, our ability to make a useful contribution to future ecosystem science through long-term studies may improve.

References

Abbott, M.; Zion, P. M. Satellite observation of phytoplankton variability during an upwelling event. Cont. Shelf Res. 4:661–680; 1985.

Abbott, M. R.; Denman, K. L.; Powell, T. M.; Richerson, P. J.; Goldman, C. R. Mixing and the dynamics of the deep chlorophyll maximum in Lake Tahoe. Limnol. Oceanogr. 29:862–878; 1984.

Aber, J. D.; Melillo, J. M. Fortnite: A computer model of organic matter and nitrogen dynamics in forest ecosystems. The Research Division of the College of Agricultural and Life Science, University of Wisconsin–Madison, R3130 Research Bulletin, 1982; 49 pp.

Aber, J. D.; Botkin, D. B.; Melillo, J. M. Predicting the effects of different harvesting regimes on forest floor dynamics in northern hardwoods. Can. J. For. Res. 8:306–315; 1978.

Adams, D. E.; Wallace, L. L. Nutrient and biomass allocation in five grass species in an Oklahoma tallgrass prairie. Am. Midl. Nat. 113:170–181; 1985.

Adey, W. H. Coral reef morphogenesis: a multidimensional model. Science 202:831–837; 1978.

Adey, W. H. The microcosm: a new tool for reef research. Coral Reefs 1:193–201; 1983.

Adey, W. H.; Steneck, R. S. Highly productive eastern Caribbean reefs: Synergistic effects of biological, chemical, physical, and geological factors. In: Reaka, M. L., ed., The ecology of coral reefs, vol. 3. Rockville, MD: NOAA Undersea Research Program; 1985:163–187.

Ajtay, G. L.; Ketner, P.; Duvigneaud, P. Terrestrial primary production and phytomass. In: Bolin, B.; Degens, E. T.; Kempe, S.; Ketner, P., eds. The global carbon cycle. Chichester: John Wiley & Sons (SCOPE 13), 1979:129–181.

Akiyama, T.; Takahashi, S.; Shiyomi, M.; Okubo, T. Energy flow at the producer level. Oikos 42:129–137; 1984.

Albrektson, A. Relations between tree biomass fractions on conventional silvicultural measurements. In: T. Persson, ed. *Structure and function of northern coniferous forests: an ecosystem study.* Stockholm: Ecological Bulletins, Swedish Natural Science Research Council (NFR), 1980:315–327.

Allaway, W. G.; Ashford, A. E. Nutrient input by seabirds to the forest on a coral island of the Great Barrier Reef. Mar. Ecol. Prog. Ser. 19:297–298; 1984.

Allen, P. M. Ecology, thermodynamics and self-organization: Towards a new understanding of complexity. In: Ulanowicz, R. E.; Platt, T., eds. *Ecosystem theory for biological oceanography.* Can. Bull. Fish. Aquat. Sci. 213:1–26; 1985.

Allen, P. M.; McGlade, J. M. The dynamics of discovery and exploitation: The case of the Scotian Shelf groundfish fisheries. Can. J. Fish. Aquat. Sci. 43:1187–1200; 1986.

Allen, T. F. H.; Hoekstra, T. W. Nested and non-nested hierarchies: A significant distinction. In: Smith, A. W., ed. *Proceedings of the society for general systems research. I. System methodologies and isomorphies.* Lewiston, NY: Intersystems Publ., Coutts Library Service, 1984:175–180.

Allen, T. F. H.; Starr, T. B. *Hierarchy: perspectives for ecological complexity.* Chicago: University of Chicago Press, 1982, 310 pp.

Allen, T. F. H.; O'Neill, R. V.; Hoekstra, T. W. Interlevel relations in ecological research and management: some working principles from hierarchy theory. Rocky Mtn. For. and Range Exp. Sta., Ft. Collins, CO: USDA Forest Serv. Gen. Tech. Rep. RM-110, 1984.

Anderson, D. W.; Coleman, D. C. The dynamics of organic matter in grassland soils. J. Soil Water Conserv. 40:211–216; 1985.

Anderson, N. H.; Cummins, K. W. The influences of diet on the life histories of aquatic insects. J. Fish. Res. Bd. Can. 36:335–342; 1979.

Andrews, J. C.; Gentien, P. Upwelling as a source of nutrients for the Great Barrier Reef ecosystems: A solution to Darwin's question? Mar. Ecol. Prog. Ser. 8:257–269; 1982.

Andrews, J. C.; Müller, H. R. Space–time variability of nutrients in a lagoonal patch reef. Limnol. Oceanogr. 28:215–227; 1983.

Andrzejewska, L.; Wojcik, Z. The influence of *Acridoidea* on the primary production of a meadow (field experiment). Ekol. Pol. 18:89–109; 1970.

Anonomyous. A comparative study of 14 estuaries. Maritimes 28:5–7; 1984.

Archer, S.; Detling, J. K. Evaluation of potential herbivore mediation of plant water status in a North American mixed-grass prairie. Oikos 47:287–291; 1986.

Archer, S.; Garrett, M. G.; Detling, J. K. Rates of vegetation change associated with prairie dog (*Cynomys ludovicianus*) grazing in North American mixed-grass prairie. Vegetatio. In press.

Arntz, W.; Landa, A.; Tarazona, J., eds. *El Niño: su impacto en la fauna marina.* Callao, Peru: Bol. Extraord., Inst. del Mar del Peru, 1985, 222 pp.

Arsuffi, T. L.; Suberkropp, K. Selective feeding by stream caddisflies (Trichoptera) detritivores on leaves with fungal colonized patches. Oikos 45:50–58; 1985.

Assmann, E. *The principles of forest yield study.* Oxford: Pergamon Press; 1970.

Atkinson, L. P.; Menzel, D. W.; Bush, K. A., eds. *Oceanography of the southeastern U. S. continental shelf.* Washington: Amer. Geophysical Union; 1985.

Atkinson, M. J. Phosphate flux as a measure of net coral reef flat productivity. Proc. Fourth Int. Coral Reef Symp. 1:417–418; 1981.

Atkinson, M. J. Phosphorus in coral reef ecosystems. In: Baker, J. T.; Carter, R. M.; Sammarco, P. W.; Stark, K. P., eds. *Proceedings: inaugural Great Barrier Reef conference* (Townsville, Aug. 28–Sept. 2, 1983) J. C. U. Press, 1983:271–274.

Atkinson, M. J.; Grigg, R. W. Model of a coral reef ecosystem. II. Gross and net primary production at French Frigate Shoals, Hawaii. Coral Reefs 3:13–22; 1984.

Atkinson, M. J.; Smith, S. V. C:N:P ratios of benthic marine plants. Limnol. Oceanogr. 28:568–574; 1983.

Azam, F.; Fenchel, T.; Field, J. G.; Meyer-Reil, L.-A.; Thingstad, F. The ecological role of water-column microbes in the sea. Mar. Ecol. Prog. Ser. 10:257–263; 1983.

Baldwin, I.T.; Schultz, J. C. Rapid changes in tree leaf chemistry induced by damage: Evidence for communication between plants. Science 221:277–279; 1983.

Banner, A. H. A fresh-water "kill" on the coral reefs of Hawaii. Hawaii Inst. Mar. Biol. Tech. Report 15:29; 1968.

Banse, K. On weight dependence of net growth efficiency and specific respiration rates among field populations of invertebrates. Oecologia 38:111–126; 1979.

Banse, K. Mass-scaled rates of respiration and intrinsic growth in very small invertebrates. Mar. Ecol. Prog. Ser. 9:281–297; 1982.

Banse, K.; Mosher, S. Adult body mass and annual production/biomass of field populations. Ecol. Monogr. 50:355–379; 1980.

Barber, R. T. The distribution of dissolved organic carbon in the Peru Current system of the Pacific Ocean. Stanford University; 1967. 132 pp. Ph. D. dissertation.

Barber, R. T. The JOINT-I expedition of the coastal upwelling ecosystem analysis program. Deep-Sea Res. 24:1–6; 1977.

Barber, R. T.; Chavez, F. P. Biological consequences of El Niño. Science 222:1203–1210; 1983.

Barber, R.T.; Chavez, F. P. Ocean variability in relation to living resources during the 1982–83 El Niño. Nature 319:279–285; 1986.

Barber, R. T.; Smith, R. L. Coastal upwelling ecosystems. In: Longhurst, A. R., ed. *Analysis of marine ecosystems.* New York: Academic Press, 1981:31–68.

Barber, R. T.; Kirby-Smith, W. W.; Parsley, P. E. Wetland alterations for agriculture. In: Greeson, P. E.; Clark, J. R.; Clark, J. E., eds. *Wetland functions and values.* American Water Resources Association, 1980:642–651.

Barber, S. A. Corn residue management and soil organic matter. Agron. J. 71:625–627; 1979.

Barmuta, L. A.; Lake, P. S. On the value of the River Continuum Concept. N. Z. J. Mar. Freshwat. Res. 16:299–231,324; 1982.

Barnes, B. T.; Ellis, F. B. Effects of different methods of cultivation and direct drilling, and disposal of straw residues, on populations of earthworms. J. Soil Sci. 30:669–679; 1979.

Barnes, D. J. Profiling coral reef productivity and calcification using pH and oxygen electrodes. J. Exp. Mar. Biol. Ecol. 66:149–161; 1983.

Barnes, J. R.; Minshall, G. W., eds. *Stream ecology: testing general ecological theory.* New York: Plenum Press, 1983, 399 pp.

Bartholomew, G. A. The role of natural history in contemporary biology. BioScience 36:324–329; 1986.

Battley, E. H. Growth-reaction equations for *Saccharomyces cerevisiae*. Physiol. Planturam 13:192–203; 1960a.

Battley, E. H. Enthalpy changes accompanying the growth of *Saccharomyces cerevisiae* (Hansen). Physiol. Planturam 13:628–640; 1960b.

Battley, E. H. A theoretical approach to the study of the thermodynamics of growth of *Saccharomyces cerevisiae* (Hansen). Physiol. Planturam 13:674–686; 1960c.

Battley, E. H. A calculation of the free-energy changes accompanying the growth of *Saccharomyces cerevisiae* on several substrates. Proc. First European Biophys. Congr.: Baden, Austria IV, 1971:299–305.

Battley, E. H. *Energetics of microbial growth.* New York: John Wiley & Sons, 1987, 450 pp.

Bauchop, T.; Eldsden, S. R. The growth of microorganisms in relation to their energy supply. J. Gen. Microbiol. 23:457–469; 1960.

Bauer, A.; Black, A. L. Soil carbon, nitrogen and bulk density comparisons in two cropland tillage systems after 25 years and in virgin grassland. Soil. Sci. Soc. Amer. J. 45:1166–1170; 1981.

Bazzaz, F. A.; Pickett, S. T. Physiological ecology of tropical succession: A comparative review. Ann. Rev. Ecol. Syst. 11:287–310; 1980.

Beaver, R. A. Geographical variation in the food web structure of *Nepenthes* pitchers. Ecol. Ent. 10: 241–248; 1985.

Beers, J. R.; Reid, F. M. H.; Stewart, G. L. Seasonal abundance of the microplankton population in the North Pacific Central Gyre. Deep-Sea Res. 29:227–245; 1982.

Belsky, A. J. Does herbivory benefit plants? A review of the evidence. Am. Nat. 127:870–892; 1986.

Belsky, A. J. The effects of grazing: Confounding of ecosystem, community, and organism scales. Am. Nat. 129:777–783; 1987.

Benecke, U.; Nordmeyer, A. H. Carbon uptake and allocation by *Nothofagus solandri var. cliffortioidies* (Hook. F.) Poole and *Pinus contorta* Dougl. ex. Loundon ssp. *contorta* at montane and subalpine altitudes. In: Waring, R. H., ed. *Carbon uptake and allocation in subalpine ecosystems as a key to management.* IUFRO Workshop. Corvallis, Oregon: Forest Research Lab, Oregon State University, 1982:9–21.

Benke, A. C.; Van Arsdall, T. C.; Gillespie, D. M.; Parrish, K. F. Invertebrate productivity in a subtropical blackwater river: the importance of habitat and life history. Ecol. Monogr. 54:25–63; 1984.

Berg, B.; Staaf, H. Decomposition rate and chemical changes of Scots pine needle litter. I. Influence of stand age. In: Persson, T., ed. *Structure and function of northern coniferous forests-an ecosystem study.* Ecological Bulletins. Stockholm: Swedish Natural Science Research Council (NFR), 1980:363–372.

Berg, B.; Jansson, P. E.; Meentemeyer, V. Litter decomposition and climate–regional and local models. In: Agren, G. I., ed. *State and change of forest ecosystems–indicators of future research.* Swedish Ecology and Environmental Research Report No. 13, 1984:389–404.

Berg, K. Biological studies on the River Susaa. Folia Limnol. Scandinavica 4:1–318; 1948.

Bernstein, R. L.; Breaker, L.; Whritner, R. California current eddy formation: ship, air and satellite results. Science 195:353–359; 1977.

Beyers, R. J. The metabolism of twelve aquatic laboratory microecosystems. Ecol. Monogr. 33:281–306; 1963.

Birch, L. C.; Clark, D. P. Forest soil as an ecological community with special reference to the fauna. Q. Rev. Biol. 28:13–36; 1953.

Bird, D. F.; Kalff, J. Bacterial grazing by planktonic lake algae. Science 231:493–495; 1986.

Birk, E. M.; Simpson, R. W. Steady state and the continuous input model of litter accumulation and decomposition in Australian eucalypt forests. Ecology 61:481–485; 1980.

Birkland, C. Terrestrial runoff as a cause of outbreaks of *Acanthaster planci* (Echinodermata, Asteroidea). Mar. Biol. 69:175–185; 1982.

Blackburn, M. Low latitude gyral regions. In: Longhurst, A. R., ed. *Analysis of marine ecosystems.* New York: Academic Press, 1981:3–29.

Blackburn, T. R. Information and the ecology of scholars. Science 181:1141–1146; 1973.

Blevins, R. L.; Smith, M. S.; Thomas, G. W. Changes in soil properties under no-tillage. In: Phillips, R. E.; Phillips, S. H., eds. *No-tillage agriculture: principles & practices.* New York: Van Nostrand Reinhold, 1984:190–230.

Bliss, L. C., ed. *Truelove Lowland, Devon Island, Canada: A high arctic ecosystem.* Edmonton: University of Alberta Press, 1977.

Blumberg, A. Y.; Crossley, D. A., Jr. Comparison of soil surface arthropod populations in conventional tillage, no-tillage and old-field systems. Agro-Ecosystems 8:247–253; 1983.

Bohlool, B. B.; Kosslak, R. M.; Woolfenden, R. The ecology of *Rhizobium* in the rhizosphere: survival, growth and competition. In: Veeger, C.; Newton, W. E., eds. *Advances in nitrogen fixation research.* The Hague: Martinus Nijhoff/W. Junk Publishers, 1984:287–293.

Boon, P. J. Uptake and release of nitrogen compounds in coral reef and seagrass, *Thalassia hemprichi* (Enrenb.) Aschers., bed sediments at Lizard Island, Queensland. Aust. J. Mar. Freshwater Res. 37:11–19; 1986.

Borchert, J. R. Climate of the central North American grasslands. Annals of the Association of American Geographers 40:1–39; 1950.

Bormann, F. H. Landscape ecology and air pollution. In: Turner, M. G., ed. *Landscape heterogeneity and disturbance.* Springer-Verlag, 1987:37–57.

Bormann, F. H.; Likens, G. E. *Pattern and process in a forested ecosystem.* Berlin: Springer-Verlag; 1979a.

Bormann, F. H.; Likens, G. E. Catastrophic disturbance and the steady state in northern hardwood forests. Am. Sci. 67:660–669; 1979b.

Bornesbusch, C. H. *The fauna of the forest soil.* Copenhagen: Neilsen and Lydiche (Axel Simmelkir), 1930, 224 pp.

Botkin, D. B.; Mellilo, J. M.; Wu, L. S. Y. How ecosystem processes are linked to large mammal population dynamics. In: Fowler, C., ed. *Dynamics of large mammal populations.* New York: John Wiley & Sons, 1981:373–387.

Braithwaite, R. B. *Scientific explanation.* Cambridge, UK: Cambridge University Press; 1953.

Briand, F. Environmental control of food web structure. Ecology 64:253–263; 1983.

Briand, F. Structural singularities of freshwater food webs. Verh. Internat. Verein. Limnol. 22:3356–3364; 1985.

Briand, F.; Cohen, J. E. Community food webs have scale-invariant structure. Nature 307:264–267; 1984.

Brock, T. D. Life at high temperatures. Science 230:132–138; 1986.

Broda, E. *The evolution of bioenergetic processes.* Oxford: Pergamon; 1975.

Brody, S. *Bioenergetics and growth* (reprinted 1968). New York: Haffner, 1945, 1023 pp.

Brougham, R. W.; Harris, W. Rapidity and extent of changes in genotypic structure induced by grazing in a ryegrass population. N. Z. J. Agric. Res. 10:56–65; 1967.

Brown, A. W. A. *Ecology of pesticides.* New York: John Wiley & Sons; 1978.

Brown, J. H. Two decades of homage to Santa Rosalia: Toward a general theory of diversity. Am. Zool. 21:877–888; 1981.

Brown, L. R. Reducing hunger. In: Stark, L., ed. *State of the world 1985.* Worldwatch Institute. New York: W. W. Norton and Company, 1985:23–41.

Bryant, J. P.; Chapin, F. S. III; Klein, D. R. Carbon/nutrient balance of boreal plants in relation to vertebrate herbivory. Oikos 40:357–368; 1983.

Brylinsky, M. Steady-state sensitivity analysis of energy flow in a marine system. In: Patten, B. C., ed. *Systems analysis and simulation in ecology.* New York: Academic Press, 1972:81–101.

Brylinsky, M. Estimating the productivity of lakes and reservoirs. In: LeCren, E. D.; Lowe-McConnell, R. H., eds. *The functioning of freshwater ecosystems,* IBP #22. Cambridge, UK: Cambridge University Press, 1980:411–447.

Bülow-Olsen, A. Net primary production and net secondary production from grazing an area dominated by *Deschampsia flexuosa* (L.) Trin. by nursing cows. Agro-ecosystems 6:51–66; 1980.

Burgis, M. J.; Darlington, J. P. E. C.; Dunn, I. G.; Gant, G. G.; Enahaba; J. J.; McGowan, L. M. The biomass and distribution of organisms in Lake George, Uganda. Proc. R. Soc. B. 184:271–298; 1973.

Burris, R. H. Nitrogen fixation by blue-green algae of the Lizard Island area of the Great Barrier Reef. Aust. J. Plant Physiol. 3:41–51; 1976.

Burris, R. H. Uptake and assimilation of $^{15}NH_{4+}$ by a variety of corals. Mar. Biol. 75:151–155; 1983.

Burtt, E. A. *The metaphysical foundations of modern physical science,* 2nd edition, Garden City, New York: Anchor Books, Doubleday; 1954.

Busalacchi, A. J.; O'Brien, J. J. Interannual variability of the equatorial Pacific in the 1960's. J. Geophys. Res. 86:10901–10907; 1981.

Cable, D. R. Influence of precipitation on perennial grass production in the semidesert Southwest. Ecology 56:981–986; 1975.

Calder, W. A. III. Size and metabolism in natural systems. In: Ulanowicz, R. E., Platt, T., eds. *Ecosystem theory for biological oceanography.* Can. Bull. Fish. Aquat. Sci. 213:65–75; 1985.

Cameron, A. M.; Endean, R. Renewed population outbreaks of a rare and specialized carnivore (the starfish *Acanthaster planci*) in a complex high-diversity system (the Great Barrier Reef). Proc. Fourth Internat. Coral Reef Symp. 2:593–596; 1982.

Cane, M. A. Oceanographic events during El Niño. Science 222:1189–1195; 1983.

Cane, M. A.; Zebiak, S. E. A theory for El Niño and the Southern Oscillation. Science 228:1085–1086; 1985.

Cane, M. A.; Zebiak, S. E.; Dolan, S. C. Experimental forecasts of El Niño. Nature 321:827–832; 1986.

Capinera, J. L.; Parton, W. J.; Detling, J. K. Application of a grassland simulation model to grasshopper pest management on the North American shortgrass prairie. In: Lauenroth, W. K.; Skogerboe, G. V.; Flug, M., eds. *Analysis of ecological systems: state-of-the-art in ecological modelling.* New York: Elsevier Scientific Publications, 1983:335–344.

Capone, D. G. Benthic nitrogen fixation. In: Carpenter, E. J.; Capone, D. G., eds. *Nitrogen in the marine environment.* New York: Academic Press, 1983:105–137.

Capone, D. G.; Carpenter, E. J. Nitrogen fixation in the marine environment. Science 217:1140–1142; 1982.

Capra, F. *The turning point: science, society and the rising culture.* Toronto: Bantam; 1982.

Cargill, A. S. III; Cummins, K. W.; Hanson, B. J.; Lowry, R. R. The role of lipids as feeding stimulants for shredding aquatic insects. Freshwat. Biol. 15:455–464; 1985a.

Cargill, A. S. III; Cummins, K. W.; Hanson, B. J.; Lowry, R. R. The role of lipids, fungi and temperature in the nutrition of a shredder Caddisfly, *Clistoronia magnifica.* Freshwat. Invert. Biol. 4:64–78; 1985b.

Carnap, R. *Introduction to the philosophy of science.* New York: Basic Books; 1966.

Carpenter, K. E. *Life in inland waters.* London: Sidgwick and Jackson; 1928.

Carr, A. Rips, FADS, and little loggerheads. BioScience 36:92–100; 1986.

Carroll, G. C. Forest canopies: complex and independent subsystems. In: Waring, R. H., ed. *Forests: fresh perspectives from ecosystem analysis.* Proceedings of the 40th Annual Biology Colloquium. Corvallis, Oregon: Oregon State University Press, 1980:87–107.

Carsin, J.-L.; Bourrouilh-Le Jan, F.; Murphy, R. C.; Taxit, R.; Niaussat, P. M. The natural eutrophication of the waters of the Clipperton lagoon: equipments, methods, results, discussions. Fifth Int. Coral Reef Symp. 3:359–364; 1985.

Cates, N.; McLaughlin, J. J. A. Differences of ammonia metabolism in symbiotic and aposymbiotic *Condylactus* and *Cassiopea spp.* J. Exp. Mar. Biol. Ecol. 21:1–5; 1976.

Caughley, G. The elephant problem: An alternative hypothesis. E. Afr. Wildl. J. 14:265–284; 1976.

Cavalieri, A. J.; Huang, A. H. C. Accumulation of proline and glycinebetaine in *Spartina alterniflora* Loisel in response to NaCl and nitrogen in the marsh. Oecologia (Berl.) 49:224–228; 1981.

Chapin, F. S.; Slack, M. Effect of defoliation upon root growth, phosphate absorption and respiration in nutrient–limited tundra graminoids. Oecologia 42:67–79; 1979.

Chapin, F. S. III; Vitousek, P. M.; Van Cleve, K. The nature of nutrient limitation in plant. Am. Nat. 127:48–57; 1986.

Chavez, F. P.; Barber, R. T. Plankton production during El Niño. In: International Conference on the TOGA Scientific Programme, World Climate Research Publications Series No 4. Geneva: World Meteorological Organization, VI, 1985:23–32.

Chew, R. M. Consumers as regulators of ecosystems: An alternative to energetics. Ohio J. Sci. 74(6):359–370; 1974.

Child, G. I.; Shugart, H. H. Frequency response analysis of magnesium cycling in a tropical forest ecosystem. In: Patten, B. C., ed. *Systems analysis and simulation in ecology,* Vol. II. New York: Academic Press, 1972:103–139.

Chilingar, G. V.; Bissell, H. J.; Wolf, K. H. Diagenesis of carbonate rocks. In: Larsen, G.; Chilingar, G. V., eds. *Diagenesis in sediments.* New York: Elsevier Science Publishers, 1967:312–322.

Christensen, B. T. Barley straw decomposition under field conditions: effect of placement and initial nitrogen content on weight loss and nitrogen dynamics. Soil Biol. Biochem. 18:523–529; 1986.

Clarholm, M. Heterotrophic, free-living protozoa: Neglected microorganisms with an important task in regulating bacterial populations. In: Klug, M. J.; Reddy, C. A., eds. *Current perspectives in microbial ecology.* Washington: Amer. Soc. Microbiol., 1984:321–326.

Clark, E. H. II; Haverkamp, J. A.; Chapman, W. *Eroding soils: the off-farm impacts.* Washington: The Conservation Foundation; 1985.

Clark, F. E. Internal cycling of 15-Nitrogen in a shortgrass prairie. Ecology 58:1322–1332; 1977.

Clements, F. E. *Plant succession: an analysis of the development of vegetation.* Washington: Carnegie Institute Publ. 242; 1916.

Codispoti, L. A., Friederich, G. E.; Packard, T. T.; Glover, H. E.; Kelly, P. J.; Spinrad, R. W.; Barber, R. T.; Elkins, J. W.; Ward, B. B.; Lipschultz, F.; Lostaunau, N. High nitrate levels off northern Peru: A signal of instability in the marine denitrification rate. Science 233:1200–1202; 1986.

Coffman, W. P.; Cummins, K. W.; Wuycheck, J. C. Energy flow in a woodland stream ecosystem. I. Tissue support trophic structure of the autumnal community. Arch. Hydrobiol. 68:232–276; 1971.

Cohen, I. B. *Revolution in science.* Cambridge: Harvard University Press; 1985.

Cohen, J. E.; Briand, F.; Newman, C. M. A stochastic theory of community food webs. II. Predicted and observed lengths of food chains. Proc. R. Soc. Lond. B. 228:317; 1986.

Coleman, D. C.; Andrews, R.; Ellis, J. E.; Singh, J. S. Energy flow and partitioning in selected man-managed and natural ecosystems. Agro-ecosystems 3:45–54; 1976.

Coleman, D. C.; Reid, C. P. P.; Cole, C. V. Biological strategies of nutrient cycling in soil systems. Adv. Ecol. Res. 13:1–55; 1983.

Coleman, D. C.; Cole, C. V.; Elliott, E. T. Decomposition, organic matter turnover, and nutrient dynamics in agroecosystems. In: House, G. J., Lowrance, R., Stinner, B., eds. *Agricultural ecosystems–unifying concepts.* New York: Wiley Interscience, 1984:83–103.

Coley, P. D.; Bryant, J. P.; Chapin, S. III. Resource availability and plant anti-herbivore defense. Science 230:895–899; 1985.

Connell, J. H. Diversity in tropical rainforests and coral reefs. Science 199:1302–1310; 1978.

Connell, J. H.; Slatyer, R. O. Mechanisms of succession in natural communities and their role in community stability and organization. Am. Nat. 111:1119–1144; 1977.

Connor, E.; Simberloff, D. The assembly of species communities: chance or competition? Ecology 60:1132–1140; 1977.

Cook, C. B. Metabolic interchange in algae–invertebrate symbioses. Int. Rev. Cytol. Suppl. 14:117–120; 1983.

Cooper, A. W. Above-ground biomass accumulation and net primary production during the first 70 years of succession in *Populus grandidentata* stands on poor sites in northern lower Michigan. In: West, D. C.; Shugart, H. H.; Botkin, D. B., eds. *Forest succession. Concepts and application.* New York: Springer-Verlag, 1981:339–360.

Cooper, W. C. The fundamentals of vegetational change. Ecology 7:391–413; 1926.

Coppock, D. L.; Detling, J. K. Alteration of bison and black-tailed prairie dog grazing interaction by prescribed burning. J. Wildl. Manage. 50:452–455; 1986.

Coppock, D. L.; Detling, J. K.; Ellis, J. E.; Dyer, M. I. Plant–herbivore interactions in a North American mixed-grass prairie I. Effects of black–tailed prairie dogs on intraseasonal aboveground plant biomass and nutrient dynamics and plant species diversity. Oecologia 56:1–9; 1983a.

Coppock, D. L.; Ellis, J. E.; Detling, J. K.; Dyer, M. I. Plant–herbivore interactions in a North American mixed-grass prairie II. Responses of bison to modification of vegetation by prairie dogs. Oecologia 56:10–15; 1983b.

Corbet, P. S. Odonata in phytotelmata. In: Frank, J. H.; Lounibos, L. P. *Phytotelmata: terrestrial plants as hosts for aquatic insect communities.* Medford, NJ: Plexus Press, 1983:29–54.

Corredor, J. E.; Capone, D. G. Studies on nitrogen diagenesis in coral reef sands. Proc. Fifth Int. Coral Reef Symp. 3:395–399; 1985.

Corredor, J. E.; Morell, J. Inorganic nitrogen in coral reef sediments. Mar. Chem. 16:379–384; 1985.

Cosgrove, D. J. Microbial transformations in the phosphorus cycle. In: Alexander, M., ed. *Advances in microbial ecology,* vol. 1. New York: Plenum Press, 1977:95–134.

Coughenour, M. B. A mechanistic simulation analysis of water use, leaf angles, and grazing in East African graminoids. Ecol. Model. 26:203–230; 1984.

Coughenour, M. B. Graminoid responses to grazing by large herbivores: adaptations, exaptations, and interacting processes. Ann. Mo. Bot. Gard. 72:852–863; 1985.

Coughenour, M. B.; Parton, W. J.; Lauenroth, W. K.; Dodd, J. L.; Woodmansee, R. G. Simulation of a grassland sulfur cycle. Ecol. Model. 9:179–213; 1980.

Coughenour, M. B.; Ellis, J. E.; Swift, D. M.; Coppock, D. L.; Galvin, K.; McCabe, J. T.; Hart, T. C. Energy extraction and use in a nomadic pastoral ecosystem. Science 230:619–625; 1985.

Cousins, S. H. A trophic continuum derived from plant structure, animal size and a detritus cascade. J. Theor. Biol. 82:607–618; 1980.

Cousins, S. H. The trophic continuum in marine ecosystems: structure and equations for a predictive model. In: Ulanowicz, R. E.; Platt, T., eds. *Ecosystem theory for biological oceanography.* Can. Bull. Fish. Aquat. Sci. 213:76–93; 1985.

Cowie, G. L.; Hedges, J. I. Carbohydrate sources in a coastal marine environment. Geochim. Cosmochim. Acta 48:2075–2087; 1984.

Cowles, T. J.; Barber, R. T.; Guillen, O. Biological consequences of the 1975 El Niño. Science 195:285–287; 1977.

Crawley, M. J. *Herbivory: the dynamics of animal–plant interactions.* Berkeley: Blackwell Scientific/University of California Press; 1983.

Crocker, R. L.; Dickson, B. A. Soil development on the recessional moraines of the Herbert and Mendenhall glaciers of south-eastern Alaska. J. Ecol. 45:169–185; 1957.

Crocker, R. L.; Major, J. Soil development in relation to vegetation and surface age at Glacier Bay, Alaska. J. Ecol. 43:427–448; 1955.

Cromack, K. Jr. Below-ground processes in forest succession. In: West, D. C.; Shugart, H. H.; Botkin, D. B., eds. *Forest succession. Concepts and application.* New York: Springer-Verlag, 1981:361–373.

Cropper, W. P. Jr.; Ewel, K. C. Carbon storage patterns in Douglas-fir ecosystems. Can. J. For. Res. 14:855–859; 1984.

Crossland, C. J. Seasonal growth of *Acropora cf. formosa* and *Pocillopora damicornis* on a high latitude reef (Houtman Abrolhos, Western Australia). Proc. Fourth Internat. Coral Reef Symp. 1:663–667; 1981.

Crossland, C. J. Dissolved nutrients in reef waters of Sesoko Island, Okinawa: a preliminary study. Galaxea 1:47–54; 1982.

Crossland, C. J.; Barnes, D. J. Acetylene reduction by coral skeletons. Limnol. Oceanogr. 21:153–156; 1976.

Crossland, C. J.; Barnes, D. J. Nitrate assimilation enzymes from two hard corals *Acropora acuminata* and *Goniastrea australensis.* Comp. Biochem. Physiol. 57B:151–157; 1977.

Crossland, C. J.; Barnes, D. J. Dissolved nutrients and organic particulates in water flowing over coral reefs at Lizard Island. Aust. J. Mar. Freshwater Res. 34:835–844; 1983.

Csirke, J. Recruitment in the Peruvian anchovy and its dependence on the adult population. Rapp. P.-V. Reun. Cons. Int. Explor. Mer 177:307–313; 1980.

Cudney, M. D.; Wallace, J. B. Life cycles, microdistribution and production dynamics of six species of net-spinning caddisflies in a large southeastern (USA) river. Holarctic Ecol. 3:169–182; 1980.

Cullen, J. J. The deep chlorophyll maximum. Comparing vertical profiles of chlorophyll a. Can. J. Fish. Aquat. Sci. 39:791–803; 1982.

Cummins, K. W. Trophic relations of aquatic insects. Ann. Rev. Ent. 18:183–206; 1973.

Cummins, K. W. Structure and function of stream ecosystems. BioScience 24:631–641; 1974.

Cummins, K. W. Macroinvertebrates. In: Whitton, B. A., ed. *River ecology.* Oxford: Blackwell Sci., 1975, 725 pp.

Cummins, K. W. The role of stream biota in watershed processes. In: DeVries, J.; Keller, E., eds. *Proc. Chaparral Ecosystem Research.* Univ. CA Davis Water Res. Report 162, 1986a:7–11.

Cummins, K. W. Riparian influence on stream ecosystems. In: Campbell, I., ed. *Stream protection.* Proc. Water Studies Cntr. Chisholm Inst.Tech. Caulfield, Australia. 1986b:45–55.

Cummins, K. W.; Klug, M. J. Feeding ecology of stream invertebrates. Ann. Rev. Ecol. Syst. 10:147–172; 1979.

Cummins, K. W.; Wilzbach, M. A. Field procedures for analysis of functional feeding groups of stream macroinvertebrates. Contrib. 1611. Appalachian Env. Lab, Frostburg: University of Maryland, 1985, 18 pp.

Cummins, K. W.; Coffman, W. P.; Roff, P. A. Trophic relations in a small woodland stream. Verh. Int. Ver. Limnol. 16:627–638; 1966.

Cummins, K. W.; Spangler, G. L.; Ward, G. M.; Speaker, R. W.; Ovink, R. W.; Mahan, D. C. Processing of confined and naturally entrained leaf litter in a woodland stream ecosystem. Limnol. Oceanogr. 25:952–957; 1980.

Cummins, K. W.; Sedell, J. R.; Swanson, F. J.; Minshall, G. W.; Fisher, S. J.; Cushing, E. E.; Petersen, R. C.; Vannote, R. L. Organic matter budgets for stream ecosystems: problems in their evaluation. In: Barnes, J. R.; Minshall, G. W., eds. *Stream ecology: testing general ecological theory.* New York: Plenum Press, 1983:299–353.

Cummins, K. W.; Minshall, G. W.; Sedell, J. R.; Cushing, C. E.; Petersen, R. C. Stream ecosystem theory. Verh. Internat. Verein. Limnol. 22:1818–1827; 1984.

Curtis, J. T. *The vegetation of Wisconsin.* Madison: University of Wisconsin Press; 1959.

Cushing, D. H. Temporal variability in production systems. In: Longhurst, A. R., ed. *Analysis of marine ecosystems.* New York: Academic Press, 1981:443–472.

Cushing, D. H. *Climate and fisheries.* New York: Academic Press, 1982, 373 pp.

Dahm, C. N. Pathways and mechanisms for removal of dissolved organic carbon from leaf leachate in streams. Can. J. Fish. Aquat. Sci. 38:68–76; 1981.

Dahm, C. N. Uptake of dissolved organic carbon in mountain streams. Verh. Internat. Verein. Limnol. 22:1842–1846; 1984.

Dahm, C. N. Uptake of dissolved organic carbon in streams: effects of source and concentration. Can. J. Fish. Aquat. Sci. In press.

Dandonneau, Y.; Donguy, J. R. Changes in sea surface chlorophyll concentration related to 1982 El Niño. Trop. Ocean-Atmos. Newsl. 21:14–15; 1983.

Darwin, C. R. *The origin of species by means of natural selection.* New York: Random House (The Modern Library); 1859.

Darwin, C. R. *The formation of vegetable mould through the action of worms, with observations on their habits.* London: John Murray; 1881.

Davis, B. D. Bacterial domestication: underlying assumptions. Science 235: 1329–1335; 1987.

Dean, R.; Ellis, J. E.; Rice, R. W.; Bement, R. E. Nutrient removal by cattle from a shortgrass prairie. J. Appl. Ecol. 12(1): 25–29; 1975.

DeAngelis, D. L. Energy flow, nutrient cycling and ecosystem resilience. Ecology 56: 238–243; 1980.

deKanel, J.; Morse, J. W. The chemistry of orthophosphate uptake from seawater onto calcite and aragonite. Geochim. Cosmochim. Acta 42:1335–1340; 1978.

Delcourt, H. R.; Delcourt, P. A.; Webb, T. III. Dynamic plant ecology: the spectrum of vegetational change in space and time. Quatern. Sci. Rev. 1:153–175; 1983.

D'Elia, C. F. Aspects of the phosphorus flux of scleractinian corals. University of Georgia; 1974. Ph.D. thesis.

D'Elia, C. F. The uptake and release of dissolved phosphorus by reef corals. Limnol. Oceanogr. 22:301–315; 1977.

D'Elia, C. F.; Webb, K. L. The dissolved nitrogen flux of reef corals. Proc. Third Int. Coral Reef Symp. 1:325–330; 1977.

D'Elia, C. F.; Wiebe, W. J. Biogeochemical nutrient cycles in coral reef ecosystems. In: Dubinsky, Z., ed. *Coral reef ecosystems*. Amsterdam: Elsevier Science Publishers. In press.

D'Elia, C. F.; Webb, K. L.; Porter, J. W. Nitrate-rich groundwater inputs to Discovery Bay, Jamaica: a significant source of N to local coral reefs? Bull. Mar. Sci. 31:903–910; 1981.

D'Elia, C. F.; Domotor, S. L.; Webb, K. L. Nutrient uptake kinetics of freshly isolated zooxanthellae. Mar. Biol. 75:157–167; 1983.

Denman, K. L. Short-term variability in vertical chlorophyll structure. Limnol. Oceanogr. 22:434–442; 1977.

Detling, J. K.; Dyer, M. I. Evidence for potential plant growth regulators in grasshoppers. Ecology 62:485–488; 1981.

Detling, J. K.; Painter, E. L. Defoliation responses of western wheatgrass populations with diverse histories of prairie dog grazing. Oecologia 57:65–71; 1983.

Detling, J. K.; Dyer, M. I.; Winn, D. T. Net photosynthesis, root respiration, and regrowth of *Bouteloua gracilis* following simulated grazing. Oecologia (Berl.) 41:127–134; 1979.

Detling, J. K.; Dyer, M. I.; Procter-Gregg, C.; Winn, D. T. Plant–herbivore interactions: examination of potential effects of bison saliva on regrowth of *Bouteloua gracilis* (H. B. K.) Lag. Oecologia 45:26–31; 1980.

Detling, J. K.; Painter, E. L.; Coppock, D. L. Ecotypic differentiation resulting from grazing pressure: evidence for a likely phenomenon. In: Joss, P. J.; Lynch, P. W.; Williams, O. B., eds. *Rangelands: a resource under seige. Proceedings of the Second International Rangeland Congress*; 1986.

De Vantier, L. M.; Reichert, R. E.; Bradbury, R. H. Does *Spirobranchus giganteus* protect host *porites* from predation by *acanthaster planci*: predation pressure as a mechanism of coevolution. Mar. Ecol. Prog. Ser. 32: 307–310; 1986.

Dickie, L. M. Food chains and fish production. Int. Comm. NW Atl. Fish. Spec. Pub. 8:201–219; 1972.

Dickie, L. M.; Valdivia, J. E. Investigacion cooperativa de la anchoveta y su ecosistema (ICANE) between Peru and Canada: a summary report. Bol. Inst. Mar. Peru Callao Vol. Extraord., 1981, pp. xiii–xxiii.

Dickie, L. M.; Kerr, S. R.; Boudreau, P. R. Size-dependent processes underlying regularities in ecosystem structure. Ecol. Monogr. 57:233–250; 1987.

Dickson, B. A.; Crocker, R. L. A chronosequence of soils and vegetation near Mt. Shasta, California. II. The development of the forest floors and the carbon and nitrogen profiles of the soils. J. Soil Sci. 4:142–154; 1953.

DiSalvo, L. H. Microbial ecology. In: Jones, O. A.; Endean, R., eds. *Biology and geology of coral reefs*, vol. 2, biology 1. New York: Academic Press, 1973:1–15.

DiSalvo, L. H. Soluble phosphorus and amino nitrogen released to seawater during recoveries of coral reef regenerative sediments. Proc. Second Int. Coral Reef Symp. 1:11–19; 1974.

Dodd, J. L.; Lauenroth, W. K. Analysis of the response of a grassland ecosystem to stress. In: French, N. R., ed. *Perspectives in grassland ecology*. New York: Springer-Verlag, Ecological Studies, Volume 32, 1979:43–58.

Domotor, S. L.; D'Elia, C. F. Nutrient uptake kinetics and growth of zooxanthellae maintained in laboratory culture. Mar. Biol. 80:93–101; 1984.

Doran, J. W. Soil microbial and biochemical changes associated with reduced tillage. Soil Sci. Soc. Am. Journal 44:761–765; 1980a.

Doran, J. W. Microbial changes associated with residue management with reduced tillage. Soil Sci. Soc. Am. Journal 44:518–524; 1980b.

Dugdale, R. C. Nutrient limitation in the sea: dynamics, identification, and significance. Limnol. Oceanogr. 12:685–695; 1967.

Dugdale, R. C.; Goering, J. J. Uptake of new and regenerated forms of nitrogen in primary productivity. Limnol. Oceanogr. 12:196–206; 1967.

Dunlap, W. C. Measurement of extra-cellular phosphatase enzyme activity in reef waters of the central Great Barrier Reef. Fifth Int. Coral Reef Symp. 3:451–456; 1985.

Dyer, M. I.; Bokhari, U. G. Plant–animal interactions: studies of the effects of grasshopper grazing in blue grama grass. Ecology 57:762–772; 1976.

Edson, M. M.; Foin, T. C.; Knapp, C. M. "Emergent properties" and ecological research. Am. Nat. 118:593–596; 1981.

Edwards, C. A. Investigations into the influence of agricultural practice on soil invertebrates. Ann. Appl. Biol. 87:515–520; 1977.

Edwards, N. T.; Shugart, H. H. Jr.; McLaughlin, S. B.; Harris, W. F.; Reichle; D. E. Carbon metabolism in terrestrial ecosystems. In: Reichle, D. E., ed. *Dynamic properties of forest ecosystems.* Cambridge, England: Cambridge University Press, 1981:499–536.

Egerton, F. N. Changing concepts of the balance of nature. Q. Rev. Biol. 48:322–350; 1973.

Elliot, E. T.; Wiegert, R. G.; Hunt, H. W. Simulation of simple food chains. In: Lauenroth, W. K.; Skogerboe, G. V.; Flug, M., eds. *Analysis of ecological systems: state-of-the-art in ecological modelling.* Amsterdam: Elsevier Science Publishers; 1983.

Elliott, E. T.; Coleman, D. C.; Ingham, R. E.; Trofymow, J. A. Carbon and energy flow through microflora and microfauna in the soil subsystem of terrestrial ecosystems. In: Klug, M. J.; Reddy, C. A., eds. *Current perspectives in microbial ecology.* Washington: Amer. Soc. Microbiol., 1984:424–433.

Ellis, J. E.; Detling, J. K. Animal effects and ecosystem change in grasslands. Bull. Ecol. Soc. Am. 64(2):52; 1983.

Ellison, L. Influence of grazing by large herbivores on nitrogen cycling in agricultural ecosystems. Bot. Rev.26(1):1–78; 1960.

Elton, C. *Animal ecology.* New York: Macmillan, 1927, 207 pp.

Elton, C. S. *The pattern of animal communities.* London: Methuen & Co. Ltd., 1966, 432 pp.

Elton, C. S.; Miller, R. S. The ecological survey of animal communities; with a practical system of classifying habitats by structural characteristics. J. Ecol. 42:460–496; 1954.

Elwood, J. W.; Newbold, J. D.; O'Neill, R. V.; VanWinkle, W. Resource spiralling: an operational paradigm for analyzing lotic ecosystems. In: Fontaine, T. D.; Bartell, S. M., eds. *The dynamics of lotic ecosystems.* Ann Arbor, Michigan: Ann Arbor Science, 1983, 494 pp.

Ember, L. M.; Williams, D. F.; Morris, J. T. Processes that influence carbon isotope variation in salt marsh sediments. Mar. Ecol. Prog. Ser. 36:33–42; 1987.

Endean, R. Population explosions of *Acanthaster planci* and associated destruction of hermatypic corals in the Indo-West Pacific region. In: Jones, O. A.; Endean, R., eds. *Biology and geology of coral reefs,* vol. 2. London: Academic Press, 1973:389–438.

Enfield, D. B.; Newberger, P. A. Peru coastal winds during 1982–83. In: Proceedings of the 9th Annual Climate Diagnostics Workshop. Washington: NOAA, 1985:147–156.

Engelmann, M. D. The role of soil arthropods in the energetics of an old field community. Ecol. Monogr. 31:221–238; 1961.

Entsch, B; Boto, K. G.; Sim, R. G.; Wellington, J. T. Phosphorus and nitrogen in coral reef sediments. Limnol. Oceangr. 28:465–476; 1983a.

Entsch, B; Sim, R. G.; Hatcher, B. G. Indications from photosynthetic components that iron is a limiting nutrient in primary producers on coral reefs. Mar. Biol. 73:17–30; 1983b.

Ertel, J. R.; Hedges, J. I. Sources of sedimentary humic substances: vascular plant debris. Geochim. Cosmochim. Acta 49: 2097–2107; 1985.

Evans, F. C. Ecosystem as the basic unit in ecology. Science 123:1127–1128; 1956.

Fahey, T. J. Nutrient dynamics of aboveground detritus in lodgepole pine (*Pinus contorta ssp. latifolia*) ecosystems, southeastern Wyoming. Ecol. Monogr. 53:51–72; 1983.

Fasham, M. J. R. Flow analysis of materials in the marine euphotic zone. In: Ulanowicz, R. E.; Platt, T., eds., *Ecosystem theory for biological oceanography.* Canad. Bull. Fish. Aquat. Sci. 213: 139–162; 1985.

Fee, E. J. The vertical and seasonal distribution of chlorophyll in lakes of the Experimental Lakes Area, northwestern Ontario: implication for primary productivity estimates. Limnol. Oceanogr. 21:767–783; 1976.

Fee, E. J. Studies of the hypolimnion chlorophyll peaks in the ELA, northwestern Ontario. Can. Fish. Mar. Ser. Tech. Rep. 757: iv + 21 pp.; 1978.

Fenchel, T. The ecology of marine microbenthos. IV. Structure and function of the benthic ecosystem, its chemical and physical factors and the microfauna communities with special reference to the ciliated protozoa. Ophelia 6:1–182; 1969.

Fenchel, T. Intrinsic rate of natural increase: the relationship with body size. Oecologia 14:317–326; 1974.

Fenchel, T.; Blackburn, T. *Bacteria and mineral cycling.* New York: Academic Press, 1979, 225 pp.

Fiedler, P. C. Some effects of El Niño 1983 on the northern anchovy. CalCOFI Rep. 25:53–58; 1984.

Fish, D. Phytotelmata: flora and fauna. In: Frank, J. H.; Lounibos, L. P., eds. *Phytotelmata: terrestrial plants as hosts for aquatic insect communities.* Medford, New Jersey: Plexus Press, 1983:173–194.

Fisher, S. G. Organic matter processing by a stream–segment ecosystem: Fort River, Massachusetts, USA. Int. Rev. Ges. Hydrobiol. 62:701–727; 1977.

Fisher, S. G. Succession in streams. In: Barnes, J. R.; Minshall, G. W., eds. *Stream ecology: application and testing of general ecological theory.* New York: Plenum Press, 1983:7–27.

Fisher, S. G.; Likens, G. E. Energy flow in Bear Brook, New Hampshire: an integrative approach to stream ecosystem metabolism. Ecol. Monogr. 43:421–439; 1973.

Fisher, S. G.; Gray, L. J.; Grimm, N. B.; Busch, D. E. Temporal succession in a desert stream ecosystem following flash flooding. Ecol. Monogr. 52:93–110; 1982.

Fittkau, E. J.; Klinge, H. On biomass and trophic structure of the central Amazonian rain forest ecosystem. Biotropica 5:2–14; 1973.

Flint, M. L.; van den Bosch, R. *Introduction to integrated pest management,* New York: Plenum Press; 1981.

Floate, M. J. S. Effects of grazing by large herbivores on nitrogen cycling in agricultural ecosystems. In: Clark, F. E.; Rosswall, T., eds. *Terrestrial nitrogen cycles.* Swed. Nat. Sci. Res. Counc. 33:585–602; 1981.

Fogel, R. Roots as primary producers in below-ground ecosystems. In: Fitter, A. H.; Atkinson, D.; Read, D. J.; Usher, M. B., eds. *Ecological interactions in soil.* Special Publ. #4, British Ecological Society. Oxford: Blackwell, 1985:23–36.

Folsome, C. E. *The origin of life*. San Francisco: Freeman and Company; 1979.

Fong, W.; Mann, K. H. Role of gut flora in the transfer of amino acids through a marine food chain. Can. J. Fish. Aquat. Sci. 37: 88–96; 1980.

Fontaine, T. E. A self-designing model for testing hypotheses of ecosystem development. Progress in Ecological Engineering and Management by Mathematical Modeling. Liege, Belgium, Editions Cebedoc Sprl: 1981:281–292.

Fontaine, T. D. III; Bartell, S. M., eds. *Dynamics of lotic ecosystems*. Ann Arbor, Michigan: Ann Arbor Science, 1983, 494 pp.

Forbes, S. A. (1887) The lake as a microcosm. Bull. Peoria Sci. Assoc.: Illinois Nat. Hist. Surv. Bull. 15:537–550; 1925.

Ford, D. E.; Stephan, H. Stratification variability in three morphometrically different lakes under identical meteorological forcing. Water Resour. Res. Bull. 16:243–247; 1980.

Forel, F. A. Le Leman: monographie limnologique. 2. Mecanique, chimie, thermique, optique, acoustique. Lausanne: F. Rouge, 1895, 651 pp.

Foreman, R. E. Benthic community modification and recovery following intensive grazing by *Strongylocentrotus droebachiensis*. Helgol. wiss. Meeresunters. 30:468–484; 1977.

Forman, R. T. T.; Godron, M. *Landscape ecology*. New York: John Wiley & Sons; 1986.

Foster, J. R. Causes of tree mortality in wave-regenerated balsam fir forests. Dartmouth College, Hanover, New Hampshire, USA. 1984. Ph.D. Thesis.

Francis, C. A.; Harwood, R. R.; Parr, J. F. The potential for regenerative agriculture in the developing world. Am. J. Alt. Agric. 1:65–74; 1986.

Franklin, J. F.; Forman, R. T. T. Creating landscape patterns by cutting: biological consequences and principles. Landscape Ecology 1:5–18; 1987.

Franklin, J. F.; Hemstrom, M. A. Aspects of succession in the coniferous forests of the Pacific Northwest. In: West, D. C.; Shugart, H. H.; Botkin, D. B., eds. *Forest succession. Concepts and application*. New York: Springer-Verlag; 1981.

Franklin, J. F.; Cromack, K. Jr.; Denison, W.; Mckee, A.; Maser, C.; Sedell, J.; Swanson, F.; Juday, G. Ecological characteristics of old-growth Douglas-fir forests. USDA Forest Service, Pacific Northwest Forest and Range Experiment Station General Technical Report PNW–118; 1981.

Franklin, R. T. Insect influences on the forest canopy. In: Reichle, D. E., ed. *Analysis of temperate forest ecosystems*. Berlin: Springer-Verlag, 1970:86–99.

Franzisket, L. Uptake and accumulation of nitrate and nitrite by reef corals. Naturwissenschaften 12:552; 1973.

Franzisket, L. Nitrate uptake by reef corals. Int. Rev. Gesamten Hydrobiol. 59:1–7; 1974.

Freivalds, J. *Grain trade, the key to world power and human survival*. Stein and Day, 1976, 248 pp.

Frempong, E. Diel variations in the abundance, vertical distribution, and species composition of phytoplankton in a eutrophic English lake. J. Ecol. 69:919–939; 1981.

French, N. R., ed. *Perspectives in grassland ecology*. New York: Springer-Verlag; 1979a.

French, N. R. Principal subsystem interactions in grasslands. In: French, N. R., ed. *Perspectives in grassland ecology. (Ecological Studies, vol. 32)*, New York: Springer-Verlag, 1979b:173–190.

Friederichs, K. Grundsätzliches über die Lebenseinheiten höheren Ordnung und den ökologischen Einheitsfaktor. Naturwissenschaften 15:153–157, 182–286; 1927.

Frissel, M. J., ed. Cycling of mineral nutrients in agricultural ecosystems. Agro-ecosystems 4:1–354; 1977.

Gage, K. S.; Reid, G. C. Response of the tropical tropopause to El Chichon and the El Niño of 1982–1983. Geophys. Res. Lett. 12:195–197; 1985.

Gander, M. V. The dynamics and trophic ecology of grasshoppers (Acridoidea) in a South African savanna. Oecologia 54:370–378; 1982.

Gaudette, H. E.; Lyons, W. B. Phosphate geochemistry in nearshore carbonate sediments: a suggestion of apatite formation. In: Bentor, Y. K., ed. *Marine phosphorites.* Soc. Econ. Paleontol. Mineral. Sp. Publ. 29, 1980:215–225.

Gebhardt, M. R.; Daniel, T. C.; Schweizer, E. E.; Allmaras, R. R. Conservation tillage. Science 230:625–630; 1985.

Gerlach, S. A.; Hahn, A. E.; Schrage, M. Size spectra of benthic biomass and metabolism. Mar. Ecol. Prog. Ser. 26:161–173; 1985.

Gholz, H. L. Environmental limits on aboveground net primary production, leaf area, and biomass in vegetation zones of the Pacific northwest. Ecology 63:469–481; 1982.

Giammatteo, P. A. On the application of chemical equilibrium models to the prediction of phosphate availability for algal growth. Clemson University, 1986. Ph. D. dissertation.

Giammatteo, P. A.; Schindler, J. E.; Waldron, M. C.; Freedman, M. L.; Speziale, B. J.; Zimmerman, M. J. Use of equilibrium programs in predicting phosphorus availability. In: Halberg, R., ed. *Environmental biogeochemistry.* Ecol. Bull. (Stockholm) 35:491–501; 1983.

Gilbert, G. N.; Mulkay, M. *Opening Pandora's box. A sociological analysis of scientists' discourse.* Cambridge, UK: Cambridge University Press; 1984.

Gill, A. E. *Atmosphere–ocean dynamics.* Academic Press, 1982, 662 pp.

Gilmour, J. S. L. The development of taxonomic theory since 1851. Nature 168:400–402; 1951.

Gladfelter, E. H. Metabolism, calification and carbon production. II. Organism-level studies. Proc. Fifth Internat. Coral Reef Congr. Tahiti 4:527–539; 1985.

Gladyshev, G. P. Classical thermodynamics, tandemism, and biological evolution. J. Theor. Biol. 94:225–239; 1982.

Glendening, G. E. Some quantitative data on the increase of mesquite and cactus on a desert grassland range in Southern Arizona. Ecology 33(3):319–328; 1952.

Glerum, C.; Balantinecz, J. J. Formation and distribution of food reserves during autumn and their subsequent utilization in jack pine. Can. J. Bot. 58:40–54; 1980.

Glynn, P. W. Ecology of a Caribbean coral reef. The Porites reef–flat biotope: Part II. Plankton community with evidence for depletion. Mar. Biol. 22:1–21; 1973.

Glynn, P. W. Extensive 'bleaching' and death of reef corals on the Pacific coast of Panama. Environ. Conser. 10:149–154; 1983.

Gochlerner, G. B. Free oxygen and evolutionary progress. J. Theor. Biol. 75:467–486; 1978.

Goldner, L. L. Nitrogen fixation (acetylene reduction) in shallow water Bahamian environments. Bull. Mar. Sci. 30:444–453; 1980.

Golley, F. B. Energy dynamics of a food chain of an old-field community. Ecol. Monogr. 30:187–206; 1960.

Golley, F. B. Energy flux in ecosystems. In: Wiens, J. A., ed. *Ecosystem structure and function.* Proceedings of the thirty-first Annual Biology Colloquium. Corvallis, Oregon: Oregon State University Press, 1972:69–90.

Gorham, E.; Vitousek, P. M.; Reiners, W. A. The regulation of chemical budgets over the course of ecosystem succession. Ann. Rev. Ecol. Syst. 10:53–84; 1979.

Gosz, J. R. Biomass distribution and production budget for a nonaggrading forest ecosystem. Ecology 61:507–514; 1980.

Gosz, J. R., Holmes, R. T.; Likens, G. E.; Bormann, F. H. The flow of energy in a forest ecosystem. Sci. Am. 238:92–102; 1978.

Gould, S. J. Is uniformitarianism necessary? Am. J. Sci. 263:223–228; 1965.

Gould, S. J. Evolution and the triumph of homology, or why history matters. Am. Sci. 74:60–69; 1986.

Graham, R. L.; Cromack, K. Jr. Mass nutrient content and decay rate of dead boles in rain forests of Olympic National Park. Can. J. For. Res. 12:511–521; 1982.

Grant, W. E.; French, N. R.; Swift, D. M. Response of a small mammal community to water and nitrogen treatments in a shortgrass prairie ecosystem. J. Mammal. 58:637–562; 1977.

Greene, J. C. *Science, ideology, and world view.* Berkeley: University of California Press; 1981.

Greenland, D. J. Soil management and soil degradation. J. Soil Sci. 32:301–322; 1981.

Gregor, J. W.; Sansome, F. W. Experiments on the genetics of wild populations. Part I. Grasses. J. Gen. 17:349–364; 1926.

Gregoric, E. G.; Anderson, D. W. Effects of cultivation and erosion on soils of 4 toposequences in the Canadian prairies. Geoderma 36:343–354; 1985.

Gregory, S. V. Plant–herbivore interactions in stream systems. In: Barnes, J. R.; Minshall, J. R., eds. *Stream ecology: application and testing of general ecological theory.* New York: Plenum Press, 1983:157–189.

Grier, C. C.; Logan, R. C. Old-growth *Pseudotsuga menziesii* communities of a western Oregon watershed: biomass distribution and production budgets. Ecol. Monogr. 47:373–400; 1977.

Grier, C. C.; Vogt, K. A.; Keyes, M. R.; Edmonds, R. L. Biomass distribution and above- and below-ground production in young and mature *Abies amabilis* zone ecosystems of the Washington Cascades. Can. J. For. Res. 11:155–167; 1981.

Grigg, R. W.; Polovina, J. J.; Atkinson, M. J. Model of a coral reef ecosystem. III. Resource limitation, community regulation, fisheries yield and resource management. Coral Reefs 3:23–27; 1984.

Grime, J. P. Evidence for the existence of three primary strategies in plants and its relevance to ecological and evolutionary theory. Am. Nat. 111:1169–1194; 1977.

Grunow, J. O.; Groeneveld, H. T.; DuToit, S. H. C. Above- ground dry matter dynamics of the grass layer of a South African tree savanna. J. Ecol. 68:877–889; 1980.

Guerinot, M. L. The association of N_2-fixing bacteria with sea urchins. Halifax, Nova Scotia: Dalhousie University; 1973. Ph. D. thesis.

Guerinot, M. L.; Fong, W.; Patriquin, D. G. Nitrogen fixation [acetylene reduction] associated with sea urchins [*Strongylacentrotus droebachiensis*] feeding on seaweeds and eel grass. J. Fish. Res. Board Can. 34:416–420; 1977.

Guither, H. D. Developing policy for a resource-conserving agriculture: The Food Security Act of 1985 in perspective. Am. J. Alt. Agric. 1:39–42; 1986.

Gurtz, M. E.; Webster, J. R.; Wallace, J. B. Seston dynamics in southern Appalachian streams: effects of clearcutting. Can. J. Fish. Aquat. Sci. 37:624–631; 1980.

Hacking, I. *The emergence of probability.* London: Cambridge University Press; 1984.

Hagen, N. T. Destructive grazing of kelp beds by sea urchins in Vestfjorden, northern Norway. Sarsia 68:177–190; 1983.

Hake, D. R.; Powell, J.; McPherson, J. K.; Claypool, P. L.; Dunn, G. L. Water stress of tallgrass prairie plants in central Oklahoma. J. Range Manage. 37:147–151; 1984.

Hallam, A. *Great geological controversies.* Oxford University Press, 1983, 182 pp.

Halvorson, H. O.; Pramer, D.; Rogul, M., eds. *Engineered organisms in the environment: scientific issues.* Washington: American Society for Microbiology; 1985.

Hanisak, M. D. F. The nitrogen relationships of marine macroalgae. In: Carpenter, E. J.; Capone, D. G. *Nitrogen in the marine environment.* New York: Academic Press, 1983:699–730.

Hanson, B. J.; Cummins, K. W.; Cargill, A. S. III; Lowry, R. W. Dietary effects of lipid and fatty acid composition of *Clistoronia magnifica* (Trichoptera: Limnephilidae). Freshwat. Biol. 2:2–15; 1983.

Hanson, B. J.; Cummins, K. W.; Barnes, J. B.; Carter, M. W. Leaf litter processing in aquatic systems: a two variable model. Hydrobiologia 111:21–29; 1984.

Hanson, B. J.; Cummins, K. W.; Cargill, A. S. III; Lowry, R. W. Composition of polyunsaturated fatty acids in aquatic insects. Comp. Biochem. Physiol. 808:257–276:1985.

Hanson, C. L.; Morris, R. P.; Wright, J. R. Using precipitation to predict range herbage production in southwestern Idaho. J. Range Manage. 36:766–770; 1983.

Hanson, N. R. *Patterns of discovery.* Cambridge: Cambridge University Press; 1965.

Hanson, N. R. *Perception and discovery.* San Francisco: Freeman, Cooper and Company; 1969.

Hanson, R. B.; Gundersen, K. Relationship between nitrogen fixation (acetylene reduction) and the C:N ratio in a polluted coral reef ecosystem, Kaneohe Bay, Hawaii. Estuarine Coast. Mar. Sci. 5:437–444; 1977.

Hardy, A. C. The herring in relation to its animate environment, Part a. The food and feeding habits of the herring. Fish. Invest. Lond. Ser. II 7, no. 3; 1924.

Harmon, M. E.; Franklin, J. F.; Swanson, F. J.; Solins, P.; Gregory, S. V.; Lattin; J. D.; Anderson, N. H.; Cline, S. P.; Aumen, N. G.; Sedell, J. R.; Lienkaemper, G. W.; Cromack, K. Jr.; Cummins, K. W. Ecology of coarse woody debris. Adv. Ecol. Res. 15:133–302; 1986.

Harris, G. P. Temporal and spatial scales in phytoplankton ecology. Mechanisms, methods, models and management. Can. J. Fish. Aquat. Sci. 37:877–900; 1980a.

Harris, G. P. The measurement of photosynthesis in natural populations of phytoplankton. In: Morris, I., ed. *The physiological ecology of phytoplankton.* Berkeley: University of California Press, 1980b:129–187.

Harris, W. F.; Santantonio, D.; McGinty, D. The dynamic belowground ecosystem. In: Waring, R. H., ed. *Forests: fresh perspectives from ecosystem analysis.* Proceedings of the 40th Annual Biology Colloquium. Corvallis, Oregon: Oregon State University Press, 1980:119–129.

Harrison, W. G. Respiration and its size-dependence in microplankton populations from surface waters of the Canadian Arctic. Polar Biol. 6:145–152; 1986.

Hart, D. D. Grazing insects mediate algal interactions in a stream benthic community. Oikos 44:40–46; 1985.

Harvey, H. W. On the production of living matter in the sea off Plymouth. J. Mar. Biol. Ass. UK 29:97–137; 1950.

Hatcher, A. I. The relationship between coral reef structure and nitrogen dynamics. Proc. Fifth Int. Coral Reef Symp. 3:407–413; 1985.

Hatcher, A. I.; Frith, C. A. The control of nitrate and ammonium concentrations in a coral reef lagoon. Coral Reefs 4:101–110; 1985.

Hatcher, A. I.; Hatcher, B. G. Seasonal and spatial variation in dissolved inorganic nitrogen in One Tree Reef Lagoon. Proc. Fourth Internat. Coral Reef Symp., Manila, vol. 1, 1981:419–424.

Hatcher, A. I.; Larkum, A. W. D. An experimental analysis of factors controlling the standing crop of the epilithic algal community on a coral reef. J. Exp. Mar. Biol. Ecol. 69:61–84; 1983.

Hattori, A. Denitrification and dissimilatory nitrate reduction. In: Carpenter, E. J.; Capone, D. G., eds. *Nitrogen in the marine environment*. New York: Academic Press, 1983:191–232.

Hauptli, H.; Goodman, R. M. Commercial priorities of small companies in agricultural biotechnology: relevance to alternative agriculture. Institute of Alternative Agriculture; 1986.

Hayes, D. C. Seasonal nitrogen translocation in big bluestem during a drought year. J. Range Manage. 38:406–410; 1985.

Heinselman, M. L. Fire and succession in the conifer forests of northern North America. In: West, D. C.; Shugart, H. H.; Botkin, D. B., eds. *Forest succession. Concepts and application*. New York: Springer-Verlag, 1981:374–405.

Heitschmidt, R. K.; Price, D. L.; Gordon, R. A.; Frasure, J. R. Short duration grazing at the Texas Experimental ranch: Effects on aboveground net primary productivity and seasonal growth dynamics. J. Range Manage. 35:367–372; 1982.

Hempel, C. G. The function of general laws in history. J. Philos. 39:5–48; 1942.

Hempel, C. G.; Oppenheim, P. Studies on the logic of explanation. Philos. Sci. 15:135–175; 1948.

Henderson, R. S. In situ and microcosm studies of diel metabolism of reef flat communities. Proc. Fourth Internat. Coral Reef Symp., Manila, vol. 1, 1981:679–686.

Hendrix, P. F.; Parmelee, R. W. Decomposition, nutrient loss and microarthropod densities in herbicide-treated grass litter in a Georgia Piedmont agroecosystem. Soil Biol. Biochem. 17:421–428; 1985.

Hendrix, P. F.; Parmelee, R. W.; Crossley, D. A. Jr.; Coleman, D. C.; Odum, E. P.; Groffman, P. M. Detritus food webs in conventional and no-tillage agroecosystems. BioScience 36:374–380; 1986.

Heywood, R. B.; Everson, I.; Priddle, J. The absence of krill from the South Georgia zone, winter 1983. Deep-Sea Res. 32:369–378; 1985.

Hilbert, D. W.; Logan, J. A. A simulation model of the migratory grasshopper (*Melanoplus sanguinipes*). In: Lauenroth, W. K.; Skogerboe, G. V.; Flug, M., eds. *Analysis of ecological systems: state-of-the-art in ecological modelling*. Amsterdam: Elsevier Scientific Publishing Company, 1983:323–334.

Hilbert, D. W.; Swift, D. M.; Detling, J. K.; Dyer, M. I. Relative growth rates and the grazing optimization hypothesis. Oecologia 51:14–18; 1981.

Hill, J.; Durham, S. L. Input, signals and control in ecosystems. Proc. IEEE Intl. Conf. on Acoustics, Speech and Signal Processing, Tulsa, Oklahoma, April 1978, 1978:1–7.

Hill, J.; Wiegert, R. G. Microcosms in ecological modeling. In: Giesy, J. P., ed. *Microcosms in ecological research*. Tech. Information Center, USDOE Symposium Series No. 52, 1980:138–163.

Hines, M. E. Microbial biogeochemistry in shallow water sediments of Bermuda. Fifth Int. Coral Reef Symp. 3:427–432; 1985.

Hines, M. E.; Lyons, W. B. Biogeochemistry of nearshore Bermuda sediments. I. Sulfate reduction rates and nutrient generation. Mar. Ecol. Prog. Ser. 8:87–94; 1982.

Hiratsuka, E. Researches on the nutrition of the silk worm. Bull. Seric. Exp. Stn., Japan 1:257–315; 1920.

Hisard, P.; Henin, C.; Houghton, R.; Piton, B.; Rual, P. Oceanic conditions in the tropical Atlantic during 1983 and 1984. Nature 322:243–245; 1986.

Hobbie, J. E.; Melillo, J. M. Role of microbes in global carbon cycling. In: Klug, M. J.; Reddy, C. A., eds. *Current perspectives in microbial ecology*. Washington: Amer. Soc. Microbiol., 1984:389–393.

Holland, E. A.; Coleman, D. C. Litter placement effects on microbial communities and organic matter dynamics in agroecosystems. Ecology. In press.

Hooghiemstra, H. *Vegetational and climatic history of the high plain of Bogota, Columbia: a continuous record of the last 3.5 million years.* Liechtenstein: Cramer; 1984.

Horne, A. J.; Fogg, G. E. Nitrogen fixation in some English lakes. Proc. R. Soc. London Ser. B. 175:351–366; 1970.

Horne, A. J.; Galat, D. L. Nitrogen fixation in an oligotrophic saline desert lake: Pyramid Lake, Nevada. Limnol. Oceanogr. 30:1229–1239; 1985.

Horne, A. J.; Goldman, C. R. Supression of N_2 fixation by blue-green algae in a eutrophic lake with trace additions of copper. Science 183:409–411; 1974.

House, G. J.; Parmelee, R. W. Comparison of soil arthropods and earthworms from conventional and no-tillage agroecosystems. Soil and Tillage Res. 5:351–360; 1985.

House, G. J.; Stinner, B. R.; Hicks, R. E.; Crossley, D. A. Jr.; Odum, E. P. Simulation models of nitrogen flux in conventional and no-tillage agroecosystems. In: Lowrance, R.; Todd, R. L.; Asmussen, L. E.; Leonard, R. A., eds. *Nutrient cycling in agricultural ecosystems.* University of Georgia, Coll. Agric. Spec. Publ. 23., 1983:569–578.

Howarth, R. W.; Cole, J. J. Molybdenum availability, nitrogen limitation, and phytoplankton growth in natural waters. Science 229:653–655; 1985.

Hulburt, E. M. The unpredictability of marine phytoplankton. Ecology 64:1157–1170; 1983.

Hulburt, E. M. Use of logical equivalence in modeling ecological relations of oceanic phytoplankton. Ecol. Model. 27: 25–43; 1985.

Humphries, W. F. Production and respiration in animal populations. J. Anim. Ecol. 48:427–453; 1979.

Humphries, W. F. Towards a simple index based on live-weight and biomass to predict assimilation in animal populations. J. Anim. Ecol. 50:543–561; 1981.

Huston, M. A. Patterns of species diversity on coral reefs. Annu. Rev. Ecol. Syst. 16:149–177; 1985.

Hutchinson, G. E. Circular causal systems in ecology. Ann. NY Acad. Sci. 50:221–246; 1948.

Hutchinson, G. E. *A treatise on limnology, vol. 1. Geography, physics, and chemistry.* New York: John Wiley & Sons, 1957, 1015 pp.

Hutchinson, G. E. The paradox of the plankton. Am. Nat. 95:137–145; 1961.

Hutchinson, G. E. The lacustrine microcosm reconsidered. Am. Sci. 52:334–341; 1964.

Hutchinson, G. E. *A treatise on limnology, vol. 2.* New York: John Wiley & Sons, 1967, 412 pp.

Hutchinson, G. E.; Loffler, H. The thermal classification of lakes. Proc. Nat. Acad. Sci. Wash. 42:84–86; 1956.

Hutchinson, K. J. Productivity and energy flow in fodder conservation/grazing systems. Herb. Abstr. 41:1–10; 1971.

Hutton, J. *Theory of the Earth, with proofs and illustrations, 2 vols.* Edinburgh. Facsimile reprint in 1959 by Wheldon and Wesley, Codicote, Herts.; 1795.

Huxley, J. S. *Problems of relative growth* (1972 reprint). New York: Dover, 1932, 312 pp.

Huxley, T. H. *Discourses: biological and geological essays.* New York: D. Appleton and Co, 1897, 388 pp.

Huyer, A.; Smith, R. L.; Hickey, B. M. Observations of a warm-core eddy off Oregon, January to March 1978. Deep-Sea Res. 31:97–117; 1984.

Hyder, D. N.; Bement, R. E.; Remmenga, E. E.; Hervey, D. F. Ecological responses of mature plants and guidelines for management of shortgrass prairie. USDA Technical Bulletin 1503; 1975.

Hynes, H. B. N. Imported organic matter and secondary productivity in streams. Proc. 14th Int. Congr. Zool. 3:324–329; 1963.

Hynes, H. B. N. *The ecology of running waters.* Toronto: University of Toronto Press, 1970a, 555 pp.

Hynes, H. B. N. The ecology of stream insects. Ann. Rev. Ent. 15:25–42; 1970b.

Hynes, H. B. N. The stream and its valley. Verh. Internat. Verein. Limnol. 19:1–15; 1975.

Hynes, H. B. N. Groundwater and stream ecology. Hydrobiologia 100:93–99; 1983.

Imberger, J. The diurnal mixed layer. Limnol. Oceanogr. 30:737–770; 1985.

Imberger, J.; Hamblin, P. F. Dynamics of lakes, reservoirs, and cooling ponds. Ann. Rev. Fluid Mech. 14:153–187; 1982.

Imberger, J.; Parker, G. Mixed layer dynamics in a lake exposed to a spatially varying wind field. Limnol. Oceanogr. 30: 473–488; 1985.

Imberger, J.; Patterson, J. A dynamic reservoir simulation model. DYRESM 5. In: Fischer, H. B., ed. *Transport models for inland and coastal waters. Proceedings of a symposium on predictive ability.* New York: Academic Press, 1981:310–361.

Imberger, J.; Patterson, J. J.; Hebbert, B.; Loh, I. Dynamics of reservoir of medium size. J. Hydrau. Div. Am. Soc. Civ. Eng. 104:725–743; 1978.

Ingham, R. E.; Detling, J. K. Plant–herbivore interactions in a North American mixed-grass prairie. III. Soil nematode populations and root biomass on *Cynomys ludovicianus* colonies and adjacent uncolonized areas. Oecologia (Berl.) 63:307–313; 1984.

Ingham, R. E.; Detling, J. K. Effects of defoliation and nematode consumption on growth and leaf gas exchange in *Bouteloua curtipendula.* Oikos 46:23–28; 1986.

Ingham, R. E.; Trofymow, J. E.; Ingam, E. R.; Coleman, D. C. Interactions of bacteria, fungi, and their nematode grazers: Effects on nutrient cycling and plant growth. Ecol. Monogr. 55:119–140; 1985.

Innis, G. S., ed. *Grassland simulation model.* Ecological Studies, vol. 26, New York: Springer-Verlag; 1978.

Jacox, A. Competing theories of science. Paper presented at a 1981 forum on Doctoral Education in Nursing. Seattle, Washington; June 25, 1981.

Jantsch, E. *The self-organizing universe: scientific and human implications of the emerging paradigm of evolution.* Oxford: Pergamon, 1980.

Jaramillo, V. The effect of grazing history and competition on defoliation of *Bouteloua gracilis.* Fort Collins: Colorado State University; 1986. M. S. thesis.

Jarvis, P. G.; Leverenz, J. W. Productivity of temperate, deciduous and evergreen forests. In: Lange, O. L.; Nobel, P. S.; Osmond, C. B.; Ziegler, H., eds. *Physiological plant ecology IV. Ecosystem processes: mineral cycling, productivity and man's influence.* Encyclopedia of Plant Physiology, New Series, vol. 12D. Berlin: Springer-Verlag, 1983:233–280.

Jeffries, M. J.; Lawton, J. H. Enemy free space and the structure of ecological communities. Biol. J. Linn. Soc. 23:269–286; 1984.

Jenkins, W. J.; Goldman, J. C. Seasonal oxygen cycling and primary production in the Sargasso Sea. J. Mar. Res. 43:465–491; 1985.

Jenkinson, D. S.; Powlson, D. S. The effects of biocidal treatments on metabolism in soil. V. A method for measuring soil biomass. Soil Biol. Biochem. 8:209–213; 1976.

Jenny, H. *Factors of soil formation.* New York: McGraw-Hill; 1941.

Johannes, R. E. Phosphorus excretion and body size in marine animals: microzooplankton and nutrient regeneration. Science 146:923–924; 1964.

Johannes, R. E. Pollution and degradation of coral reef communities. In: Wood, E. J. F.; Johannes, R. E., eds. *Tropical marine pollution.* Elsevier Oceanography Series, #12. New York: Elsevier, 1975:13–51.

Johannes, R. E.; Betzer, S. B. Introduction: marine communities respond differently to pollution in the tropics than at higher latitudes. In: Wood, E. J. F.; Johannes, R. E., eds. *Tropical marine pollution.* Elsevier Oceanography Series. #12. New York: Elsevier, 1975:1–12.

Johannes, R. E.; Alberts, J.; D'Elia, C. F.; Kinzie, R. A.; Pomeroy, L. R.; Sottile, W.; Wiebe, W.; Marsh, J. A. Jr.; Helfrich, P.; Maragos, J.; Meyer, J.; Smith, S.; Crabtree, D.; Roth, A.; McCloskey, L. R.; Betzer, S.; Marshall, N.; Pilson, M. E. Q.; Telek, G.; Clutter, R. I.; DuPaul, W. D.; Webb, K. L.; Wells, J. M. Jr. The metabolism of some coral reef communities: A team study of nutrient and energy flux at Eniwetok. BioScience 22:541–543; 1972.

Johannes, R. E.; Wiebe, W. J.; Crossland, C. J. Three patterns of nutrient flux in a coral reef community. Mar. Ecol. Prog. Ser. 12:131–136; 1983a.

Johannes, R. E.; Wiebe, W. J.; Crossland, C. J.; Rimmer, D. W.; Smith, S. V. Latitudinal limits of coral reef growth. Mar. Ecol. Prog. Ser. 11:105–111; 1983b.

Johanssen, P. M., ed. *Ecology of eutrophic, subarctic Lake Myvatn and the River Laxa.* Oikos 32:1–308; 1979.

Johns, P. M.; Mann, K. H. An experimental investigation of juvenile lobster habitat preference and mortality among habitats of varying structural complexity. J. Exp. Mar. Biol. Ecol. 109:275–285; 1987.

Johnson, A. H.; Siccama, T. G. Acid deposition and forest decline. Environ. Sci. Tech. 17:294A–305A; 1983.

Johnson, D. W.; Cole, D. W.; Bledsoe, C. S.; Cromack, K.; Edmonds, R. L.; Gessell, S. P.; Grier, C. C.; Richards, B. N.; Vogt, K. A. Nutrient cycling in the forests of the Pacific Northwest. In: Edmonds, R. L., ed. *Analysis of coniferous forest ecosystems in the Western United States.* Stroudsburg, Pennsylvania: Hutchinson Ross Publishing Company (US/IBP Synthesis Series 14), 1982:186–232.

Jones, N. S.; Kain, J. M. Subtidal algal recolonization following the removal of *Echinus.* Helgol. wiss. Meeresunters. 15:460–466; 1967.

Jordan, C. F. A world pattern in plant energetics. Am. Sci. 59:425–433; 1971.

Jordan, C. F.; Murphy, P. G. A latitudinal gradient of wood and litter production, and its implications regarding competition and species diversity in trees. Am. Midl. Nat. 99:415–434; 1978.

Jordan, C. F.; Kline, J. R.; Sasscaer, D. S. Effective stability of mineral cycles in forest ecosystems. Am. Nat. 106: 237–253; 1972.

Kaplan, L. A.; Bott, T. L. Diel fluctuations of DOC generated by algae in a piedmont stream. Limnol. Oceanogr. 27:1091–1100; 1982.

Kawaguti, S. Ammonium metabolism of the reef corals. Biol. J. Okayama Univ. 1:171–176; 1953.

Kaye, H. L. *The social meaning of modern biology.* New Haven: Yale University Press; 1986.

Kelly, E. F. Long term erosional effects on cropland vs. rangeland in semi–arid agroecosystems. Colorado State University; 1984. M. S. thesis.

Kelly, R. Aspectos generales de El Niño 1982–83. Invest. Pesq. (Chile) 32:5–7; 1985.

Kemp, W. B. Natural selection within plant species as exemplified in a permanent pasture. J. Hered. 28:329–333; 1937.

Keyes, M. R.; Grier, C. C. Above- and below-ground net production in 40-year old Douglas-fir stands on low and high productivity sites. Can. J. For. Res. 11:599–605; 1981.

Kilgore, B. M. Fire in ecosystem distribution and structure: western forests and scrublands. In: Mooney, H. A.; Bonnicksen, T. M.; Christensen, N. L.; Lotan, J. E.; Reiners, W. A., eds. *Proceedings of the conference. Fire regimes and ecosystem properties.* USDA Forest Service General Technical Report WO–26, 1981:58–89.

Kilham, P.; Kilham, S. S. The evolutionary ecology of phytoplankton. In: Morris, I., ed. *The physiological ecology of phytoplankton.* Berkeley: University of California Press, 1980, 625 pp.

Kilham, S. S.; Kilham, P. The importance of resource supply rates in determining phytoplankton community structure. In: Meyers, D. G.; Strickeler, J. R., eds. *Trophic interactions within aquatic ecosystems.* A. A. A. S. Selected Symposium Series. Boulder, Colorado: Westview Press; 1984.

Kimmerer, W. J.; Walsh, T. W. Tarawa Atoll Lagoon: circulation, nutrient fluxes, and the impact of human waste. Micronesica 17:161–179; 1981.

Kimmins, J. P.; Scoullar, K. A.; Feller, M. C. FORCYTE-9: A computer simulation approach to evaluating the effect of whole tree harvesting on nutrient budgets and future forest productivity. Mitteilungen der Forstlichen Bundesversuchsanst 140:189–205; 1981.

Kingsland, S. E. *Modeling nature. Episodes in the history of population ecology.* Chicago: University of Chicago Press, 1985, 267 pp.

Kinsey, D. W. Seasonality and zonation in coral reef productivity and calcification. Proc. Third Int. Coral Reef Symp. 2:383–388; 1977.

Kinsey, D. W. Carbon turnover and accumulation by coral reefs. University of Hawaii; 1979, 248 pp. Ph.D. dissertation.

Kinsey, D. W. Johnston Atoll–A modern analogy of early holocene reefs? (abstract). Fifteenth Congress of Pac. Sci. Assn. 1:128; 1983a.

Kinsey, D. W. Short-term indicators of gross material flux in coral reefs–how far have we come and how much further can we go? In: Baker, J. T.; Carter, R. M.; Sammarco, P. W.; Stark, K. P., eds. *Proceedings: inaugural Great Barrier Reef conference,* Townsville, Aug 28–Sept. 2, 1983. J.C.U. Press, 1983b:333–339.

Kinsey, D. W. Metabolism, calcification and carbon production–I. System level studies. Proc. Fifth Internat. Coral Reef Congress 4:505–526; 1985.

Kinsey, D. W.; Davies, P. J. Carbon turnover, calcification and growth in coral reefs. In: Trudinger, P. A.; Swaine, D. J., eds. *Biogeochemical cycling of mineral forming elements.* Elsevier, 1979a:131–162.

Kinsey, D. W.; Davies, P. J. Effects of elevated nitrogen and phosphorus on coral reef growth. Limnol. Oceanogr. 24:935–940; 1979b.

Kinsey, D. W.; Domm, A. Effects of fertilization on a coral reef environment–primary production studies. Proc. Second Int. Coral Reef Symp., 1974:49–66.

Kitching, R. L. Spatial and temporal variation in food webs in water-filled tree holes. Oikos 48: 280–288; 1987.

Kitching, R. L.; Callaghan, C. The fauna of water-filled tree holes in box forest in southeast Queensland. Aust. Ent. Mag. 8:61–70; 1981.

Kitching, R. L.; Pimm, S. L. The length of food chains: phytotelmata in Australia and elsewhere. Proc. Ecol. Soc. Aust. 14:123–139; 1985.

Kitts, D. B. *The structure of geology.* Dallas: Southern Methodist University Press; 1977.

Kleiber, M. *The fire of life.* New York: John Wiley & Sons, 1961, 453 pp.

Klug, M. J.; Kotarski, S. Bacteria associated with the gut tract of larval stages of the aquatic cranefly *Tipula abdominalis* (Diptera: Tipulidae). Appl. Environ. Microbiol. 40:408–416; 1980.

Knight, D. H. Parasites, lightning, and the vegetation mosaic in wilderness landscapes. In: Turner, M. G., ed. *Disturbance and landscape heterogeniety.* New York: Springer-Verlag, 1987:59–83.

Koblenz-Mishke, O. J.; Vovkovinsky, V. V.; Kabanova, J. J. Plankton primary productivity of the world ocean. In: Wooster, W. S., ed. *Scientific exploration of the South Pacific.* Washington: National Academy of Sciences; 1970.

Kohn, A. J. ; Helfrich, P. Primary organic productivity of a Hawaiian coral reef. Limnol. Oceanogr. 2:241–251; 1957.

Krebs, C. J. *Ecology.* New York: Harper and Row; 1985.

Krueger, K. Feeding relationships among bison, pronghorn and prairie dogs: an experimental analysis. Ecology 67:760–770; 1986.

Kruger, F. J.; Mitchell, D. T.; Jarvis, J. U. M. *Mediterranean-type ecosystems.* Berlin: Springer-Verlag, 1983, 552 pp.

Kuhn, T. S. *The structure of scientific revolutions.* Chicago: University of Chicago Press; 1962.

Lacey, J. R.; Van Poolen, H. W. Comparison of herbage, production on moderately grazed and ungrazed western ranges. J. Range Manage. 34:210–212; 1981.

Lakatos, I. Falsification and the methodology of scientific research programs. In: Lakatos, I.; Musgrave, A., eds. *Criticism and the growth of knowledge.* London: Cambridge University Press; 1970.

Lal, R. Conversion of tropical rainforest: agronomic potential and ecological consequences. Adv. Agron. 39:173–264; 1986.

Lamb, J. R.; Peterson, G. A.; Fenster, C. R. Wheat fallow tillage systems' effect on a newly cultivated grassland soil's nitrogen budget. Soil Sci. Soc. Am. Journal 49:352–356; 1985.

Lamotte, M. The structure and function of a tropical savannah ecosystem. In: Golley, F. R.; Medina, E., eds. *Tropical ecological systems: Trends in terrestrial and aquatic research.* New York: Springer-Verlag; 1979.

Lamprey, H. F. Estimation of the large mammal densities, biomass, and energy exchange in the Tarangire Game Reserve and the Masai Steppe in Tanganyika. E. Afr. Wildl. J. 2:1–46; 1964.

Lang, G. E. Forest turnover and the dynamics of bole wood litter in subalpine balsam fir forest. Can. J. For. Res. 15:262–268; 1985.

Langbien, W. B.; Leopold, L. B. Quasi-equilibrium states in channel morphology. Amer. J. Sci. 262:782–794; 1964.

Langdale, G. W.; Leonard, R. A. Nutrient and sediment losses associated with conventional and reduced-tillage agricultural practices. In: Lowrance, R.; Todd, R. L.; Asmussen, L. E.; Leonard, R. A., eds. *Nutrient cycling in agricultural ecosystems.* University of Georgia, Coll. Agric. Spec. Publ. 23., 1983:457–468.

Langdale, G. W.; Hargrove, W. L.; Giddens, J. Residue management in double-crop conservation tillage systems. Agron. J. 76:689–694; 1984.

Larson, R. A. Dissolved organic matter of a low-colored stream. Freshwat. Biol. 8:91–104; 1978.

Larson, W. E. Protecting the soil resource base. J. Soil Water Conserv. 36:13–16; 1981.

Larsson, P.; Brittain, J. E.; Lein, L.; Lillehammer, A.; Tangen, K. The lake ecosystem of Ovre Heimdalsvatn. Holarctic Ecology 1:304–320; 1978.

Lau, K.-M.; Chan, P. H. Aspects of the 40–50 day oscillation during the northern winter as inferred from outgoing longwave radiation. Mon. Wea. Rev. 113:1889–1909.

Laudan, L. *Progress and its problems: toward a theory of scientific growth.* Berkeley: University of California Press; 1977.

Laudan, L. *Science and hypothesis: essays on scientific methodology.* Boston: Dordrecht, Holland, 1981, 258 pp.

Lauenroth, W. K. Grassland primary production: North American grasslands in perspective. In: French, N. R., ed. *Perspectives in grassland ecology.* New York: Springer-Verlag, 1979:3–24.

Lauenroth, W. K.; Sims, P. L. Evapotranspiration from a shortgrass prairie subjected to water and nitrogen treatments. Water Resour. Res. 12:437–442; 1976.

Lauenroth, W. K.; Hunt, H. W.; Swift, D. M.; Singh, J. S. Reply to Vogt et al. Ecology 67:580–582; 1986.

Launchbaugh, J. L.; Owensby, C. E. Kansas rangelands. Their management based on a half century of research. Kansas Ag. Exper. Sta. Bull. 622. Manhattan; 1978.

Lavoisier, A. L. Expériences sur la respiration des animaux et sur les changements qui arrivent l'air en passant par leurs poumons. Mem. Acad. Sci., 1777, 1985 pp.

Laws, R. M. Elephants as agents of habitat and landscape change in East Africa. Oikos 21:1–15; 1970.

Laws, E. A. Man's impact on the marine nitrogen cycle. In: Carpenter, E. J.; Capone, D. G., eds. *Nitrogen in the marine environment.* New York: Academic Press, 1983: 459–485.

Lee, K. E. *Earthworms, their ecology and relationships with soils and land use.* Orlando, Florida: Academic Press; 1985.

Leege, T. A.; Herman, D. J.; Zamora, B. Effects of cattle grazing on mountain meadows in Idaho. J. Range Manage. 34:324–328; 1981.

Legendre, L. Hydrodynamic control of marine phytoplankton production. The paradox of stability. In: Nihoul, J. C. J., ed. *Ecohydrodynamics.* Amsterdam: Elsevier, 1981, 358 pp.

Legendre, L.; Demers, S. Towards dynamic biological oceanography and limnology. Can. J. Fish. Aquat. Sci. 41:2–19; 1984.

Leggett, W. C.; Frank, K. T.; Carscadden, J. E. Meteorological and hydrographic regulation of year-class strength in Capelin (*Mallotus villosus*). Can. J. Fish. Aquat. Sci. 41:1193–1201; 1984.

Leigh, E. G. Jr. Population fluctuations, community stability, and environmental variability. In: Cody, M. L.; Diamond, J. M., eds. *Ecology and evolution of communities.* Cambridge: Belknap Press; 1976.

Leighton, D. L. Grazing activities of invertebrates in Southern California kelp beds. In: North, W. J., ed. *The biology of giant kelp beds (Macrosystis) in California.* Nova Hedwigia Suppl. 32: 421–453; 1971.

Leopold, A. C. Aging, senescence, and turnover in plants. BioScience 25:659–662; 1975.

Leopold, L. B.; Wolman, M. G.; Miller, J. P. Fluvial processes in geomorphology. San Francisco: Freeman, 1964, 522 pp.

Lessios, H. A.; Robertson, D. R.; Cubit, J. D. Spread of *Diadema* mass mortality through the Caribbean. Science 226:335–337; 1984.

Levasseur, M.; Therriault, J. C.; Legendre, L. Hierarchical control of phytoplankton succession by physical factors. Mar. Ecol. Prog. Ser. 19:211–222; 1984.

Levins, R. Evolution in communities near equilibrium. In: Cody, M. J.; Diamond, J. M., eds. *Ecology and evolution of communities.* Cambridge: Belknap Press; 1976.

Lewis, J. B. Process of production on coral reefs. Biol. Rev. 52:305–347; 1977.

Lewis, J. B. Groundwater discharge onto coral reefs, Barbados (West Indies). Proc. Fifth Int. Coral Reef Symp. 6:477–481; 1985.

Lewis, J. B. Measurements of groundwater seepage flux onto a coral reef: spatial and temporal variations. Limnol. Oceanogr. 32:1165–1169; 1987.

Lewis, M. R.; Cullen, J. J.; Platt, T. Phytoplankton and thermal structure in the upper ocean: consequences of nonuniformity in chlorophyll profile. J. Geophys. Res. 88: 2565–2570; 1983.

Lewis, R. W. Evolution: a system of theories. Persp. Biol. Med 23:551–572; 1980.

Lewis, S. M.; Wainwright, P. C. Herbivore abundance and grazing intensity on a Caribbean coral reef. J. Exp. Mar. Biol. Ecol. 87:215–228; 1985.

Lewis, W. M. The thermal regime of Lake Lanao (Philippines) and its theoretical implications for tropical lakes. Limnol. Oceanogr. 18:200–217; 1973.

Lieberman, D.; Lieberman, M.; Peralta, R.; Hartshorn, G. S. Mortality patterns and stand turnover rates in a wet tropical forest in Costa Rica. J. Ecol. 73:915–925; 1985.

Liebig, J. Principles of agricultural chemistry with special reference to the late researches made in England. London: Walton and Maberly; 1855.

Lieth, H. Modeling the primary productivity of the world. In: Lieth, H.; Whittaker, R. H., eds. *Primary productivity of the biosphere*, Ecological Studies, vol. 14, New York: Springer-Verlag, 1975:237–263.

Likens, G. E. Primary productivity of inland aquatic ecosystems. In: Lieth, H.; Whittaker, R. H., eds. *Primary productivity of the biosphere*, Ecological Studies, vol. 14, New York: Springer-Verlag, 1975.

Likens, G. E. Beyond the shoreline: a watershed–ecosystem approach. Verh. Internat. Verein. Limnol. 22:1–22; 1984.

Likens, G. E.; Bormann, F. H.; Pierce, R. S.; Eaton, J. S.; Johnson, N. M. *Biogeochemistry of a forested ecosystem*. Berlin: Springer-Verlag, 1977, 146 pp.

LIMER 1975 Expedition Team: Barnes, D. J.; Caperon, J.; Cox, T. J.; Crossland, C. J.; Davies, P. J.; Devereux, M.; Hamner, W. M.; Jitts, H. R.; Kinsey, B. E.; Kinsey, D. W.; Knauer, G. A.; Lundgren, J. A.; Olafson, R.; Skyring, G. W.; Smith, D. F.; Webb, K. L.; Wiebe, W. J. Metabolic processes of coral reef communities at Lizard Island, Queensland. Search 7:11–12, 463–468; 1976.

Lindeman, R. L. The trophic–dynamic aspect of ecology. Ecology 23:399–418; 1942.

Lindstedt, S. L.; Calder, W. A. III. Body size, physiological time, and longevity of homeothermic animals. Q. Rev. Biol. 56:1–16; 1981.

Linley, E. A. S.; Newell, R. C. Estimates of bacterial growth yields based on plant detritus. Bull. Mar. Sci. 35:409–425; 1984.

Lock, M. A.; Williams, D. D., eds. *Perspectives in running water ecology*. New York: Plenum, 1981, 430 pp.

Long, J. N. Productivity of western coniferous forests. In: Edmonds, R. L., ed. *Analysis of coniferous forest ecosystems in the Western United States*. US/IBP synthesis Series 14. Stroudsburg, Pennsylvania: Hutchinson Ross, 1982:89–125.

Longhurst, A. R. Heterogeneity in the ocean: implication for fisheries. Rapp. P.-v. Reun. Cons., int. Explor. Mer 185:268–282; 1984.

Longstreth, D. J.; Strain, B. R. Effects of salinity and illumination on photosynthesis and water balance of *Spartina alterniflora* Loisel. Oecologia (Berl.) 31:191–199; 1977.

Lorenzen, C. J. A method for the continuous measurement of in vivo chlorophyll concentration. Deep-Sea Res. 13:223–227; 1966.

Lotka, A. J. *Elements of physical biology*. Baltimore: Williams and Wilkins; 1925.

Loucks, O. Evolution of diversity, efficiency, and community stability. Am. Zool. 10:17–25; 1970.

Loucks, O. L.; Ek, A. R.; Johnson, W. C.; Monserud, R. A. Growth, aging and succession. In: Reichle, D. E., ed. *Dynamic properties of forest ecosystems. International Biological Programme 23*. Cambridge, UK: Cambridge University Press, 1981:37–85.

Lovelock, J. E. *Gaia*. Oxford: Oxford University Press; 1979.

Lovett, G. M.; Reiners, W. A. Canopy structure and cloud water deposition in subalpine coniferous forests. Tellus. In press.

Lowrance, R.; House, J. G.; Stinner, B., eds. *Agricultural ecosystems–unifying concepts.* New York: Wiley/Interscience; 1984a.

Lowrance, R.; Todd, R. L.; Asmussen, L. E. Nutrient cycling in an agricultural watershed: I. Phreatic movement. J. Environ. Qual. 13:22–27; 1984b.

Lyell, C. *Principles of geology, being an attempt to explain the former changes of the Earth's surface by reference to causes now in operation.*, 3 vol., London: J. Murray; 1830–1833.

Lynch, J. M. *Soil biotechnology: microbial factors in crop productivity.* Oxford: Blackwell; 1983.

Macan, T. T. *Freshwater ecology, 2nd edn.* New York: John Wiley & Sons, 1974, 343 pp.

MacFadyen, A. The contribution of the microfauna to total soil metabolism. In: Doeksen, J.; van der Drift, J., eds. *Soil organisms.* Amsterdam: North–Holland Publishing Company, 1963:3–16.

MacIsaac, J. J.; Dugdale, R. C.; Barber, R. T.; Blasco, D.; Packard, T. T. Primary production cycle in an upwelling center. Deep-Sea Res. 32:503–529; 1985.

Mague, T. H.; Holm-Hansen, O. Nitrogen fixation on a coral reef. Phycologia 14:87–92; 1975.

Mann, K. H. *Ecology of coastal waters.* Studies in Ecology, vol. 8. Berkeley: University of California Press, 1982a, 322 pp.

Mann, K. H. Kelp, sea urchins and predators: A review of strong interactions in rocky subtidal systems of eastern Canada, 1970–1980. Netherlands J. Sea Res. 16:414–423; 1982b.

Mann, K. H. Invertebrate behaviour and the structure of marine benthic communities. In: Sibly, R. M.; Smith, R. H., eds. *Behavioural ecology.* Oxford: Blackwell Scientific Publishers, 1985:227–246.

Mann, K. H.; Britton, R. H.; Kowalczewski, A.; Lack, T. J.; Mathews, C. P.; McDonald, I. Productivity and energy flow at all trophic levels in the River Thames, England. In: Kajak, Z.; Hillbricht-Ilkowska, A. *Productivity problems of freshwater.* Warsawa-Krakow: PWN Polish Scientific Publishers, 1972:579–596.

Maragos, J. E.; Evans, C.; Holthus, P. Reef corals in Kaneohe Bay six years before and after termination of sewage discharges (Oahu, Hawaiian Archipelago). Proc. Fifth Internat. Coral Reef Congr. Tahiti 4:189–194; 1985.

Markus, E. Naturkomplekse. Sitzber. Naturforsch-Ges. Univ. Tartu. 32:79–94; 1926.

Marra, J.; Boardman, D. C. Late winter chlorophyll *a* distributions in the Weddell Sea. Mar. Ecol. Prog. Ser. 19:197–205; 1984.

Marsh, J. A. Jr. Terrestrial inputs of nitrogen and phosphorus on fringing reefs of Guam. Proc. Fifth Int. Coral Reef Symp. 1:331–336; 1977.

Marsh, J. A. Jr.; Smith, S. V. Productivity measurements of coral reefs in flowing water. In: Stoddart, D. R.; Johannes, R. E., eds. *Coral reefs: research methods.* Monogr. Oceanogr. Methodol. 5. Paris: SCOR/UNESCO, 1978:361–377.

Marshall, J. D.; Waring, R. H. Predicting fine root production and turnover by monitoring root starch and soil temperature. Can. J. For. Res. 15:791–800; 1985.

Martin, M. M.; Martin, J. S.; Kukor, J. J.; Merritt, R. W. The digestion of protein and carbohydrate by the stream detritivore, *Tipula abdominalis* (Diptera, Tipulidae). Oecologia 46:360–364; 1980.

Martin, M. M.; Kukor, J. J.; Martin, J. S.; Lawson, D. L.; Merritt, R. W. Digestive enzymes of larvae of the three species of caddisflies (Trichoptera). Insect Biochem. 11:501–505; 1981a.

Martin, M. M.; Martin, J. S.; Kukor, J. J.; Merritt, R. W. The digestive enzymes of detritus-feeding stone-fly nymphs (Plecoptera:Pteronarcidae). Can. J. Zool. 59:1947–1951; 1981b.

Marzolf, G. R. The potential effects of clearing and snagging on stream ecosystems. U. S. Fish and Wildlife Service FWS/OBS–78/14, 1978, 31 pp.

Matson, P. A.; Waring, R. H. Effects of nutrient and light limitation on mountain hemlock: susceptibility to laminated root rot. Ecology 65:1517–1524; 1984.

Mattson, W. J.; Addy, N. D. Phytophagous insects as regulators of forest primary production. Science 190(4214):515–522; 1975.

May, R. M. Biological populations with nonoverlapping generations: stable points, stable cycles and chaos. Science 186:645–647; 1974.

May, R. M. Models for single populations. In: May, R. M., ed. *Theoretical ecology: principles and applications.* Philadelphia: W.B. Saunders; 1976.

May, R. M.; Oster, G. F. Bifurcations and dynamic complexity in simple ecological models. Am. Nat. 110:573–599; 1976.

Mayor, A. G. Structure and ecology of Samoan reefs. Washington: Carnegie Inst. Dept. Marine Biol. Papers 19:1–25, 51–72; 1924.

McClure, J. W. The physiology of phenolic compounds in plants. Recent Adv. Phytochem. 12:525–556; 1979.

McEwen, W. M. Some aspects of seed development and seedling growth of Rimu, *Dacrydium cupressinum,* Lamb. Hamilton, New Zealand: University of Waikate; 1983. Ph.D. thesis.

McFarland, D.; Houston, A. *Quantitative ethology. The state space approach.* Boston: Pitman Publishing Inc., 1981, 204 pp.

McGowan, J. A. Oceanic biogeography of the Pacific. In: Funnel, B. M.; Riedel, W. R., eds. *The micropaleontology of oceans.* Cambridge, UK: Cambridge University Press, 1971:3–74.

McGowan, J. A. The nature of oceanic ecosystems. In: Miller, C., ed. *The biology of the oceanic Pacific.* Oregon State University Press, 1974:9–28.

McGowan, J. A.; Hayward, T. L. Mixing and oceanic productivity. Deep-Sea Res. 25:771–793; 1978.

McGowan, J. A.; Williams, P. M. Oceanic habitat differences in the North Pacific. J. Exp. Mar. Biol. Ecol. 12:187–217; 1973.

McIntosh, R. P. Succession and ecological theory. In: West, D. C.; Shugart, H. H; Botkin, D. B., eds. *Forest succession. Concepts and application.* New York: Springer-Verlag, 1981:10–23.

McIntosh, R. P. *The background of ecology: concept and theory.* Cambridge, UK: Cambridge University Press; 1985.

McLaughlin, S. B.; Edwards, N. T.; Beauchamp, J. J. Spatial and temporal patterns in transport and respiratory allocation of [14C] sucrose by white oak (*Quercus alba*) roots. Can. J. Bot. 55:2971–2980; 1977.

McLaughlin, S. B.; McConathry, R. K.; Barnes, R. L.; Edwards, N. T. Seasonal changes in energy allocation by white oak (*Quercus alba*). Can. J. For. Res. 10:379–388; 1980.

McNaughton, S. J. Serengeti migratory wildebeest: Facilitation of energy flow by grazing. Science 191:92–94; 1976.

McNaughton, S. J. Grassland-herbivore dynamics. In: Sinclair, A. R. E.; Norton-Griffiths, M., eds. *Serengeti, dynamics of an ecosystem.* Chicago: University of Chicago Press, 1979a:46–81.

McNaughton, S. J. Grazing as an optimization process: Grass–ungulate relationships in the Serengeti. Am. Nat. 113:691–703; 1979b.

McNaughton, S. J. Serengeti grassland ecology: The role of composite environmental factors and contingency in community organization. Ecol. Monogr. 53(3):291–320; 1983.

McNaughton, S. J. Grazing lawns: animals in herds, plant form, and coevolution. Am. Nat. 124:863–886; 1984.

McNaughton, S. J. Ecology of a grazing ecosystem: the Serengeti. Ecol. Monogr. 55:259–294; 1985.

McNaughton, S. J. On plants and herbivores. Am. Nat. 128:765–770; 1986.

McNeill, S.; Lawton, J. H. Annual production and respiration in animal populations. Nature 225:427–424; 1970.

Meentemeyer, V. The geography of organic decomposition rates. Ann. Assoc. Am. Geog. 74:551–560; 1984.

Meentemeyer, V.; Box, E. O.; Thompson, R. World patterns and amounts of terrestrial plant litter production. BioScience 32:125–128; 1982.

Meentemeyer, V.; Gardner, J.; Box, E. O. World patterns and amounts of detrital soil carbon. Earth Surface Processes and Landforms 10:557–567; 1985.

Melillo, J. M.; Aber, J. D. Nutrient immobilization in decaying litter: an example of carbon–nutrient interactions. In: Cooley, J. H.; Golley, F. B., eds. *Trends in ecological research for the 1980's.* New York: Plenum, 1984:193–214.

Mercer, M., ed. Multispecies approaches to fisheries management advice. Can. Spec. Publ. Fish. Aquat. Sci. 59:1–169; 1982.

Merritt, R. W.; Cummins, K. W., eds. *An introduction to the aquatic insects of North America.* Dubuque, Iowa: Kendall/Hunt, 1984, 722 pp.

Merritt, R. W.; Cummins, K. W.; Barnes, J. R. Demonstration of stream watershed community processes with some simple bioassay techniques. Fish. Mar. Serv. Tech. Rept. Can. Dept. Fish. Environ. 43:101–113; 1979.

Meyer, J. L.; O'Hop, J. Leaf shredding insects as a source of dissolved organic carbon in headwater streams. Amer. Midl. Nat. 109:175–183; 1983.

Meyer, J. L.; Schultz, E. T. Migrating haemulid fishes as a source of nutrients and organic matter on coral reefs. Limnol. Oceanogr. 30:146–156; 1985a.

Meyer, J. L.; Schultz, E. T. Tissue condition and growth rate of corals associated with schooling fish. Limnol. Oceanogr. 30:157–166; 1985b.

Meyer, J. L.; Tate, C. M. The effects of watershed disturbance on dissolved organic matter carbon dynamics of a stream. Ecology 64:33–44; 1983.

Meyer, J. L.; Schultz, E. T.; Helfman, G. S. Fish schools: an asset to corals. Science 220:1047–1049; 1983.

Miller, C. B.; Frost, B. W.; Batchelder, H. P.; Clemons, M. J.; Conway, R. E. Life histories of large, grazing copepods in a subarctic ocean gyre: *Neocalanus plumchrus, Neocalanus cristatus,* and *Eucalanus bungii* in the Northeast Pacific. Prog. Oceanogr. 13:201–243; 1984.

Miller, R. J.; Colodey, A. G. Widespread mass mortalities of the green sea urchin in Nova Scotia, Canada. Mar. Biol. 73:263–267; 1983.

Miller, R. J.; Mann, K. H.; Scarratt, D. J. Production potential of a seaweed–lobster community in eastern Canada. J. Fish. Res. Bd. Can. 28:1733–1738; 1971.

Minderman, G. Addition, decomposition and accumulation of organic matter in forests. J. Ecol. 56:355–362; 1968.

Minshall, G. W. Autotrophy in stream ecosystems. BioScience 28:767–771; 1978.

Minshall, G. W.; Petersen, R. C.; Cummins, K. W.; Bott, T. L.; Sedell, J. R.; Cushing, C. E.; Vannote, R. L. Interbiome comparison of stream ecosystem dynamics. Ecol. Monogr. 53:1–25; 1983.

Minshall, G. W.; Cummins, K. W.; Petersen, R. C.; Cushing, C. E.; Bruns, D. A.; Sedell, J. R.; Vannote, R. L. Developments in stream ecology. Can. J. Fish. Aquat. Sci. 42:1045–1055; 1985.

Missimer, C. L. Nonseasonal mixing: Its influence on the distribution, abundance, and succession of phytoplankton. Clemson University, 1986. Ph.D. dissertation.

Mitchell, J. M. An overview of climate variability and its causal mechanisms. Quat. Res. 6:481–493; 1976.

Möbius, K. *Die Auster und die Austernwirtschaft*. Berlin: Wiegundt Hempel Parey; 1877.

Moll, R. A.; Stormer, E. F. A hypothesis relating trophic status and subsurface chlorophyll maxima of lakes. Arch. Hydrobiol. 94:425–440; 1982.

Moller, C. M.; Muller, D.; Nielsen, J. Graphic presentation of dry matter production in European beech. Forstlige Forsogsuaesen (Danmark) 21:327–335; 1954.

Moloney, C. L.; Field, J. G. Use of particle-size data to predict potential pelagic fish yields of some southern African areas. S. Afr. J. Mar. Sci. 3:119–128; 1985.

Morgan, N. C.; Backiel, T.; Bretschko, G.; Duncan, A.; Hillbricht-Ilkowska, A.; Kajak, Z.; Kitchell, J. F.; Larson, P.; Leveque, C.; Nauwerck, A.; Schiemer, F.; Thorpe, J. E. Secondary production. In: LeCren, E. D.; Lowe-McConnell, R. H., eds. *The functioning of freshwater ecosystems*. IBP #22. Cambridge, UK: Cambridge University Press, 1980:247–338.

Mori, S.; Yamamoto, G. Productivity of communities in Japanese inland waters. Japanese International Biological Programme Synthesis 10. Tokyo: University of Tokyo Press; 1975.

Moriarty, D. J. W.; Pollard, P. C.; Alongi, D. M.; Wilkinson, C. R.; Gray, J. S. Bacterial productivity and trophic relationships with consumers on a coral reef (MECOR I). Proc. Fifth Int. Coral Reef Symp. 3:457–462; 1985.

Morisawa, M. *Streams, their dynamics and morphology*. New York: McGraw-Hill, 1968, 175 pp.

Morris, B.; Barnes, J.; Brown, F.; Markham, J. The Bermuda marine environment. Vol. 1. Bermuda Biol. Sta. for Res., Inc. Spec. Pub. 15; 1977, 120 pp.

Moser, L. E.; Anderson, K. L. Nitrogen and phosphorus fertilization of bluestem range. Trans. Kans. Acad. Sci. 67:613–616; 1965.

Mueller-Dombois, D.; Canfield, J. E.; Hold, R. A.; Buelow, G. P. Tree-group death in North American and Hawaiian forests: a pathological problem or a new problem for vegetation ecology? Phytocoenologia 11:117–137; 1983.

Mulholland, P. J. Organic carbon flow in a swamp–stream ecosystem. Ecol. Monogr. 51:307–322; 1981.

Murdoch, W. W. Community structure, population control and competition–a critique. Am. Nat. 100:219–236; 1966.

Murphy, P. G. Net primary productivity of tropical, terrestrial ecosystems. In: Lieth, H.; Whittaker, R. H., eds. *Primary productivity of the biosphere*, Ecological Studies, vol. 14, New York: Springer-Verlag, 1975.

Muscatine, L. Productivity of zooxanthellae. In: Falkowski, P. G., ed. *Primary productivity in the sea*. New York: Plenum Press, 1980a:381–402.

Muscatine, L. Uptake, retention, and release of dissolved inorganic nutrients by marine alga–invertebrate associations. In: Cook, C. B., et al., eds. *Cellular interactions in symbiosis and parasitism*. Ohio State University Press, 1980b:229–244.

Muscatine, L.; D'Elia, C. F. The uptake, retention, and release of ammonium by reef corals. Limnol. Oceanogr. 23:725–734; 1978.

Muscatine, L.; Marian, R. E. Dissolved inorganic nitrogen flux in symbiotic and nonsymbiotic medusae. Limnol. Oceanogr. 27:910–918; 1982.

Muscatine, L.; Masuda, T. H.; Burnap, R. Ammonium uptake by symbiotic and apo-symbiotic reef corals. Bull. Mar. Sci. 29:572–575; 1979.

Muttkowski, R. A. The ecology of trout streams in Yellowstone National Park. Roosevelt Wildlife Ann. 2:155–240; 1929.

Nadelhoffer, K. J.; Aber, J. D.; Melillo, J. M. Fine roots, net primary production, and soil nitrogen availability: a new hypothesis. Ecology 66(4):1377–1389; 1985.

Naiman, R. J. Characteristics of sediment and organic carbon export from pristine boreal forest watersheds. J. Fish. Aquat. Sci. 39:1699–1718; 1982.

Naiman, R. J. The annual pattern and spatial distribution of aquatic oxygen metabolism in boreal forest watersheds. Ecol. Monogr. 53:73–94; 1983.

Nanney, D. Cytoplasmic inheritance in protozoa. In: Burdett, W. J., ed. *Methodology in basic genetics*. San Francisco: Holden-Day, 1963:355–374.

National Research Council. *Global change in the geosphere–biosphere: initial priorities for an IGBP*. Washington: National Academy of Sciences, Commission on Physical Sciences, Mathematics and Resources, 1986, 91 pp.

Naveh, Z.; Lieberman, A. S. *Landscape ecology: theory and application*. New York: Springer-Verlag; 1984.

Needham, P. R. *Chemical embryology, vol. 2*. New York: Cambridge University Press; 1931.

Needham, P. R. Quantitative studies of stream bottom foods. Trans. Amer. Fish. Soc. 64:238–247; 1934.

Needham, P. R. *Trout streams*. Ithaca, New York: Comstock; 1938.

Nelson, S. G. Immediate enhancement of photosynthesis by coral reef macrophytes in response to ammonia enrichment. Fifth Int. Coral Reef Symp. 5:35–46; 1985.

Nemoto, T.; Harrison, G. High latitude ecosystems. In: Longhurst, A. R., ed. *Analysis of marine ecosystems*. New York: Academic Press, 1981:95–126.

Nestler, J. Interstitial salinity as a cause of ecophenic variation in *Spartina alterniflora*. Est. Coast. Mar. Sci. 5:717–714; 1977.

Newbold, J. D.; Elwood, J. W.; O'Neill, R. V.; Winkle, W. V. Measuring nutrient spiralling in streams. Can. J. Fish. Aquat. Sci. 38:860–863; 1981.

Newbold, J. D.; Mulholland, P. J.; Elwood, J. W.; O'Neill, R. V. Organic carbon spiralling in stream ecosystems. Oikos 38:266–272; 1982.

Newbold, J. D.; Elwood, J. W.; O'Neill, R. V.; Sheldon, A. L. Phosphorus dynamics in a woodland stream ecosystem: a study of nutrient spiralling. Ecology 64:1249–1265; 1983.

Newell, R. E.; Gould-Steward, S. A stratospheric fountain? J. Atmos. Sci. 38:2789–2796; 1981.

Newman, E. I. The rhizosphere: carbon sources and microbial populations. In: Fitter, A. H.; Atkinson, D.; Read, D. J.; Usher, M. B., eds. *Ecological interactions in soil*. Special Publ. #4, British Ecological Society. Oxford: Blackwell, 1985:107–121.

Nicholis, G.; Prigogine, I. *Self-organization in nonequilibrium systems*. New York: John Wiley & Sons, 1977.

Nixon, S. W.; Oviatt, C. A.; Frithsen, J.; Sullivan, B. Nutrients and the productivity of estuarine and coastal marine systems. J. Limnol. Soc. Sth. Afr. 12:43–71; 1986.

Norton-Griffin, M. Influence of grazing, browsing, and fire on the vegetation dynamics of the Serengeti. In: Sinclair, A. R. E.; Norton-Griffin, M., eds. *Serengeti: dynamics of an ecosystem*. Chicago: University of Chicago Press, 1979:310–352.

Nowak, R. S.; Caldwell, M. M. A test of compensatory photosynthesis in the field: implications for herbivory tolerance. Oecologia (Berl.) 61:311–318; 1984.

Ochoa, N.; Rojas de Mendiola, B.; Gomez, O. Identification of the "El Niño" phenomenon through phytoplankton organisms. Bol. Extraord., Inst. del Mar del Peru, Callao, Peru, 1985:23–31.

Odum, E. P. The strategy of ecosystem development. Science 164:262–270; 1969.

Odum, E. P. *Ecology.* New York: Holt, Rinehart and Winston, 1971a.

Odum, E. P. *Fundamentals of ecology, 3rd edn.* Philadelphia: Saunders; 1971b.

Odum, E. P. Halophytes, energetics and ecosystems. In: Reimold, R. J.; Queen, W. H., eds. *Ecology of halophytes.* New York: Academic Press, 1974:509–602.

Odum, E. P. *Ecology: The link between the natural and the social sciences, 2nd edn.* New York: Holt, Rinehart and Winston; 1975.

Odum, E. P. The emergence of ecology as a new integrative discipline. Science 195:1289–1293; 1977.

Odum, E. P. Keynote address: dedication of new wing of Narragansett EPA Laboratory. In: Jacoff, F. S., ed. *Advances in marine environment research: proceedings of a symposium.* Publ. EPA–600/9–79–035, 1979:iv–viii.

Odum, E. P. *Basic ecology.* Philadelphia: Saunders, 1983, 613 pp.

Odum, E. P. Properties of agroecosystems. In: *Agricultural ecosystems–unifying concepts.* New York: Wiley/Interscience, 1984a:5–11.

Odum, E. P. The mesocosm. BioScience 34:558–562; 1984b.

Odum, E. P.; de la Cruz, A. A. Detritus as a major component of ecosystems. Am. Inst. Biol. Sci. Bull. 13(3):39–40; 1963.

Odum, E. P.; de la Cruz, A. A. Particulate organic detritus in a Georgia salt-marsh-estuarine ecosystem. In: Lauff, G. H., ed. *Estuaries.* Washington: Amer. Assoc. for the Adv. Sci., 1967:383–388.

Odum, E. P.; Fanning, M. E. Comparisons of the productivity of *Spartina alterniflora* and *S. cynosuroides* in Georgia coastal marshes. Bull. Georgia Acad. Sci. 31:1–12; 1973.

Odum, E. P.; Smalley, A. E. Comparison of population energy flow of a herbivorous and deposit-feeding invertebrate in a salt–marsh ecosystem. Proc. Nat. Acad. Sci. USA 45:617–622; 1959.

Odum, E. P.; Marshall, S. C.; Marples, T. G. The caloric content of migrating birds. Ecology 46:901–904; 1965.

Odum, E. P.; Finn, J. T.; Franz, E. The subsidy–stress gradient. Bioscience 29:349–352; 1979.

Odum, H. T. Trophic structure and productivity of Silver Springs, Florida. Ecol. Monogr. 27:55–112; 1957.

Odum, H. T. Summary: an emerging view of the ecological system at El Verde. In: Odum, H. T.; Pigeon, R. F., eds. *A tropical rain forest. A study of irradiation and ecology at El Verde, Puerto Rico.* Springfield, Virginia: US Atomic Energy Commission, Division of Technical Information TID–2427 (PRNC–138), 1970, pp. I-191-I–281.

Odum, H. T. *Environment, power and society.* New York: John Wiley & Sons, 1971, 331 pp.

Odum, H. T. *Systems ecology.* New York: John Wiley & Sons, 1983, 644 pp.

Odum, H. T.; Odum, E. P. Trophic structure and productivity of a windward coral reef community on Eniwetok Atoll. Ecol. Monogr. 25:291–320; 1955.

Odum, H. T.; Pigeon, R. F., eds. *A tropical rain forest. A study of irradiation and ecology at El Verde, Puerto Rico.* Springfield, Virginia: US Atomic Energy Commission, Division of Technical Information TID–2427 (PRNC- 138); 1970.

Odum, H. T.; Pinkerton, R. Time's speed regulator: the optimum efficiency for maximum power in physical and biological systems. Am. Sci. 43:331–343; 1955.

Odum, W. E.; Johannes, R. E. The response of mangroves to man-induced environmental stress. In: Wood, E. J. F.; Johannes, R. E., eds. *Tropical marine pollution.* Oceanography Series, #12. New York: Elsevier, 1975:52–62.

Ogawa, H.; Yoda, K.; Ogino, K.; Kira, T. Comparative ecological studies on three main types of forest vegetation in Thailand. II. Plant biomass. Nature and Life in SE Asia 4:50–80; 1965.

Ogden, J. C.; Lobel, P. S. The role of herbivorous fishes and urchins in coral reef communities. Env. Biol. Fish. 3:49–63; 1978.

Ohiagu, C. E. A quantitative study of seasonal foraging by the grass harvesting termite, *Trinervitermes geminatus* (Wasmann) (Isoptera, Nasutitermitinae) in Southern Guinea Savanna, Mokwa, Nigeria. Oecologia 40:179–188; 1979.

O'Keefe, P.; Kristoferson, L. The uncertain energy path–energy and third world development. Ambio 13:168–170; 1984.

Olson, J. S. Energy storage and the balance of producers and decomposers in ecological systems. Ecology 44:322–331; 1963.

O'Neill, R. V. Ecosystem persistence and heterotrophic regulation. Ecology 57:1244–1253; 1976.

O'Neill, R. V.; De Angelis, D. L. Comparative productivity and biomass relations of forest ecosystems. In: Reichle, D. E., ed. *Dynamic properties of forest ecosystems.* Cambridge, UK: Cambridge University Press, 1981:411–449.

O'Neill, R. V.; DeAngelis, D. L.; Waide, J. B.; Allen, T. F. H. *A hierarchical concept of the ecosystem.* Princeton, New Jersey: Princeton University Press; 1986.

Oren, R.; Thies, W. G.; Waring, R. H. Tree vigor and stand growth of Douglas-fir as influenced by laminated root rot. Can. J. For. Res. 15:985–988; 1985.

Orndorff, K. A.; Lang, G. E. Leaf litter redistribution in a West Virginia hardwood forest. J. Ecol. 69:225–235; 1981.

Ovington, J. D. Quantitative ecology and the woodland ecosystem concept. Adv. Ecol. Res. 1:103–192; 1962.

Ovington, J. D. Organic production, turnover and mineral cycling in woodlands. Biol. Rev. Cambridge Philos. Soc. 40:295–336; 1965.

Ovington, J. D.; Heitkamp, D.; Lawrence, D. B. Plant biomass and productivity of prairie, savanna, oakwood and maize field ecosystems in central Minnesota. Ecology 44:52–63; 1963.

Owen, D. R.; Wiegert, R. G. Do consumers maximize plant fitness? Oikos 27:488–492; 1976.

Owensby, C. E.; Hyde, R. M.; Anderson, K. L. Effects of clipping and supplemental nitrogen and water on loamy upland bluestem range. J. Range Manage. 23:341–346; 1970.

Pace, M. L.; Glasser, J. E.; Pomeroy, L. R. A simulation analysis of continental shelf food webs. Mar. Biol. 82:47–63; 1984.

Paerl, H. W.; Webb, K. L.; Baker, J.; Wiebe, W. J. Nitrogen fixation in waters. In: Broughton, W. J., ed. *Nitrogen fixation.* Ecology. Oxford: Oxford University Press 1:193–240; 1981.

Paine, R. T. Controlled manipulations in the marine intertidal zone, and their contributions to ecological theory. In: Goulden, C. E., ed. *The changing scenes in natural sciences, 1776–1976.* Philadelphia: Acad. Nat. Sci. Spec. Publ. 12:245–270; 1977.

Palm, C. A.; Houghtan, R. A.; Melillo, J. M.; Skole, D.; Woodwell, G. M. The effect of tropical deforestation on atmospheric CO_2. In: Lal, R., Sanchez, P. A.; Cummings, R. W. Jr., eds. *Land clearing and development in the tropics.* Rotterdam: A. A. Balkema, 1986:181–194.

Pares-Sierra, A. F.; Inoue, M.; O'Brien, J. J. Estimates of oceanic horizontal heat transport in the tropical Pacific. J. Geophys. Res. 90:3293–3303; 1985.

Parkin, T. B.; Brock, T. D. Photosynthetic bacterial production and carbon mineralization in a meromictic lake. Arch. Hydrobiol. 91:366–382; 1981.

Parmelee, R. W.; Alston, D. G. Nematode trophic structure in conventional and no-tillage agroecosystems. J. Nematol. 18:403–407; 1986.

Parton, W. J.; Risser, P. G. Impact of management on the tallgrass prairie. Oecologia 46:223–234; 1980.

Parton, W. J.; Lauenroth, W. K.; Smith, F. W. Water loss from a shortgrass steppe. Agric. Meteorol. 24:97–109; 1981.

Patrick, R.; Reimer, C. W. The diatoms of the United States. I. Monogr. Acad. Nat. Sci. Phila. 13, 1966, 688 pp.

Patriquin, D. G. The origin of nitrogen and phosphorus for growth of marine angiosperm *Thalassia testudinum*. Mar. Biol. 15:35–46; 1972.

Patriquin, D. G. Nitrogen fixation (acetylene reduction) associated with cord grass *Spartina alterniflora* Loisel. Ecol. Bull. (Stockholm) 26:20–27; 1978.

Patriquin, D. G.; McLung, C. R. Nitrogen accretion and the nature and possible significance of N_2 fixation (acetylene reduction) in a Nova Scotian *Spartina alterniflora* stand. Mar. Biol. 47:227–242; 1978.

Patten, B. C. An introduction to the cybernetics of the ecosystem: the trophic dynamic aspect. Ecology 40:221–231; 1959.

Patten, B. C. *Systems analysis and simulation in ecology, vols. I–IV.* New York: Academic Press; 1971–1976.

Patten, B. C. Energy cycling in the ecosystem. Ecol. Modelling 28:1–71; 1985.

Patten, B. C.; Odum, E. P. The cybernetic nature of ecosystems. Am. Nat. 118:886–895; 1981.

Patterson, J. C.; Hamblin, P. H.; Imberger, J. Classification and dynamic simulation of vertical density of lakes. Limnol. Oceanogr. 29:845–861; 1984.

Pearson, H. W.; Taylor, R. Nitrogen fixation in a polluted canal system. In: Granhall, U., ed. *Environmental role of nitrogen-fixing blue-green algae and asymbiotic bacteria.* Ecol. Bull. (Stockholm) 26:69–82; 1978.

Pearson, J. A.; Knight, D. H.; Fahey, T. J. Biomass and nutrient accumulation during stand development in Wyoming lodgepole pine forests. Ecology 68: 1966–1987; 1987.

Pearson, R. G. Recovery and recolonization of coral reefs. Mar. Ecol. Prog. Ser. 4:105–122; 1981.

Peckarsky, B. L.; Dodson, S. I. Do stonefly predators influence benthic distributions in streams? Ecology 61:1275–1282; 1980.

Peet, R. K. Changes in biomass and production during secondary succession. In: West, D. C.; Shugart, H. H; Botkin, D. B., eds. *Forest succession. Concepts and application.* New York: Springer-Verlag, 1981:324–338.

Perry, M. J. Phosphate utilization by an oceanic diatom in phosphorus-limited chemostat culture and in the oligotrophic waters of the central north Pacific. Limnol. Oceanogr. 21:88–107; 1976.

Persson, H. Fine-root production, mortality and decomposition in forest ecosystems. Vegetatio 41:101–109; 1979.

Persson, H. A. The distribution and productivity of fine roots in boreal forests. Plant Soil 71:87–101; 1983.

Persson, T.; Baath, E.; Clarholm, M.; Lundkvist, H.; Soderstrom, B. E.; Sohlenius, B. Trophic structure, biomass dynamics and carbon metabolism of soil organisms in a Scots pine forest. In: Persson, T., ed. *Structure and function of northern coniferous forests–an ecosystem study.* Stockholm: Ecological Bulletins, Swedish Natural Science Research Council (NFR), 1980:419–459.

Peterjohn, W. T.; Correll, D. L. Nutrient dynamics in an agricultural watershed: observations on the role of a riparian forest. Ecology 65:1466–1475; 1984.

Peterman, R. M. A simple mechanism that causes collapsing stability regions in exploited salmonid populations. J. Fish. Res. Board Can. 34:1130–1142; 1977.

Peters, R. H. *The ecological implications of body size.* Cambridge, Massachusetts: Cambridge University Press, 1983, 329 pp.

Petersen, R. C. Life history and bionomics of *Nigronia serricornis* (Say) (Megaloptera: Corydalidae). Michigan State Univ., 1974. 210 pp. Ph.D. dissertation.

Petersen, R. The paradox of the plankton: an equilibrium hypothesis. Am. Nat. 109:35–49; 1975.

Petersen, R. C.; Cummins, K. W. Leaf processing in a woodland stream. Freshwat. Biol. 4:343–368; 1974.

Petersen, R. C.; Cummins, K. W.; Ward, G. M. An annual equilibrium model for microbial–animal processing of detritus in a first order woodland stream. Ecol. Monogr. In press.

Peterson, B. J.; Howarth, R. W.; Garritt, R. H. Multiple stable isotopes used to trace the flow of organic matter in estuarine food webs. Science 227: 1361–1363; 1985.

Petrides, G. A.; Swank, W. G. Estimating the productivity and energy relations of an African elephant population. Proc., Ninth International Grasslands Congress, San Paulo, Brazil; 1965.

Philander, S. G. H. El Niño Southern Oscillation phenomena. Nature 302:295–301; 1983.

Philander, S. G. H. Predictability of El Niño. Nature 321:810–811; 1986a.

Philander, S. G. H. Unusual conditions in the tropical Atlantic Ocean in 1984. Nature 322:236–238; 1986b.

Phillips, J. Succession, development, the climax and the complex organism: an analysis of concepts. I. J. Ecol. 22:554–571; 1934.

Phillips, J. Succession, development, the climax and the complex organism: an analysis of concepts. II. and III. J. Ecol. 23:210–246, 488–508; 1935.

Phillips, R. E.; Blevins, R. L.; Thomas, G. W.; Frye, W. W.; Phillips, S. H. No-tillage agriculture. Science 208:1108–1113; 1980.

Phillips, R. E.; Phillips, S. H., eds. *No-tillage agriculture.* New York: Van Nostrand Reinhold, 1984.

Phillipson, J. A miniature bomb calorimeter for small biological samples. Oikos 15:130–139; 1964.

Pick, F. R.; Nalewajko, C.; Lean, D. R. The origin of a metalimnetic peak. Limnol. Oceanogr. 29:125–134; 1984.

Pickett, S. T. A.; White, P. S., eds. The ecology of natural disturbance and patch dynamics. Orlando, Florida: Academic Press; 1985.

Pieper, R. D. Consumption rates of desert grassland herbivores. In: Smith, J. A.; Hayes, V. W., eds. *Proceedings of the XIV international grassland congress, Lexington, Kentucky.* Boulder, Colorado: Westview Press, 1981:465–467.

Pilson, M. E. Q.; Betzer, S. B. Phosphorus flux across a coral reef. Ecology 54:581–588; 1973.

Pimm, S. L. *Food webs.* London: Chapman and Hall, 1982, 219 pp.

Pimm, S. L. The complexity and stability of ecosystems. Nature 307:321–326; 1984.

Pimm, S. L.; Kitching, R. L. Experimental studies of food chain lengths. Oikos 50: 302–307; 1987.

Pimm, S. L.; Lawton, J. H. The number of trophic levels in ecological communities. Nature 268: 329–331; 1977.

Platt, J. R. Strong inference. Science 146:347–353; 1964.

Platt, T. Structure of the marine ecosystem. In: Ulanowicz, R. E.; Platt, T., eds. *Ecosystem theory for biological oceanography*. Can. Bull. Fish. Aquat. Sci. 213:55–64; 1985.

Platt, T.; Denman, K. L. Organisation in the pelagic ecosystem. Helgol. Wiss. Meeresunters. 30:575–581; 1977.

Platt, T.; Denman, K. L. The structure of pelagic marine ecosystems. Rapp. Reun. Cons. Int. Explor. Mer 173:60–65; 1978.

Platt, T.; Harrison, W. G. Biogenic fluxes of carbon and oxygen in the ocean. Nature 318:55–58; 1985.

Platt, T.; Mann, K. H.; Ulanowicz, R. E. *Mathematical models in biological oceanography*. Paris: The UNESCO Press, 1981, 157 pp.

Platt, T.; Lewis, M.; Geider, R. Thermodynamics of the pelagic ecosystem: Elementary closure conditions for biological production in the open ocean. In: Fasham, M. J. R., ed. *Flows of energy and materials in marine ecosystems: theory and practice*. London: Plenum Press, 1984, 733 pp.

Playfair, J. *Illustrations of Huttonian theory*. Edinburgh: Creech; 1802.

Plowman, K. P. Litter and soil fauna of two Australian subtropical forests. Aust. J. Ecol. 4:87–104; 1979.

Pomeroy, L. R. Residence time of dissolved phosphate in natural waters. Science 131:1731–1734; 1960.

Pomeroy, L. R. The strategy of mineral cycling. Annu. Rev. Ecol. Syst. 1:171–190; 1970.

Pomeroy, L. R. Mineral cycling in marine ecosystems. In: Howell, F. G.; Gentry, J. B.; Smith, M. H., eds. *Mineral cycling in southeastern ecosystems*. ERDA Symp. Series (CONF–740513), 1975:209–223.

Pomeroy, L. R.; Kuenzler, E. J. Phosphorus turnover by coral reef animals. In: Nelson, D. J.; Evans, F. S., eds. Proc. Second National Symp. Radioecology 2:474–482; 1969.

Pomeroy, L. R.; Wiebe, W. J. Energetics of microbial food webs. Hydrobiologia. In press.

Pomeroy, L. R.; Pilson, M. E. Q.; Wiebe, W. J. Tracer studies of the exchange of phosphorus between reef water and organisms on the windward reef of Eniwetok Atoll. Proc. Second Int. Coral Reef Symp. 1:87–96; 1974.

Pomeroy, L. R.; Hanson, R. B.; McGillivary, P. A.; Sherr, B. F.; Kirchman, D.; Deibel, D. Microbiology and chemistry of fecal products of pelagic tunicates: rates and fates. Bull. Mar. Sci. 35:426–439; 1984.

Popper, K. R. *The poverty of historicism*. Boston: Beacon Press; 1957.

Popper, K. R. *Objective knowledge, an evolutionary approach*. Oxford: Clarendon Press; 1972.

Porter, J. W. Autotrophy, heterotrophy, and resource partitioning in Caribbean reef-building corals. Am. Nat. 110:731–741; 1976.

Porter, J. W.; Battey, J. F.; Smith, G. J. Perturbation and change in coral reef communities. Proc. Natl. Acad. Sci. USA 79:1678–1681; 1982.

Post, W. M.; Emmanuel, W. R.; Zinke, P. J.; Stangenberger, A. G. Soil carbon pools in world life zones. Nature 298:156–159; 1982.

Potts, M.; Whitton, B. A. Nitrogen fixation by blue-green algal communities in the intertidal zone of the lagoon of Aldabra Atoll. Oecologia 27:275–283; 1977.

Price, D. de S. *Little science, big science, and beyond*. New York: Columbia University Press; 1986.

Prigogine, I. Time, structure, and fluctuations. Science 201:777–785; 1978.

Prigogine, I.; Stengers, I. *Order out of chaos: Man's new dialogue with nature.* New York: Bantam; 1984.

Pugsley, C. W.; Hynes, H. B. N. Three dimensional distribution of winter stonefly nymphs, *Albcapnia pygmaea,* within the substrate of a Southern Ontario river. Can. J. Fish. Aquat. Sci. 43:1812–1817; 1986.

Quinn, J. A.; Miller, R. V. A biotic selection study utilizing *Muhlenbergia montana.* Bull. Torrey Bot. Club 94:423–432; 1967.

Quinn, J. F.; Dunham, A. E. On hypothesis testing in ecology and evolution. Am. Nat. 122:602–617; 1983.

Quinn, W. H.; Neal, V. T. Long-term variations in the Southern Oscillation, El Niño, and Chilean subtropical rainfall. Fishery Bull. 81:363–373; 1983.

Rafes, P. M. Estimation of the effects of phytophagous insects on forest production. In: Reichle, D. E., ed. *Analysis of temperate forest ecosystems.* Berlin: Springer-Verlag, 1970:100–106.

Ragotzkie, R. A. Heat budgets of lakes. In: Lerman, A., ed. *Lakes: chemistry, geology, physics.* New York: Springer-Verlag, 1978:1–18.

Ramage, C. S. Role of a tropical "maritime continent" in the atmospheric circulation. Mon. Wea. Rev. 96:365–370; 1969.

Rapport, D. J.; Regier, H. A.; Thorpe, C. Diagnosis, prognosis, and treatment of ecosystems under stress. In: Barrett, G. W.; Rosenberg, R., eds. *Stress effects on natural ecosystems.* Chichester: John Wiley & Sons, 1981:269–280.

Rasmusson, E. M.; Wallace, J. M. Meteorological aspects of the El Niño/Southern Oscillation. Science 222:1195–1202; 1983.

Rauner, J. L. Deciduous forests. In: Monteith, J. L., ed. *Vegetation and the atmosphere, vol. 2. Case studies.* London: Academic Press, 1976:241–264.

Rayner, R. F.; Drew, E. A. Nutrient concentrations and primary productivity at the Peros Banhos and Salomon Atolls in the Chagos Archipelago. Est. Coast. Shelf Sci. 18:121–132; 1984.

Read, D. J.; Francis, R.; Finlay, R. D. Mycorrhizal mycelia and nutrient cycling in plant communities. In: Fitter, A. H.; Atkinson, D.; Read, D. J.; Usher, M. B., eds. *Ecological interactions in soil.* Special Publ. #4, British Ecological Society. Oxford: Blackwell, 1985:193–217.

Reardon, P. O.; Huss, D. L. Effects of fertilization on little bluestem community. J. Range Manage. 18:238–241; 1965.

Redfield, A. C. On the proportions of organic derivatives in sea water and their relation to the composition of plankton. In: James Johnstone Memorial Volume. Liverpool: Liverpool University Press, 1934:176–192.

Reichenbach, H. *The theory of probability, 2nd ed.* Berkeley: University of California Press; 1949.

Reichenbach, H. Verifiability theory of meaning. Amer. Acad. Arts and Sci. 80:52; 1951.

Reichle, D. E., ed. *Analysis of temperate forest ecosystems.* New York: Springer-Verlag; 1970.

Reid, J. L., Jr. *Intermediate waters of the Pacific Ocean.* Baltimore: Johns Hopkins Press, 1965, 85 pp.

Reid, J. L.; Brinton, E.; Fleminger, A.; Venrick, E. L.; McGowan, J. A. Ocean circulation and marine life. In: Charnock, H.; Deacon, G., eds. *Advances in oceanography.* New York: Plenum Press, 1978:65–130.

Reiners, W. A. Structure and energetics of three Minnesota forests. Ecol. Monogr. 42:71–94; 1972.

Reiners, W. A. Terrestrial detritus and the carbon cycle. In: Woodwell, G. M.; Pecan, E. V., eds. *Carbon in the biosphere.* 24th Brookhaven Symposium of Biology. Springfield, Virginia: National Technical Information Service, 1973:303–327.

Reiners, W. A. Disturbance and basic properties of ecosystem energetics. In: Mooney, H. A.; Godron, M., eds. *Disturbance and ecosystem-components of response.* New York: Springer-Verlag, 1983:83–98.

Reiners, W. A. Complementary models for ecosystems. Am. Nat. 127:59–73; 1986.

Reiners, W. A.; Reiners, N. M. Energy and nutrient dynamics of forest floors in three Minnesota forests. J. Ecol. 58:497–519; 1970.

Reiners, W. A.; Worley, I. A.; Lawrence, D. B. Plant diversity in a chronosequence at Glacier Bay, Alaska. Ecology 52(1):55–69; 1971.

Resh, V. H.; Rosenberg, D. M., eds. *The ecology of aquatic insects.* New York: Praeger Sci., 1984, 625 pp.

Revsbech, N. P.; Jorgensen, B. P. Microelectrodes: their use in microbial ecology. In: Marshall, K. C., ed. *Advances in microbial ecology, vol. 9.* Plenum, 1986:293–352.

Revsbech, N. P.; Ward, D. M. Oxygen microelectrode that is insensitive to medium chemical composition: use in an acid microbial mat dominated by *Cyanidium caladrium.* Appl. Environ. Microbiol. 45:755–759; 1983.

Reynolds, C. S. Phytoplankton periodicity: The interaction of the form, function, and environmental variability. Freshwat. Biol. 14:111–142; 1984.

Ricard, M.; Delesalle, B. Phytoplankton and primary production of the Scilly Lagoon waters. Proc. Fourth Int. Coral Reef Symp., Manila, vol. 1, 1981:425–429.

Rice, C. W.; Smith, M. S.; Blevins, R. L. Soil nitrogen availability after long-term continuous no-tillage and conventional tillage corn production. Soil Sci. Soc. Am. J. 50:1206–1210; 1986.

Rich, P. H. Trophic–detrital interactions: vestiges of ecosystem evolution. Am. Nat. 123:20–29; 1984a.

Rich, P. H. Trophic vs. detrital energetics: is detritus productive? Bull. Mar. Sci. 35:312–317;1984b.

Richerson, P.; Armstrong, R.; Goldman, C. R. Contemporaneous disequilibrium, a new hypothesis to explain the "paradox of the plankton." Proc. Nat. Acad. Sci. USA 67:1710–1714; 1970.

Richerson, P. J.; Lopez, M.; Coon, T. The deep chlorophyll layer of Lake Tahoe. Verh. Internat. Verein. Limnol. 20:426–433; 1978.

Richey, J. E. The phosphorus cycle. In: Bolin, B.; Cook, R. B., eds. *The major biogeochemical cycles and their interactions.* SCOPE 21. Chichester: John Wiley & Sons, 1983:51–55.

Richman, S. The transformation of energy by *Daphnia pulex.* Ecol. Monogr. 28:273–291; 1958.

Ricker, W. E. Stock and recruitment. J. Fish. Res. Bd. Can. 11:559–623; 1954.

Ricklefs, R. E. Review of O'Neill et al. (1986). Science 236: 206–207; 1987.

Rigler, F. H. A tracer study of the phosphorus cycle in lake water. Ecology 37: 550–562; 1956.

Rigler, F. H. The concept of energy flow between trophic levels. In: van Dobben, W. H.; Lowe-McConnell, R. H., eds. *Unifying concepts of ecology.* The Hague: W. Junk, 1975a:15–26.

Rigler, F. H. Nutrient kinetics and the new typology. Verh. Internat. Verein. Limnol. 19:197–210; 1975b.

Risk, M. J.; Müller, H. R. Porewater in coral heads: evidence for nutrient regeneration. Limnol. Oceanogr. 28:1004–1008; 1983.

Risser, P. G.; Mankin, J. B. Simplified simulation model of the plant producer function in shortgrass steppe. Am. Midl. Nat. 115:348–360; 1986.

Risser, P. G.; Parton, W. J. Ecological analysis of a tallgrass prairie: nitrogen cycle. Ecology 63:1342–1351; 1982.

Risser, P. G.; Birney, E. C.; Blocker, H. D.; May, S. W.; Parton, W. J.; Wiens, J. A. *The true prairie ecosystem.* Stroudsburg, Pennsylvania: Hutchinson Ross Publishing Company; 1981.

Roberts, R. J.; Morton, R. Biomass of larval Scarabaeidae (Coleoptera) in relation to grazing pressures in temperate, sown pastures. J. Appl. Ecol. 22:863–874; 1985.

Rodin, L. E.; Bazilevich, N. J. Production and mineral cycling in terrestrial vegetation. Edinburgh: Oliver and Boyd; 1967.

Roff, D. A. Predicting body size with life history models. BioScience 36:316–323; 1986.

Rogler, G. A.; Haas, H. J. Range production as related to soil moisture and precipitation on the northern Great Plains. J. Am. Soc. Agron. 39:378–389; 1974.

Rojas de Mendiola, B.; Gomez, O.; Ochoa, N. Effects of "El Niño" on phytoplankton. Bol. Extraord., Inst. del Mar del Peru, Callao, Peru, 1985:34–40.

Romme, W. H.; Knight, D. H.; Yavitt, J. B. Mountain pine beetle outbreaks in the Rocky Mountains: regulators of primary productivity? Am. Nat. 127:484–494; 1986.

Rosenzweig, M. L. Net primary productivity of terrestrial communities: prediction from climatological data. Am. Nat. 102:67–74; 1968.

Ross, H. H. Stream communities and terrestrial biomes. Arch. Hydrobiol. 59:235–242; 1963.

Rosswall, T. The nitrogen cycle. In: Bolin, B.; Cook, R. B., eds. *The major biogeochemical cycles and their interactions.* SCOPE 21. Chichester: John Wiley & Sons, 1983:46–50.

Rougerie, F.; Wauthy, B. Le concept d'endo-upwelling dans le fonctionnement des atolls–oasis. Oceanol. Acta 9:133–148; 1986.

Rowe, F. W. E.; Vail, L. Crown-of-Thorns: GBR not under threat. Search 15:211–213; 1984.

Royama, T. Population dynamics of the spruce budworm *Choristoneura fumiferana.* Ecol. Monogr. 54:429–462; 1984.

Ruess, R. W. Nutrient movement and grazing: Experimental effects of clipping and nitrogen source on nutrient uptake in *Kyllinga nervosa.* Oikos 43:183–188; 1984.

Ruess, R. W.; McNaughton, S. J.; Coughenour, M. B. The effects of clipping, nitrogen source and nitrogen concentration on the growth responses and nitrogen uptake of an East African sedge. Oecologia 59:253–261; 1983.

Runkle, J. R. Disturbance regimes in temperate forests. In: Pickett, S. T. A.; White, P. S., eds. *The ecology of natural disturbance and patch dynamics.* Orlando, Florida: Academic Press, 1985:17–33.

Russell, E. W. *Soil conditions and plant growth, 10th edn.* London: Longmans; 1973.

Russell, F. S. A summary of the observations on the occurrence of planktonic stages of fish off Plymouth 1924–1972. J. Mar. Biol. Ass. UK 53:347–355; 1973.

Russell-Hunter, W. D. *Aquatic productivity.* New York: Macmillan, 1970, 306 pp.

Ruttner, F. *Fundamentals of limnology.* Toronto: University of Toronto Press, 1963, 295 pp.

Ryder, R. A.; Kerr, S. R.; Loftus, K. H.; Regier, H. A. The morphoedaphic index, a fish yield estimator–review and evaluation. J. Fish. Res. Bd. Can. 31:663–688; 1974.

Rymer, L. The use of uniformitarianism and analogy in paleoecology, particularly pollen analysis. In: Walker, D.; Guppy, J. C., eds. *Biology and quaternary environments.* Canberra City: Australian Academy of Science; 1978.

Ryther, J. H. Photosynthesis and fish production in the sea. Science 166:72–76; 1969.

Sala, O. E.; Lauenroth, W. K. Small rainfall events: an ecological role in semiarid regions. Oecologia 53:301–304; 1982.

Salt, G. W. A comment on the use of the term emergent properites. Am. Nat. 113:145–148; 1979.

Salt, G. W. Roles: their limits and responsibilities in ecological and evolutionary research. Am. Nat. 122:697–705; 1983.

Sammarco, P. W. Echinoid grazing as a structuring force in coral communities: whole reef manipulation. J. Exp. Mar. Biol. Ecol. 61:31–55; 1982.

Sano, M.; Shimiza, M.; Nose, Y. Changes in structure of coral reef fish communities by destruction of hermatypic corals: observational and experimental views. Pac. Sci. 38:51–79; 1984.

Santander, H.; Zuzunaga, J. Cambios en algunos componentes del ecosistema marino frente al Peru durante el fenomeno El Niño 1982–1983. Rev. Com. Perm. Pacifico Sur 15:311–331; 1984.

Santantonio, D.; Mermann, R. K.; Overton, W. S. Root biomass studies in forest ecosystems. Pedobiologia 17:1–31; 1977.

Sargent, M. C.; Austin, T. S. Organic productivity of an atoll. Trans. Am. Geophys. Union 30:245–249; 1949.

Sargent, M. C.; Austin, T. S. Biologic economy of coral reefs. Bikini and nearby atolls. Part 2. Oceanography. U.S. Geol. Survey Prof. Paper 260:293–300; 1954.

Satoo, T. A synthesis of studies by the harvest method: primary production relations in the temperate deciduous forests of Japan. In: Reichle, D. E., ed. *Analysis of temperate forest ecosystems.* New York: Springer-Verlag, 1970:55–72.

Scheibling, R. E.; Stephenson, R. L. Disease-related mortality of *Strongylocentrotus droebachiensis* (Echinodermata: Echinoidea) off Nova Scotia, Canada. Mar. Biol. 78:153–164; 1984.

Schimel, D.; Stillwell, M. A.; Woodmansee, R. G. Biogeochemistry of C, N, and P in a soil catena of the shortgrass steppe. Ecology 66:276–282; 1985a.

Schimel, D. S.; Coleman, D. C.; Horton, K. A. Soil organic matter dynamics in paired rangeland and cropland toposequences in North Dakota. Geoderma 36:201–214; 1985b.

Schimel, D. S.; Parton, W. J.; Adamsen, F. J.; Woodmansee, R. G.; Senft, R. L.; Stillwell, M. A. The role of cattle in the volatile loss of nitrogen from a shortgrass steppe. Biogeochemistry 2:39–52; 1986.

Schindler, D. W. Evolution of phosphorus limitation in lakes. Science 195:260–262; 1977.

Schindler, D. W.; Mills, K. H.; Malley, D. F.; Findlay, D. L.; Shearer, J. A.; Davies, I. J.; Turner, M. A.; Linsey, G. A.; Cruikshank, D. R. Long-term ecosystem stress: the effects of years of experimental acidification on a small lake. Science 228:1395–1401; 1985.

Schindler, J. E.; Missimer, C. L.; Schreiner, S. P. The use of CE-Therm-R1 and real-time meteorological data for the prediction of nonseasonal mixing; the development of event-related sampling. International Symposium on Applied Lake and Watershed Management, 13–16 November 1985. Lake Geneva, Wisconsin: North American Lake Management Society; 1986.

Schlesinger, W. H. Carbon balance in terrestrial detritus. Annu. Rev. Ecol. Syst. 8:51–81; 1977.

Schopf, J. W., ed. *Earth's earliest biosphere.* Princeton: Princeton University Press; 1983.

Schultz, J. C.; Baldwin, I. T. Oak leaf quality declines in response to defoliation by gypsy moth larvae. Science 217:149–151; 1982.

Schuster, J. L. Root development of native plants under three grazing intensities. Ecology 45:63–70; 1964.

Schwinghamer, P. Characterisitic size distributions of integral benthic communities. Can. J. Fish. Aquat. Sci. 38:1255–1263; 1981.

Schwinghamer, P. Generating ecological hypotheses from biomass spectra using causal analysis: a benthic example. Mar. Ecol. Prog. Ser. 13:151–166; 1983.

Schwinghamer, P. Observations on size-structure and pelagic coupling of some shelf and abyssal benthic communities. In: Gibbs, P. E., ed. *Proc. 19th European Mar. Biol. Symp.* Cambridge, UK: Cambridge University Press, 1985:347–359.

Scott, D. The determination and use of thermodynamic data in ecology. Ecology 46:673–680; 1965.

Scott, D. A microcalorimeter with a range of 0.1–1.0 calories. Limnol. Oceanogr. 27(3):585–590; 1982.

Scott, J. A.; French, N. R.; Leetham, J. W. Patterns of consumption in grasslands. In: French, N. R., ed. *Perspectives in grassland ecology. Results and applications of the US/IBP grassland biome study.* New York: Springer-Verlag, 1979:89–105.

Scriven, M. Truisms as the grounds for historical explanations. In: Gardiner, P., ed. *Theories of history.* Glencoe, Illinois: The Free Press, 1959a:443–475.

Scriven, M. Explanation and prediction in evolutionary theory. Science 130: 477–482; 1959b.

Seastedt, T. R. Maximization of primary and secondary productivity by grazers. Am. Nat. 126:559–564; 1985a.

Seastedt, T. R. Canopy interception of nitrogen in bulk precipitation by annually burned and unburned tallgrass prairie. Oecologia 66:88–92; 1985b.

Seastedt, T. R.; Crossley, D. A. The influence of arthropods on ecosystems. BioScience 34:157–161; 1984.

Sedell, J. R.; Frogett, J. L. Importance of streamside forests to large rivers: the isolation of the Willamette River, Oregon, USA, from its floodplain by snagging and streamside forest removal. Verh. Internat. Verein. Limnol. 22:1828–1834; 1984.

Seitzinger, S. P.; D'Elia, C. F. Preliminary studies of denitrification on a coral reef. In: Reaka, M. L. *The ecology of coral reefs.* Symp. Ser. Undersea Res. NOAA Undersea Research Program, Rockville, Maryland 3:199–208; 1985.

Senft, R. L. The redistribution of nitrogen by cattle. Fort Collins: Colorado State University; 1983. Ph. D. dissertation.

Shah, D.; Horsch, R.; Klee, H.; Kishore, G.; Winter, J.; Turner, N.; Hironaka, C.; Sanders, P.; Gasser, C.; Aykent, S.; Siegel, N.; Rogers, S.; Fraley, R. Engineering herbicide tolerance in transgenic plants. Science 233:478–481; 1986.

Sheldon, R. W. A universal grade scale for particulate materials. Proc. Geol. Soc. Lond. 1659:293–295; 1969.

Sheldon, R. W.; Prakash, A.; Sutcliffe, W. H. The size distribution of particles in the ocean. Limnol. Oceanogr. 17:327–340; 1972.

Sheldon, R. W.; Sutcliffe, W. H.; Prakash, A. The production of particles in the surface waters of the ocean with particular reference to the Sargasso Sea. Limnol. Oceanogr. 18:719–733; 1973.

Sheldon, R. W.; Sutcliffe, W. H.; Paranjape, M. A. Structure of pelagic food chains and relationship between plankton and fish production. J. Fish. Res. Bd. Can. 34:2344–2353; 1977.

Shelford, V. E. *Animal communities in temperate North America.* Chicago: University of Chicago Press; 1913.

Shelford, V. E. An experimental study of the behavior agreement among animals of an animal community. Biol. Bull. 26:294–315; 1914.

Short, F. T.; Davis, M. W.; Gibson, R. A.; Zimmermann, C. F. Evidence for phosphorus limitation in carbonate sediments of the seagrass *Syringodium filiforme*. Est. Coast. Shelf Sci. 20:419–430; 1985.

Shugart, H. H. *A theory of forest dynamics*. New York: Springer-Verlag; 1984.

Shugart, H. H.; O'Neill, R. V., eds. *Systems ecology*. Stroudsburg, Pennsylvania: Dowden, Hutchinson and Ross; 1978.

Shugart, H. H.; Seagle, S. W. Modeling forest landscapes and the role of disturbance in ecosystems and communities. In: Pickett, S. T. A.; White, P. S., eds. *The ecology of natural disturbance and patch dynamics*. Orlando, Florida: Academic Press, 1985:353–368.

Shugart, H. H.; West, D. C. Size and pattern of simulated forest stands. For. Sci. 25:120–122; 1979.

Shugart, H. H. Jr.; West, D. C. Forest succession models. BioScience 30:308–313; 1980.

Shugart, H. H.; West, D. C. The long-term dynamics of forest ecosystems. Am. Sci. 69:647–652; 1981.

Shugart, H. H.; Reichle, D. E.; Edwards, N. T.; Kercher, J. R. A model of calcium-cycling in an east Tennessee *Liriodendron* forest: Model structure, parameters and frequency response analysis. Ecology 57:99–109; 1976.

Shumway, S. E. Cucci, T. L.; Newell, R. C.; Yentsch, C. M. Particle selection, ingestion and absorption in filter-feeding bivalves. J. Exp. Mar. Biol. Ecol. 91:77–92; 1985.

Simkiss, K. Phosphates as crystal poisons of calcification. Biol. Rev. 39:487–505; 1964.

Simpson, G. G. Uniformitarianism. An inquiry into principle, theory, and method in geohistory and biohistory. In: Hecht, M. K.; Steere, W. C., eds. *Essays in evolution and genetics in honor of Theodosius Dobzhansky*. New York: Appleton Century Crofts, 1970, 594 pp.

Sims, P. L.; Singh, J. S. The structure and function of ten western North American grasslands. III. Net primary production, turnover, and efficiencies of energy capture and water use. J. Ecol. 66:573–597; 1978.

Sims, P. L.; Singh, J. S.; Lauenroth; W. K. The structure and function of ten western North American grasslands. I. Abiotic and vegetational characteristics. J. Ecol. 66:251–285; 1978.

Sinclair, A. R. E. The resource limitation of trophic levels in tropical grassland ecosystems. J. Anim. Ecol. 44:497–520; 1975.

Sinclair, M. M.; Tremblay, M. J.; Bernal, P. El Niño events and variability in Pacific mackerel (*Scomber japonicus*) survival index: Support for Hjort's second hypothesis. Can. J. Fish. Aquat. Sci. 42:602–608; 1985.

Singh, J. S.; Lauenroth, W. K.; Milchunas, D. G. Geography of grassland ecosystems. Prog. Phys. Geogr. 7:46–80; 1983.

Slobodkin, L. B. Energetics in *Daphnia pulex* populations. Ecology 40:232–243; 1959.

Slobodkin, L. B.; Richman, S. The availability of a miniature bomb calorimeter for ecology. Ecology 41:784; 1960.

Smayda, T. J. Succession of phytoplankton, and the ocean as an holocoenotic environment. In: Oppenheimer, C. H., ed. *Symposium on marine microbiology*. Springfield: Thomas, 1963:260–274.

Smayda, T. J. Phytoplankton species succession. In: Morris, I., ed. *The physiological ecology of phytoplankton*. Boston: Blackwell Science Publishers, 1980:493–570.

Smeck, N. E. Phosphorus dynamics in soils and landscapes. Geoderma 36:185–199; 1985.

Smetacek, V.; Pollehne, F. Nutrient cycling in pelagic systems: A reappraisal of the conceptual framework. Ophelia 26:401–428; 1986.

Smith, F. E. Effects of enrichment in mathematical models. In: *Eutrophication: causes, consequences, correctives.* Proceedings of a symposium. Washington: Nat. Acad. Sci.; 1969.

Smith, R. C. An analysis of 100 years of grasshopper populations in Kansas (1854 to 1954). Trans. Kansas Acad. Sci. 57:397–437; 1954.

Smith, R. L. Peru coastal currents during El Niño: 1976 and 1982. Science 221:1397–1399; 1983.

Smith, S. B.; Young, L. B.; Miller, C. E. Evaluation of soil nitrogen mineralization potentials under modified field conditions. Soil Sci. Soc. of Am. J. 40:74–76; 1977.

Smith, S. V. Carbon dioxide dynamics: record of organic production, respiration and calcification in the Eniwetok reef flat community. Limnol. Oceanogr. 18:106–120; 1973.

Smith, S. V. Kaneohe Bay: Nutrient mass balance, sewage diversion, and ecosystem responses. In: Jacoff, F. S., ed. *Advances in marine environment research: proceedings of a symposium.* Publ. EPA-600/9-79-035, 1979:344–358.

Smith, S. V. The Houtman Abrolhos Islands: Carbon metabolism of coral reefs at high latitude. Limnol. Oceanogr. 26:612–621; 1981.

Smith, S. V. Coral reef calcification. In: Barnes, D. J., ed. *Perspectives on coral reefs.* Aust. Inst. Mar. Sci., 1983:240–247.

Smith, S. V. Phosphorus versus nitrogen limitation in the marine environment. Limnol. Oceanogr. 29:1149–1160; 1984.

Smith, S. V. Mass balance in coral reef dominated areas. Proc. SCOR/UNESCO/IABO Workshop on Coastal–Offshore Ecosystem Couplings. In press.

Smith, S. V.; Harrison, J. T. Calcium carbonate production of the mare incognitum, the upper windward reef slope, at Enewetak Atoll. Science 197:556–559; 1977.

Smith, S. V.; Jokiel, P. L. Water composition and biogeochemical gradients in the Canton Atoll lagoon. 2. Budgets of phosphorus, nitrogen, carbon dioxide and particulate materials. Mar. Sci. Comm. 1:162–207; 1975.

Smith, S. V.; Jokiel, P. L. Water composition and biogeochemical gradients in the Canton Atoll Lagoon. In: Smith, S. V., Henderson, R. S., eds. *An environmental survey of Canton Atoll Lagoon.* San Diego: Naval Undersea Research and Development Center, 1976:15–53.

Smith, S. V.; Pesret, F. Processes of carbon dioxide flux in the Fanning Island lagoon. Pac. Sci. 28:225–245; 1974.

Smith, S. V.; Kimmerer, W. J.; Laws, E. A.; Brock, R. E.; Walsh, T. W. Kaneohe Bay sewage diversion experiment: Perspectives on ecosystem responses to nutritional perturbation. Pac. Sci. 35:279–402; 1981.

Smith, S. V.; Chandra, S.; Kwitko, L.; Schneider, R. C.; Schoonmaker, J.; Seelo, J.; Tebano, T.; Tribble, G. W. Chemical stoichiometry of lagoonal metabolism: Preliminary report on an environmental chemistry survey of Christmas Island, Kiribati. Technical Report, Joint UH/USF International Sea Grant Programme, 1984, 30 pp.

Smith, W. O.; Nelson, D. M. Phytoplankton bloom produced by a receding ice edge in the Ross Sea: spatial coherence with the density field. Science 227:163–166; 1985.

Smoliak, S. Influence of climatic conditions on forage production of shortgrass rangeland. J. Range Manage. 9:89–91; 1956.

Smoliak, S. Influence of climate conditions on production of *Stipa–Bouteloua* prairie over a 50-year period. J. Range Manage. 39:100–103; 1986.

Smolik, J. D. Nematode studies at the Cottonwood site. US/IBP Biome Tech. Rep. 251. Fort Collins: Colorado State University, 1974, 80 pp.

Smolik, J. D.; Dodd, J. L. Effect of water and nitrogen, and grazing on nematodes in a shortgrass prairie. J. Range Manage. 36:744–748; 1983.

Sneva, F. A.; Hyder, D. N. Estimating herbage production on semiarid ranges in the intermountain region. J. Range Manage. 15:88–93; 1962.

Sorokin, Y. I. Decomposition of organic matter and nutrient regeneration. In: Kinne, O., ed. *Marine ecology*, vol. 4. New York: John Wiley & Sons; 1978.

Sournia, A.; Richard, M. Données sur l'hydrologie et la productivité du lagoon d'un atoll ferme (Takapoto, Iles Tumaotui). Vie Milieu 26:243–279; 1976.

Southwood, T. R. E. Bionomic strategies and population parameters. In: May, R. M., ed. *Theoretical ecology: principles and applications.* Philadelphia: Saunders; 1976.

Southwood, T. R. E. Habitat, the template for ecological strategies? J. Anim. Ecol. 46:337–365; 1977.

Spanner, D. C. The green leaf as a heat engine. Nature 198:934–937; 1963.

Speaker, R. M.; Moore, K.; Gregory, S. V. Analysis of the process of retention of organic matter in stream ecosystems. Verh. Internat. Verein. Limnol. 22:1835–1841; 1984.

Spencer, H. *First principles.* London; 1861.

Spencer, H. *The principles of biology, Rev. edn, 2 vols.* London; 1899.

Speziale, B. J. Consequences of the formation of calcium phosphate solids for algal growth and phosphorus accumulation. Clemson University; 1985. Ph.D. dissertation.

Spigel, R. H.; Imberger, J. The classification of mixed-layer dynamics in lakes of small to medium size. J. Phys. Oceanogr. 10:1104–1121; 1980.

Sprugel, D. G. Density, biomass, productivity, and nutrient-cycling changes during stand development in wave-regenerated balsam fir forests. Ecol. Monogr. 54(2):165–186; 1984.

Sprugel, D. G. Natural disturbance and ecosystem energetics. In: Pickett, S. T. A.; White, P. S., eds. *The ecology of natural disturbance and patch dynamics.* Orlando, Florida: Academic Press, 1985:335–352.

Stanton, N. L. The effect of clipping and phytophagous nematodes on net primary production of blue grama, *Bouteloua gracilis.* Oikos 40:249–257; 1983.

Stanton, N. L.; Allen, M.; Campion, M. The effect of the pesticide carbofuran on soil organisms and root and shoot production in shortgrass prairie. J. Appl. Ecol. 18:417–431; 1981.

Stapledon, R. G. Cocksfoot grass (*Dactylis glomerata* L.): Ecotypes in relation to the biotic factor. J. Ecol. 16:71–104; 1929.

Stavn, R. H. The three-parameter model of the submarine light field: radiant energy absorption and energy trapping in nepheloid layers. J. Geophys. Res. 87:2079–2082; 1982.

Stavn, R. H. Light attenuation in natural waters: The effects of optically discrete layers. Applied Optics 22: 649; 1983.

Stearns, S. C. On measuring fluctuating environments: predictability, constancy, and contingency. Ecology 62:185–199; 1981.

Steele, J. H. A comparison of terrestrial and marine ecological systems. Nature 313:355–358; 1985.

Steele, J. H.; Henderson, E. W. Modeling long-term fluctuations in fish stocks. Science 224:985–987; 1984.

Steinmann, P. Die Tierwelt der Gebirgsbäche. Ein faunistischbiologische studie. Ann. Biol. Lacustre. 2:30–150; 1907.

Sterner, R. W. Herbivores' direct and indirect effects on algal populations. Science 231:493–495; 1986.

Stewart, J. W. B.; Cole, C. V.; Maynard, D. G. Interactions of biogeochemical cycles in grassland ecosystems. In: Bolin, B.; Cook, R. B., eds. *The major biogeochemical cycles and their interactions.* SCOPE 21. Chichester: John Wiley and Sons, 1983:247–270.

Stewart, W. D. P.; Mague, T.; Fitzgerald, G. P.; Burris, R. H. Nitrogenase activity in Wisconsin lakes of differing degrees of eutrophication. New Phytol. 70:497–509; 1971.

Stillwell, M. A. The effect of bovine urine on the nitrogen cycle of a shortgrass prairie. Colorado State University; 1983. Ph.D. dissertation.

Stout, J. D. Organic matter turnover by earthworms. In: Satchell, J. E., ed. *Earthworm ecology.* New York: Chapman and Hall, 1983:5–48.

Strayer, D. An essay on long-term ecological studies. Bull. Ecol. Soc. Amer. 67: 271–274; 1986.

Strong, W. L.; LaRoi, G. H. Rooting depths and successional development of selected boreal forest communities. Can. J. For. Res. 13:577–588; 1983.

Strugnell, R. G.; Pigott, C. D. Biomass, shoot-production, and grazing of two grasslands in the Rwenzori National Park, Uganda. J. Ecol. 66:73–96; 1978.

Summons, R. E.; Osmund, C. B. Nitrogen assimilation in the symbiotic marine alga *Gymnodinum microadriaticum:* Direct analysis of ^{15}N incorporation by GC–MS methods. Phytochemistry 20:575–578; 1981.

Suppe, F. Criticism of the received view. In: Suppe, F. *The structure of scientific theories, 2nd edn.* Urbana: University Illinois Press, 1977:62–118.

Swank, W. T.; Schreuder, H. T. Comparison on three methods of estimating surface area and biomass for forests of young eastern white pine. For. Sci. 20:91–100; 1974.

Swanson, F. J.; Fredriksen, R. L.; McCorison, F. M. Material transfer in a western Oregon forested watershed. In: Edmonds, R. L., ed. *Analysis of coniferous forest ecosystems in the western United States.* US/IBP Synthesis Series 14. Stroudsburg, Pennsylvania: Hutchinson Ross Publishing Company, 1982a:233–266.

Swanson, F. J.; Gregory, S. V.; Sedell, J. R.; Campbell, A. G. Land–water interactions: the riparian zone. In: Edmonds, R. L., ed. *Analysis of coniferous forest ecosystems in the western United States.* US/IBP Synthesis Series 14. Stroudsburg, Pennsylvania: Hutchinson Ross Publishing Company, 1982b:267–291.

Swift, M. C., Cummins, K. W.; Smucker, R. A. Effects of Dimilin on stream leaf-litter processing rates. Verh. Internat. Verein. Limnol., 1987, 23 pp.

Swift, M. J. Overview: Management of tropical soils for sustainable production. Presentation of IV International Congress of Ecology. Syracuse, New York August 10–16, 1986.

Swift, M. J.; Heal, O. W.; Anderson, J. M. *Decomposition in terrestrial ecosystems.* Studies in Ecology, vol. 5. Oxford: Blackwell Science Publishers; 1979.

Szmant-Froelich, A.; Pilson, M. E. Q. Nitrogen excretion by colonies of the temperate coral *Astrangia danae* with and without zooxanthellae. Proc. Third Int. Coral Reef Symp. 1:417–423; 1977.

Szmant-Froelich, A.; Johnson, V.; Hoehn, T.; Battey, J.; Smith, G. J.; Fleischmann, E.; Porter, J.; Dallmeyer, D. The physiological effects of oil-drilling muds on the Caribbean coral *Montastrea annularis.* Proc. Fourth Internat. Coral Reef Symp., Manila, vol. 1, 1981:163–168.

Tadaki, Y.; Sato, A.; Sakurai, S.; Takeuchi, I.; Kawahara, T. Studies on the production structure of forest. XVIII. Structure and primary production in subalpine "dead trees strips" Abies forest near Mt. Asahi. Jpn. J. Ecol. (Nippon Seitai Gakkaishi) 27:83–90; 1977.

Tangley, L. Crop productivity revisited. BioScience 36:142–147; 1986a.

Tangley, L. Biotechnology on the farm. BioScience 36:590–593; 1986b.

Tansley, A. G. The use and abuse of vegetational concepts and terms. Ecology 16:284–307; 1935.

Tate, C. M.; Meyer, J. L. The influence of hydrologic conditions and successional state on dissolved organic carbon export from forested watersheds. Ecology 64:25–32; 1983.

Taylor, R. J. *Predation.* London: Chapman and Hall, 1984, 166 pp.

Teal, J. M. Community metabolism in a temperate cold spring. Ecol. Monogr. 27:283–302; 1957.

Teal, J. M. Energy flow in the salt marsh ecosystem of Georgia. Ecology 43:614–624; 1962.

Thienemann, A. Die Binnengewässer Mitteleuropas. Stuttgart, I:1–255, 1925.

Thienemann, A. Grundzuge einer allgemeinen Ökologie. Arch. Hydrobiol. 35:267–285; 1939.

Tieszen, L. L.; Archer, S. Physiological responses of plants in tundra grazing systems. In: Johnson, D. A., ed. *Special management needs of alpine ecosystems.* Range Science Series. Denver: Society for Range Management 5:22–42; 1979.

Tilman, D. Resource competition between planktonic algae: an experimental and theoretical approach. Ecology 58:338–348; 1977.

Tilman, D. *Resource competition and community structure.* Princeton, New Jersey: Princeton University Press, 1982, 296 pp.

Tilman, D.; Kilham, S. S. Phosphate and silicate growth and uptake kinetics of the diatoms *Asterionella formosa* and *Cyclotella meneghiniana* in batch and semicontinuous culture. J. Phytol. 12:375–383; 1976.

Toulmin, S. *The philosophy of science.* London: Hutchinson; 1953.

Transeau, E. N. The prairie peninsula. Ecology 16:423–437; 1935.

Triska, F. J. Role of wood debris in modifying channel geomorphology and riparian areas of a large lowland river under pristine conditions: an historical case study. Verh. Internat. Verein. Limnol. 22:1876–1892; 1984.

Ulanowicz, R. E. *Growth and development: Ecosystems phenomenology.* New York: Springer-Verlag, 1986, 203 pp.

Ulanowicz, R. E.; Platt, T., eds. Ecosystem theory for biological oceanography. Can. Bull. Fish. Aquat. Sci. 213: 1985.

Ulrich, B.; Benecke, P.; Harris, W. F.; Khanna, P. K.; Mayer, R. Soil processes. In: Reichle, D. E., ed. *Dynamic properties of forest ecosystems.* International Biological Programme 23. Cambridge, UK: Cambridge University Press, 1981:265–339.

Urban, D. L.; O'Neill, R. V.; Shugart, H. H. Jr. Landscape ecology. BioScience 37:119–127; 1987.

US Committee for an International Geosphere–Biosphere Program. Global change in the geosphere–biosphere. Initial priorities for an IGBP. National Research Council. Washington: National Academy Press; 1986.

Valiela, I.; Teal, J. M. Nutrient limitation in salt-marsh vegetation. In: Reimold, R. J.; Queen, W. H., eds. *Ecology of halophytes.* New York: Academic Press, 1974:547–563.

Van Cleve, K.; Dyrness, C. T.; Viereck, L. A.; Fox, J.; Chapin, F. S. III; Oechel, W. Characteristics of tiaga ecosystems in interior Alaska. BioScience 33:39–44; 1983.

Vanderhoef, L. N.; Dana, B.; Emerich, D.; Burris, R. H. Acetylene reduction in relation to levels of phosphate and fixed nitrogen in Green Bay. New Phytol. 71:1097–1105; 1972.

Vanderhoef, L. N.; Huang, C.; Musil, R.; Williams, J. Nitrogen fixation (acetylene reduction) by phytoplankton in Green Bay, Lake Michigan, in relation to nutrient concentrations. Limnol. Oceanogr. 19:119–125; 1974.

Van Hook, R. I. Energy and nutrient dynamics of spider and orthopteran populations in a grassland ecosystem. Ecol. Monogr. 14:1–26; 1971.

Vannote, R. L.; Minshall, G. W.; Cummins, K. W.; Sedell, J. R.; Cushing, C. E. The river continuum concept. Can. J. Fish. Aquat. Sci. 37:130–137; 1980.

Veeh, H. H. Uranium series dating applied to phosphate deposits on coral reef islands. Proc. Fifth Int. Coral Reef Symp. 3:463–469; 1985.

Vernadskii, V. I. *Biosfera.* Leningrad; 1926.

Vernberg, F. J. Comparative studies of tropical and temperate zone coastal systems. Bull. Mar. Sci. 31:801–808; 1981.

Vickery, P. J. Grazing and net primary production of a temperate grassland. J. Appl. Ecol. 9:307–314; 1972.

Vinogradov, M. E. Ecosystems of equatorial upwelling. In: Longhurst, A. R., ed. *Analysis of marine ecosystems.* New York: Academic Press, 1981:69–93.

Visser, S. Role of the soil invertebrates in determining the composition of soil microbial commmunities. In: Fitter, A. H.; Atkinson, D.; Read, D. J.; Usher, M. B., eds. *Ecological interactions in soil.* Special Publ #4, British Ecological Society. Oxford: Blackwell, 1985:297–317.

Vitousek, P. M.; Reiners, W. A. Ecosystem succession and nutrient retention: an hypothesis. BioScience 25:376–381; 1975.

Vitousek, P.; Ehrlich, P. R.; Ehrlich, A. H.; Matson, P. A. Human appropriation of the products of photosynthesis. BioScience 36:368–373; 1986.

Vogt, K. A.; Moore, E. E.; Vogt, D. J.; Redlin, M. J.; Edmonds, R. L. Conifer fine root biomass within the forest floors of Douglas-fir stands of different ages and site productivities. Can. J. For. Res. 13:429–437; 1983.

Vogt, K. A.; Grier, C. C.; Vogt, D. J. Production, turnover, and nutrient dynamics of above- and belowground detritus of world forests. Adv. Ecol. Res. 15:303–377; 1986a.

Vogt, K. A.; Grier, C. C.; Gower, S. T.; Sprugel, D. G.; Vogt, D. J. Overestimation of net root production: a real or imaginary problem? Ecology 67:577–579; 1986b.

Wafar, M. V. M.; Devassy, V. P.; Slawyk, G.; Goes, J.; Jayakumar, D. A.; Rajendran, A. Nitrogen uptake by phytoplankton and zooxanthellae in a coral atoll. Proc. Fifth Int. Coral Reef Symp. 6:29–37; 1985.

Waldron, M. C. Zonation of biochemical processes in stratifying lakes: the importance of particulate layers as sites of protein decomposition and ammonification. Clemson University; 1985. Ph. D. dissertation.

Walker, K. F.; Likens, G. E. Meromixes and a reconsidered typology of lake circulation patterns. Verh. Internat. Verein. Limnol. 19:442–458; 1975.

Walker, T. W.; Syers, J. K. The fate of phosphorus during pedogenesis. Geoderma 15:1–19; 1976.

Wallace, C. C.; Watt, A.; Bull, G. D. Recruitment of juvenile corals onto coral tables preyed upon by *Acanthaster planci.* Mar. Ecol. Prog. Ser. 32:299–306; 1986.

Wallace, L. L.; McNaughton, S. J.; Coughenour, M. B. Compensatory photosynthetic responses of three African graminoids to different fertilization, watering, and clipping regimes. Bot. Gaz. 145:151–156; 1984.

Wallace, L. L.; McNaughton, S. J.; Coughenour, M. B. Effect of clipping and four levels of nitrogen on the gas exchange, growth, and production of two East African graminoids. Am. J. Bot. 72:222–230; 1985.

Walsh, J. J. Herbivory as a factor in patterns of nutrient utilization in the sea. Limnol. Oceanogr. 21:1–13; 1976.

Walsh, J. J. Shelf sea ecosystems. In: Longhurst, A. R., ed. *Analysis of marine ecosystems.* London: Academic Press, 1981:159–198.

Wang, M.; Harleman, D. R. F. Hydrothermal–biological coupling of lake eutrophication models. Technical report No. 270. Cambridge, Mass: Ralph M. Parsons Laboratory, Massachusetts Institute of Technology; 1982.

Ward, G. M.; Cummins, K. W. Effects of food quality on growth of a stream detritivore, *Paratendipes albimanus* (Meigen) (Diptera: Chironomidae). Ecology 60:57–64; 1979.

Ward, J. V.; Stanford, J. A. The serial discontinuity concept of lotic ecosystems. In: Fontaine, T. D.; Bartell, S. M., eds. *Dynamics of lotic ecosystems.* Ann Arbor, Michigan: Ann Arbor Sci. Publ., 1983, 494 pp.

Ware, D. M.; Tsukayama, T. A possible recruitment model for the Peruvian anchovy. Bol. Inst. Mar. Peru, Callao. Vol. Extraord. 1981:55–61.

Wargo, P. M. Defoliation: induced chemical changes in sugar maple roots stimulate growth of *Armillaria mellea.* Phytopathology 62:1278–1283; 1972.

Waring, R. H. Imbalanced forest ecosystems: assessments and consequences. For. Ecol. Manage. 12:93–112; 1985.

Waring, R. H.; Franklin, J. F. Evergreen coniferous forest of the Pacific Northwest. Science 204:1380–1386; 1979.

Waring, R. H.; Pitman, G. B. Modifying lodgepole pine stands to change susceptibility to mountain pine beetle attack. Ecology 66:889–897; 1985.

Waring, R. H.; Schlesinger, W. H. *Forest ecosystems. Concepts and management.* Orlando, Florida: Academic Press; 1985.

Waring, T. H.; McDonald, A. J. S.; Larsson, S.; Ericsson, T.; Wiren, A.; Arwidsson, E.; Ericsson, A.; Lohammar, T. Differences in chemical composition of plants grown at constant relative growth rates with stable mineral nutrition. Oecologia 66:157–160; 1985.

Warwick, R. M. Species size distributions in marine benthic communities. Oecologia 61:32–41; 1984.

Watt, A. S. On the ecology of the British beechwoods with special reference to their regeneration. II. The development and structure of beech communities on the Sussex Downs. J. Ecol. 13:27–73; 1925.

Watt, A. S. Pattern and process in the plant community. J. Ecol. 35:1–22; 1947.

Weaver, J. E. Effects of different intensities of grazing on depth and quantity of roots of grasses. J. Range Manage. 3:100–113; 1950.

Weaver, J. E. *North American prairie.* Lincoln, Nebraska: Johnsen Publishing Company; 1954.

Weaver, J. E.; Albertson, F. W. *Grasslands of the great plains.* Lincoln, Nebraska: Johnsen Publishing Company; 1956.

Weaver, J. E.; Bruner, W. E. Nature and place of transition from true prairie to mixed prairie. Ecology 35:117–126; 1954.

Weaver, J. E.; Stoddart, L. A.; Noll, W. Response of the prairie to the great drought of 1934. Ecology 16:612–629; 1935.

Webb, K. L.; Wiebe, W. J. Nitrification on a coral reef. Can. J. Microbiol 21:1427–1431; 1975.

Webb, K. L.; Wiebe, W. J. The kinetics and possible significance of nitrate uptake by several algal–invertebrate symbioses. Mar. Biol. 47:21–27; 1978.

Webb, K. L.; DuPaul, W. D.; Wiebe, W. J.; Sottile, W.; Johannes, R. E. Enewetak (Eniwetok) Atoll: aspects of the nitrogen cycle on a coral reef. Limnol. Oceanogr. 20:198–210; 1975.

Webb, K. L.; D'Elia, C. F.; DuPaul, W. D. Biomass and nutrient flux measurements on *Holothuria atra* populations on windward reef flats at Enewetak, Marshall Islands. Proc. Third Int. Coral Reef Symp. 1:409–415; 1977.

Webb, W. L. Relation of starch content to conifer mortality and growth loss after defoliation by the Douglas-fir tussock moth. For. Sci. 27:224–232; 1981.

Webb, W.; Szarek, S.; Lauenroth, W.; Kinerson, R.; Smith, M. Primary productivity and water use in native forest, grassland, and desert ecosystems. Ecology 59:1239–1247; 1978.

Webster, J. R.; Benfield, E. F. Vascular plant breakdown in freshwater ecosystems. Ann. Rev. Ecol. Syst. 17:567–594 + 22 pp. appendix; 1986.

Weickmann, K. M. Intraseasonal circulation and outgoing long wave radiation modes during northern hemisphere winter. Mon. Wea. Rev. 111:1838; 1983.

Weinstein, D. A.; Shugart, H. H. Ecological modeling of landscape dynamics. In: Mooney, H. A.; Godron, M., eds. *Disturbance and ecosystems. Components of response.* Berlin: Springer-Verlag, 1983:29–45.

Weisberg, R. H.; Colin, C. Equatorial Atlantic Ocean temperature and current variatins during 1983 and 1984. Nature 322: 240–243; 1986.

Welbourn, M. L.; Stone, E. L.; Lassoie, J. P. Distribution of net litter inputs with respect to slope position and wind direction. For. Sci. 27:651–659; 1981.

Welch, D. W. A study of the effects of density-dependence and age-structure on the dynamics of marine fish populations. Halifax: Dalhousie University; 1984. Ph. D. thesis.

Welch, P. S. *Limnology.* New York: McGraw-Hill, 1952, 538 pp.

Wells, J. W. Corals. Mem. Geol. Soc. Am. 67:1087–1104; 1957.

Welsh, B. L.; Bessette, D.; Herring, J. P.; Read, L. M. Mechanisms for detrital cycling in nearshore waters at Bermuda. Bull. Mar. Sci. 29:125–139; 1979.

Wesenberg-Lund, C. Biologie der Susswasserinsekten. Berlin: Springer-Verlag, Gylderdalske Boghandel; 1943.

West, D. C.; Shugart, H. H.; Botkin, D. B., eds. *Forest succession. Concepts and application.* New York: Springer-Verlag; 1981.

Wetzel, R. G. *Limnology.* Philadelphia: Saunders Publications, 1983, 767 pp.

Wheeler, C. T.; Hoker, J. E.; Crowe, A.; Berrie, A. M. M. The improvement and utilization in forestry nitrogen fixation by actinorhizal plants with special reference to Alnus in Scotland. Plant Soil 90:393–406; 1986.

White, P. S.; Pickett, S. T. A. Natural disturbance and patch dynamics. In: Pickett, S. T. A., White, P. S., eds. *The ecology of natural disturbance and patch dynamics.* Orlando, Florida: Academic Press, 1985:3–13.

White, T. C. R. Weather, food, and plagues of locusts. Oecologia 22:119–134; 1976.

White, W. B.; Meyers, G.; Donguy, J. R.; Pazan, S. Short-term climatic variability in the thermal structure of the Pacific Ocean during 1979–1982. J. Phys. Oceanogr. 15:917–935; 1985.

Whittaker, R. H. *Vegetation of the Siskiyou Mountains, Oregon and California.* Ecol. Monogr. 30:279–338; 1960.

Whittaker, R. H.; Likens, G. E. The biosphere and man. In: Lieth, H.; Whittaker, R. H., eds. *Primary productivity of the biosphere.* New York: Springer-Verlag, 1975:305–328.

Whitton, B. A., ed. *River ecology.* Oxford: Blackwell Science, 1975, 725 pp.

Wicken, J. S. Entropy and evolution: a philosophic review. Perspect. Biol. Med. 22:285–300; 1979.

Wicken, J. S. Thermodynamics and the conceptual structure of evolutionary theory. J. Theor. Biol. 117:363–383; 1985.

Wiebe, W. J. Nitrogen cycle on a coral reef. Micronesica 12:23–26; 1976.

Wiebe, W. J. Aquatic microbial ecology–research questions and opportunities. In: Cooley, J. H.; Golley, F. B. *Trends in ecological research for the 1980's.* New York: Plenum Press, 1984:35–49.

Wiebe, W. J. Nitrogen dynamics on coral reefs. Fifth Int. Coral Reef Symp. 3:401–406; 1985.

Wiebe, W. J.; Johannes, R. E.; Webb, K. L. Nitrogen fixation in a coral reef community. Science 188:257–259; 1975.

Wieder, R. K.; Lang, G. E. A critique of the analytical methods used in examining decomposition data obtained from litter bags. Ecology 63:1636–1642; 1982.

Wiegert, R. G. Population energetics of meadow spittlebugs (*Philaenus spumarius L.*) as affected by migration and habitat. Ecol. Monogr. 34:217–241; 1964.

Wiegert, R. G. Thermodynamic considerations in animal nutrition. Am. Zool. 8:71–81; 1968.

Wiegert, R. G. *Ecological energetics.* Benchmark papers in ecology, vol. 4. Dowden, Hutchinson and Ross, Inc.; 1976.

Wiegert, R. G. Modeling spatial and temporal variability in a salt marsh: sensitivity to rates of primary production, tidal migration and microbial degradation. In: Wolfe, D. A., ed. *Estuarine variability,* 1986:405–426.

Wiegert, R. G.; Evans, F. C. Investigations of secondary productivity in grasslands. In: Petrusewicz, K. *Secondary productivity of terrestrial ecosystems: Principles and methods.* Warsaw: Pánstwowe Wydawnictwo Naukowe 2:499–518; 1976.

Wiegert, R. G.; Owen, D. F. Trophic structure, available resources, and population density in terrestrial vs. aquatic ecosystems. J. Theor Biol. 30:69–81; 1971.

Wight, J. R.; Hanks, R. J. A water-balance, climate model for range herbage production. J. Range Manage. 34:307–311; 1981.

Wight, J. R.; Neff, E. L. *Soil-vegetation-hydrology studies, vol. II. User manual for ERHYM: The Ekalaka rangeland hydrology and yield model.* U.S. Department of Agriculture, ARS, Agriculture Research Series, ARR–W–29; 1983.

Wight, J. R.; Hanson, C. L.; Whitmer, D. Using weather records with a forage production model to forecast range forage production. J. Range Manage. 37:3–6; 1984.

Wilkerson, F. P.; Muscatine, L. Uptake and assimilation of dissolved inorganic nitrogen by a symbiotic sea anemone. Proc. R. Soc. Lond. B 221:71–86; 1984.

Wilkerson, F. P.; Trench, R. K. Nitrate assimilation by zooxanthellae maintained in laboratory culture. Mar. Chem. 16:385–393; 1985.

Wilkinson, C. R.; Sammarco, P. W. Effects of fish grazing and damselfish territoriality on coral reef algae. II. Nitrogen fixation. Mar. Ecol. Prog. Ser. 13:15–19; 1983.

Wilkinson, C. R.; Williams, D. McB.; Sammarco, P. W.; Hogg, R. W.; Trott, L. A. Relationships between fish grazing and nitrogen fixation rates on reefs across the central Great Barrier Reef. In: Baker, J. T.; Carter, R. M.; Sammarco, P. W.; Stark, K. P., eds. *Proceedings: inaugural Great Barrier Reef Conference,* Townsville, Aug. 28–Sept. 2, 1983. J.C.U. Press, 1983, 375 pp.

Wilkinson, C. R.; Williams, D. McB.; Sammarco, P. W.; Hogg, R. W.; Trott, L. A. Rates of nitrogen fixation on coral reefs across the continental shelf of the central Great Barrier Reef. Mar. Biol. 80:255–262; 1984.

Williams, D. D. Migrations and distributions of stream benthos. In: Lock, M. A.; Williams, D. D., eds. *Perspectives in running water ecology.* New York: Plenum Press, 1981:155–207.

Williams, D. D.; Hynes, H. B. N. The occurrence of benthos deep in the substratum of a stream. Freshwat. Biol. 4:233–256; 1974.

Williams, P. J.; Jenkinson, N. W. A transportable microprocessor-controlled precise Winkler titration suitable for field station and shipboard use. Limnol. Oceanogr. 27:576–584; 1982.

Williams, S. L.; Gill, I. P.; Yarish, S. M. Nitrogen cycling in backreef sediments. Fifth Int. Coral Reef Symp. 3:389–394, 1985a.

Williams, S. L.; Yarish, S. M.; Gill, I. P. Ammonium distributions, production, and efflux from backreef sediments, St. Croix, U.S. Virgin Islands. Mar. Ecol. Prog. Ser. 24:57–64; 1985b.

Wilson, A. T. Pioneer agriculture explosion and CO_2 levels in the atmosphere. Nature 273:40–41; 1978.

Winterbourn, M. J. The River Continuum Concept–reply to Barmuta and Lake. N. Z. J. Mar. Freshwat. Res. 16:229–231; 1982.

Winterbourn, M. J.; Rounick, J. S.; Cowie, B. Are New Zealand streams really different? N. Z. J. Mar. Freshwat. Res. 15:321–328; 1981.

Wittgenstein, L. *Philosophical investigations.* Anscombe, G. E. M., translator. New York: Macmillan; 1953.

Woodmansee, R. G. Additions and losses of nitrogen in grassland ecosystems. BioScience 28(7):448–453; 1978.

Woods, L. E.; Todd, R. L.; Leonard, R. A.; Asmussen, L. E. Nutrient cycling in a Southeastern United States agricultural watershed. In: Lowrance, R.; Todd, R. L.; Asmussen, L. E.; Leonard, R. A. *Nutrient cycling in agricultural ecosystems.* Athens, Georgia: University of Georgia. Coll. Agric. Spec. Publ. 23, 1983:301–312.

Woodwell, G. M.; Whittaker, R. H. Primary production in terrestrial ecosystems. Am. Zool. 8:19–30; 1968.

Wooster, W. S.; Fluharty, D. L., eds. El Niño North: Niño effects in the eastern subarctic Pacific Ocean. Washington Sea Grant; 1985.

Worster, D. *Nature's economy.* San Francisco: Sierra Club Books; 1977.

Wyrtki, K. El Niño–the dynamic response of the equatorial Pacific Ocean to atmospheric forcing. J. Phys. Oceanogr 5:572–584; 1975.

Wyrtki, K. Water displacements in the Pacific and the genesis of El Niño cycles. J. Geophys. Res. 90:7129–7132; 1985.

Wyrtki, K.; Myers, G. The trade wind field over the Pacific Ocean. Hawaii Inst. Geophys. Rep., HIG–75–1; 1975.

Wyrtki, K.; Myers, G. The trade wind field over the Pacific Ocean. J. Appl. Meteorol. 15:698–704; 1976.

Yamazato, K. Calcification in a solitary coral, *Fungia scutaria* L., in relation to environmental factors. Bull. Sci. Eng. Div., University of the Ryukyus, Mathematics and Natural Sciences, Naha, Okinawa 13:57–122; 1970.

Yeates, G. W. Soil nematode populations depressed in the presence of earthworms. Pedobiologia 22:191–195; 1981.

Yeates, G. W.; Coleman, D. C. Role of nematodes in decomposition. In: Freckman, D. W., ed. *Nematodes in soil ecosystems.* Austin: University of Texas Press, 1982:55–80.

Yonge, C. M. Studies on the physiology of corals. I. Feeding mechanisms and food. Sci. Reports Great Barrier Reef Exped. 1:13–57; 1930.

Yonge, C. M.; Nichols, A. G. Studies on the physiology of corals. IV. The structure, distribution and physiology of zooxanthellae. Sci. Reports Great Barrier Reef Exped. 1:135–176; 1931.

Young, S. A.; Kovalak, W. P.; Del Signore, K. A. Distances traveled by autumn-shed leaves introduced into a woodland stream. Am. Midl. Nat. 100:217–222; 1978.

Zaret, T. M. *Predation and freshwater communities.* New Haven: Yale University Press, 1980, 187 pp.

Zieman, J. C. Tropical sea grass ecosystems and pollution. In: Wood, E. J. F.; Johannes, R. E., eds. *Tropical marine pollution.* Elsevier Oceanography Series, #12, New York: Elsevier, 1975:63–74.

Zimmerman, M. J.; Waldron, M. C.; Schreiner, S. P.; Freedman, M. L.; Giammatteo, P. A.; Hains, J. J.; Nestler, J. M.; Speziale, B. J.; Schindler, J. E. High frequency mixing dynamics of lakes. Verh. Internat. Verein. Limnol. 21:88–93; 1981.

Index